African Perspectives on Agroecology

Praise for this book

'Wynberg's comprehensively curated volume offers a rich and stimulating selection of perspectives on the challenges to be overcome and the opportunities to be embraced if Africans are to feed themselves justly and sustainably. These essays are especially valuable for the insights they provide on how sovereignty over the use and control of seeds will shape the struggle over a hoped-for transition to an agroecological agriculture for Africa.'

Jack Kloppenburg, Professor Emeritus, University of Wisconsin-Madison, USA;
and author of First the Seed: The Political Economy of Plant Biotechnology

'Despite minimal state and donor support, the 'silent revolution' of agroecological practices is taking root across Africa. Farmers, NGOs, and research teams are innovating and organizing to fight climate change, inequality, and hunger. This fine collection includes contributions from the frontlines, assembling an array of reflections on the possibilities and constraints facing the wider adoption of agroecology. This terrific anthology is a rejoinder to Afropessimism, and an inspiring call to action.'

Raj Patel, Research Professor, University of Texas, USA
and author of Stuffed and Starved

'Rachel Wynberg's very useful, and well written, book is packed with relevant overviews, analyses, vignettes of lived experiences and more. It's an upbeat tour de force, providing thoughtful historical perspectives and current realities about overcoming barriers and building on local knowledge and the growing seed and food sovereignty movements in the continent. She weaves all these together in her brilliant opening and closing chapters. These demonstrate the imperative of having smallholder farmer seed systems at the heart of viable agroecological, food systems that are biodiverse, nutritious and environmentally sustainable, and need to be scaled-out across all of Africa. This well referenced book is a joy to read.'

Patrick Mulvany, Food Ethics Council, UK

'*African Perspectives on Agroecology* brings together a vast range of experience from diverse communities and countries in Africa. The contributing authors illustrate the complexities and unique contexts in which agroecological transitions based on local seed biodiversity and indigenous knowledge are occurring. These are stories of hope and resistance for freedom and the renewal of life. As such, this is an inspiring book – a 'must read' for all who care about the future of Africa and its people.'

Michel Pimbert, Professor at the Centre for Agroecology, Water and Resilience,
Coventry University, UK

'*African Perspectives on Agroecology* narrates the story on how a coalition of farmers, NGOs and academics engage in a process of restoration of traditional seed and knowledge as the pillars for re-creating a biodiverse, resilient and socially just agriculture capable of fulfilling food, water and seed sovereignty and adaptation to climate change in a planet in polycrisis.'

Miguel A Altieri, Professor Emeritus of Agroecology,
University of California, Berkeley

'An agroecological transformation of our food systems is urgently needed. This timely book takes the reader through the steps that are already being taken to achieve this, highlighting the importance of farmer-led seed and knowledge systems that are often overlooked. This book is an essential companion for anyone working to transform food systems in Africa.'

Emile Frison, Senior Advisor, Agroecology Coalition and
Former Director General of Bioversity International

'A response to the biodiversity crisis … *African Perspectives on Agroecology: Why Farmer-led Seed and Knowledge Systems Matter* reminds us of the need to deepen the seed and knowledge work with farmers to revive, enhance and create pockets of resilience for hope and learning.'

Gertrude Pswarayi-Jabson, Country Coordinator, Participatory Ecological
Land Use Management (PELUM) Zimbabwe

'This book is a must for all stakeholders engaged in improving seed and food security in Africa. It provides central building blocks for the transition required to boost farmers' own seed and agricultural systems across Africa and shows with compelling clarity that such an approach is vital for achieving sustained seed and food security on the continent.'

Regine Andersen, Research Director, Fridtjof Nansen Institute, Norway and author
of Governing Agrobiodiversity: plant genetics and developing countries

'*African Perspectives on Agroecology: Why Farmer-led Seed and Knowledge Systems Matter* shines a spotlight on the agroecological transition pathway, and the importance of farmers' access to genetic resources for sustainable food systems in Africa and beyond. This book is a holistic exploration of the seed biodiversity crises created by the corporate capture of seed systems in Africa.'

Mamadou Goïta, Executive Director, Institute for Research and Promotion of
Alternatives in Development

African Perspectives on Agroecology
Why farmer-led seed and knowledge systems matter

Edited by
Rachel Wynberg

Practical
ACTION
PUBLISHING

Practical Action Publishing Ltd
25 Albert Street, Rugby, Warwickshire, CV21 2SD, UK
www.practicalactionpublishing.com

A catalogue record for this book is available from the British Library.

A catalogue record for this book has been requested from the Library of Congress.

ISBN 978-1-78853-022-4 Paperback
ISBN 978-1-78853-021-7 Hardback
ISBN 978-1-78044-744-5 Electronic book

Citation: Wynberg, R., (2024) *African Perspectives on Agroecology: Why farmer-led seed and knowledge systems matter*, Rugby, UK: Practical Action Publishing
http://doi.org/10.3362/ 9781780447445

Since 1974, Practical Action Publishing has been publishing and disseminating books and information in support of international development work throughout the world. Practical Action Publishing is a trading name of Practical Action Publishing Ltd (Company Reg. No. 1159018), the wholly owned publishing company of Practical Action. Practical Action Publishing trades only in support of its parent charity objectives and any profits are covenanted back to Practical Action (Charity Reg. No. 247257, Group VAT Registration No. 880 9924 76).

The views and opinions in this publication are those of the editors and do not represent those of Practical Action Publishing Ltd or its parent charity Practical Action. Reasonable efforts have been made to publish reliable data and information, but the editors and publisher cannot assume responsibility for the validity of all materials or for the consequences of their use.

Cover design by Katarzyna Markowska, Practical Action Publishing
Cover photo credit: SKI_Toddy Sibanda, Zambia_Photo by Xavier Vahed.
Typeset by vPrompt eServices, India

In memory and deep gratitude for two Biodiversity Giants who laid the foundation for a socially just, genetically diverse and agroecological future for Africa and her farming communities

Tewolde-Berhan Gebre Egziabher
1940–2023

Dr Melaku Worede
1936–2023

Contents

List of boxes, figures, photos, and tables

Boxes

Figures

Photos

Tables

About the authors

Million Belay coordinates the Alliance for Food Sovereignty for Africa (AFSA) and is a member of the International Panel of Experts on Sustainable Food Systems (IPES-Food). He is the founder of MELCA–Ethiopia, an indigenous NGO working on issues of agroecology, intergenerational learning, advocacy, and livelihood improvement of local and Indigenous peoples. Million has been working for more than two decades on issues of intergenerational learning of biocultural diversity, sustainable agriculture, and the rights of local communities to seed and food sovereignty. His current interests involve advocacy on food sovereignty, knowledge dialogues, and the use of participatory mapping for social learning, identity building, and the mobilization of memory for resilience. He has a PhD in environmental learning and a master's degree in tourism and conservation, and studied biology at undergraduate level.

Rachel Bezner Kerr is a professor in global development at Cornell University. She does participatory research on agroecology, gender, climate change adaptation, and food and nutrition security, in collaboration with the non-profit organization Soils, Food and Healthy Communities (SFHC) in Malawi. She has published over 100 scientific articles in journals such as *Proceedings of the National Academy of Sciences* and *Agriculture, Ecosystems & Environment*. She was a coordinating lead author of the 'food chapter' for the report on climate change impacts, vulnerabilities, and adaptation of the Intergovernmental Panel on Climate Change. She has also written a report on agroecology for the United Nations Committee on World Food Security.

Johannesburg-born **Vanessa Black** completed a Bachelor of Architecture Degree at the University of the Witwatersrand, briefly working as an architect before following her passion for environmental justice. Vanessa joined Earthlife Africa as a student, and after graduating became the Gauteng coordinator of the Environmental Justice Networking Forum. Earthlife Africa Johannesburg conceptualized the GreenHouse People's Environmental Centre Project, which Vanessa coordinated from 1997 to 2003. Relocating to Durban in 2004, she worked with various initiatives, including the African Centre for Biodiversity, the Institute for Zero Waste in Africa, Young Insights in Planning, the Agricultural Management Unit of eThekwini local government, and the South Durban Community Environmental Alliance. She joined Biowatch South Africa in February 2017 as the Advocacy and Research Coordinator.

Audrey Carlson was born and raised in Colorado Springs in the United States. After studying abroad in Denmark in her junior year of high school, she decided to pursue an undergraduate degree in economics with a double minor in political science and geography at the University of Denver. While attending that university, she conducted research to understand food security and the acculturation of diets within the Ghanaian population in the Denver metropolitan area. She has since worked in the nonprofit sector and plans to pursue a master's in economics.

Laifolo Dakishoni, or 'Dak' as he is affection-ately known, is the deputy director of the Malawian non-profit organization Soils, Food and Healthy Communities (SFHC). Responsible for research and project finance at SFHC, Dak studied at the College of Accountancy in Blantyre, Malawi. His passion for community involvement and participation drove him to become involved in the organization and he has contributed to the successes of SFHC in food security and nutrition over the years. Since 2001, he has enjoyed being part of the organization's evolution, and has been greatly motivated by the farmers' eagerness to learn and try new things. He is also a country manager for the InnovAfrica project.

Richard Dimba Kiaka is a lecturer on the human dimensions of conservation at the School for Field Studies, Kenya. Richard earned a PhD in anthro-pology from the University of Hamburg. Since 2019, he has been part of an ongoing longitudinal research programme on maize transformation in western Kenya (Luoland). Being native to Luoland, Richard has engaged personally with the socio-technical changes affecting maize seed in western Kenya. He is also part of a collaborative research project – *Digital Tech in African Agriculture*. At the time of contributing to this volume, Richard worked as a post-doctoral researcher at Jaramogi Oginga Odinga University of Science and Technology in Kenya, in a collaborative project – *Grassroots Financial Innovations for Inclusive*

Economic Growth, funded by the Danish Ministry of Foreign Affairs, Grant No. GFIIEG 18-11-CBS.

Vanessa Farr specializes in gender and crisis, including the gendered impacts of climate disruption. She works on women, peace, and security, and disaster response across Africa, particularly in the Islamic world. Her fieldwork helps her identify patterns in the intersection of patriarchal and religious extremism with climate crisis, resulting in soil degradation, food insecurity, militarization, and gendered violence. She edits the dispatches section of the online journal *Studies in Social Justice*, supporting the publication of reports or commentaries from the non-academic and academic spaces of social justice practice, discourse, and contestation. She is the co-editor of two books: *Back to the Roots: Security Sector Reform and Development* (2012) and *Sexed Pistols: The Gendered Impacts of Small Arms and Light Weapons* (2009). She holds a PhD from the School of Women's Studies at York University, Toronto.

David Fig is a Johannesburg-based academic, writer and activist. He writes about the agrofood chain, corporate behaviour, and risky technologies (for example, nuclear, fracking, asbestos, agrofuels, pesticides, GMOs). In addition to South African environmental policy, he has analyzed the eucalyptus industry in the Atlantic Forest of Brazil, Mozambican and Vietnamese energy options, and uranium in Namibia. He is linked with the Transnational Institute in Amsterdam and with the SARChI Chair on the Bio-economy at the University of Cape Town. He is a co-founder of the food sovereignty NGO Biowatch which promotes agroecological farming methods.

Stephen Greenberg is a researcher who has worked on food systems, land, agriculture, seed, and smallholder farming for the past 25 years. With a PhD in development studies from the University of Sussex, Stephen is currently focusing on agroecology and food systems transitions in South Africa. Between 2013 and 2020 he was actively involved with the African Centre for Biodiversity (ACB) in continental networks on farmer seed systems and agroecology, during which time he wrote his contribution to this book, which also benefited from the shared knowledge and experience of the ACB team. Stephen has researched and critiqued corporate concentration in the South African food system and workshopped proposed alternatives based on the democratic and active agency of food system actors who are marginalized in the dominant system.

Paul Hebinck, who passed away in 2022, before the publication of this book, was a rural sociologist and Emeritus Associate Professor in the Sociology of Development and Change group at Wageningen University in the Netherlands. He was also an adjunct professor at the University of Fort Hare, South Africa, and senior research associate in the Department of Environmental Science at Rhodes University in Grahamstown, South Africa. His research was longitudinal, with a focus on socio-material transformations in Luoland, West Kenya, the Eastern Cape province of South Africa, north-eastern Zimbabwe, and communal areas of Namibia.

Angelika Hilbeck is a senior scientist and lecturer at the Swiss Federal Institute of Technology in Zurich, where she leads the Environmental Biosafety and Agroecology group at the Institute of Integrative Biology. Her research focuses on biosafety and risk assessment of GMOs in the context of agroecology and biodiversity, and she has been engaged in these issues through numerous projects in Africa, South America, and Asia. Through her international work, she has become involved in broader issues around the development of technology, working towards a democratically legitimated, sustainable global future, and actively contributing to the debate on biosafety, digitalization, sustainability, agroecological transformation, hunger, and poverty alleviation. She is a co-founder of the European Network of Scientists for Social and Environmental Responsibility and Critical Scientists Switzerland and a former member of the board of directors of both the Swiss development organization Bread for All and the Federation of German Scientists.

Witness Kozanayi was born and raised in Chimanimani District, Zimbabwe. He has worked in academia and NGOs, focusing on rural development and livelihoods, and local-level institutions and institutional arrangements for natural resources. A holder of a PhD on the governance of non-timber forest products from the University of Cape Town, South Africa, he has a master's degree in environmental management for business from Cranfield University in the United Kingdom and a BSc degree and diploma in agriculture. Witness is currently a lecturer at Marondera University of Agricultural Sciences and Technology (MUAST), Zimbabwe. Before joining MUAST he was a postdoctoral fellow in the Bio-economy Chair at UCT, where he was involved in research linked to the Seed and Knowledge Initiative. Witness continues to be involved in research in the Chimanimani District and elsewhere in Southern Africa, including participatory action research linked to SKI. Soon after the Cyclone Idai disaster in Chimanimani, Witness joined the interdisciplinary team which was commissioned to research the multidimensional impacts of the cyclone on communities.

Kudzai Kusena is a PhD graduate from the Department of Environmental and Geographical Science at the University of Cape Town, where his study focused on the resilience of local seed systems in Zimbabwe. He completed a master's degree in biodiversity management at the Swedish University of Agricultural Sciences, and holds a bachelor's degree in agriculture from the University of Zimbabwe. He works as a policy and programmes specialist with the United Nations Food and Agriculture Organization, coordinating a global project on Capacity Building on Multilateral Environmental Agreements, with a special focus on the Convention on Biological Diversity. He has worked as a curator at the National Genebank of Zimbabwe and was appointed as an agrobiodiversity officer at the Community Technology Development Trust in Harare, where he worked with rural smallholder farmers. Previously he was the Zimbabwean National Focal Point for the International Treaty on Plant Genetic Resources for Food and Agriculture (ITPGRFA), an expert member of the African Union technical committee responsible for developing the African Seed and Biotechnology Programme, and chief African negotiator for the ITPGRFA's Multilateral Access and Benefit-sharing Mechanism.

Noelle LaDue graduated from Cornell University in 2019 with a Bachelor of Science degree in development sociology. She was a member of Professor Rachel Bezner Kerr's research team from 2018–2019. Her research focuses on seed saving, food security, and agricultural biodiversity in Malawi. She has also participated in research with the Cornell Farmworker Program and Cornell's Plant Breeding Department.

Morgan Lee is a PhD candidate in the Bio-economy Research Chair in the Environmental and Geographical Science Department at the University of Cape Town. She completed her undergraduate and honours degree with distinction at Rhodes University. Morgan is passionate about sustainable land management and food production, and has a particular interest in the dynamics of socio-technical systems in transition. Her PhD research explores the complexities of sustainable agricultural transitions on commercial grain farms (wheat and maize) in South Africa and investigates lock-in mechanisms restricting change.

Esther Lupafya has worked as Project Director and Gender Coordinator at Soils, Food and Healthy Communities (SFHC) in Malawi since 2017. Her duties include coordinating participatory farmer research and development projects, overseeing the Farmer Research and Training Centre, and carrying out gender sensitization campaigns and training. Prior to joining SFHC she spent 16 years at Ekwendeni Hospital as Deputy Primary Health Care Director. In this role she was responsible for coordinating and running a range of programmes including building sustainable livelihoods, home-based care for HIV/AIDS-affected households, support groups for people living with HIV/AIDS, prevention of mother-to-child transmission, orphans and vulnerable children, early childhood growth development centres, and running a community HIV and counselling youth centre.

Sidney Madsen is a PhD student in the Department of Global Development at Cornell University. She grew up on a small farm in rural upstate New York and this background has influenced her interest in transitioning agricultural livelihoods and the dynamics of globalization on food security. After graduating from the undergraduate Biology and Society programme at Cornell, she worked on community-based efforts to address food insecurity in Guatemala, Bolivia, Mexico, and Malawi. These experiences informed her thesis work, in which she used a participatory methodology to examine how agroecology affects conditions of farmer well-being and community vitality.

Maya Marshak holds a PhD in environmental and geographical science from the University of Cape Town. Her doctoral research was linked to the Seed and Knowledge Initiative and the Bio-economy Research Chair and focused on the impacts of seed technologies on social-ecological relationships and knowledge. She has worked in the field of food systems research for over 10 years and, during this time, has been involved in various food and agricultural projects in Cape Town and in Eswatini where she currently co-runs an agroecological farm. She is interested in the transformative power of eating and growing food locally and in more socially and ecologically sensitive ways. She enjoys collaborative environmental work that brings in many ways of knowing and working, including visual and other creative methods.

Fakazile Mthethwa (also known as **Gogo Qho**) sadly passed away at the end of 2020. She was a permaculturalist based in Mtubatuba in northern KwaZulu-Natal province, South Africa. For more than two decades, her farming practice centred on conserving indigenous knowledge around health and indigenous herbs. Her unique approach to farming attracted attention from students and researchers from various universities around the world.

Shepherd Mudzingwa is an agroecology extension and education specialist with more than a decade's experience working towards the development and implementation of agroecology programmes at Fambidzanai Training Centre in Harare. He has been at the forefront of the development of agroecology education programmes and their integration into the formal education system of Zimbabwe. He has a master's in food security and sustainable agriculture and is currently considering taking up a PhD to further his research interests in indigenous knowledge science systems and participatory action research.

With a PhD from UCT in environmental and geographical sciences, focusing on seed and food security, **Bulisani Ncube** is a researcher and development practitioner with over 20 years' experience in the field of smallholder agriculture, rural livelihoods, seed systems, and food security. This experience was gained by working with research organizations, local and international NGOs, and the donor community. He has worked across countries in southern and east Africa, including South Sudan. Key competencies include project and programme management, monitoring and evaluation, and conducting research. He is experienced in the application of transversal themes such as gender equality, HIV mainstreaming, and climate change.

Mvuselelo Ngcoya is a farmer and a lecturer. Together with his family, he practises agroecological farming on a slope in the undulating hills of Richmond, KwaZulu-Natal, South Africa. An associate professor in development studies at the University of KwaZulu-Natal in Durban, his teaching focuses on development theory, sustainable agriculture, and rural development. Topics ranging from indigenous plants, indigenous knowledge systems, and the political economy of food to land reform inform much of his research.

Kristof Nordin is a co-founder of Never Ending Food in Malawi. This community-based initiative implements, teaches, and demonstrates sustainable solutions for food and nutrition security. Originally from the United States, Kristof and his wife, Stacia, have lived and worked in Malawi for over two decades. Kristof has a degree in social work and his wife is a registered dietician; both hold diplomas in permaculture design. Their daughter, Khalidwe, was born and raised in Malawi and is also certified in permaculture design.

Hanson Nyantakyi-Frimpong is an associate professor of geography at the University of Denver, Colorado, in the United States. His research focuses on the political ecology of rural development, the human dimensions of global environmental change, and sustainable agriculture and food systems. His work has been published in *Global Environmental Change*, the *Journal of Peasant Studies*, *Land Use Policy*, *Ecology & Society*, *Geoforum*, *The Professional Geographer*, *Applied Geography*, and other interdisciplinary journals. He is active in various national and international organizations involved with sustainability, social-environmental change, and food security.

Mugove Walter Nyika is the founding Coordinator of the Regional Schools and Colleges Permaculture (ReSCOPE) Programme and is currently the General Coordinator of the ReSCOPE network. He would like to see resilient and food sovereign communities living in natural abundance. To this end he has facilitated the development of the Integrated Land Use Design process as a tool for school communities in east and southern Africa to adopt holistic and sustainable land use systems. Walter serves communities in multiple ways, including as a member of the Global Ecovillage Network Africa Advisory Council. His mission is to use his life skills, land-use design skills, and passion for the environment to listen, encourage, and share with all, especially children, so that they can be empowered to look after themselves and the environment for the common good.

Elfrieda Pschorn-Strauss has been advocating, writing, and organizing towards defending the diversity of our planet since 1990. After completing a master's degree in town and regional planning at the University of Cape Town, she was drawn into the emerging environmental justice movement of the early 1990s. Since then, she has been involved in starting up new initiatives, including Biowatch South Africa, the African

Biodiversity Network, and the Alliance for Food Sovereignty in Africa. From 2013 to 2019 she acted as the regional coordinator of the Seed and Knowledge Initiative, and while working for GRAIN, she contributed to its publications and the work of environmental rights organizations in anglophone Africa. She never ceases to marvel at seeds.

Lizzie Shumba has worked with Soils, Food and Healthy Communities since 2003. Based in northern Malawi, she works with farmers as an extension officer, focused on agroecology, nutrition, and gender. Although she is from Lilongwe, she enjoys working in Ekwendeni, where she lives with her family. Lizzie holds a BSc in agriculture extension and nutrition which she obtained from Lilongwe University of Agriculture and Natural Resources. She likes interacting with the farmers in the communities and looks forward to sharing what she has learnt in the field with her colleagues when she returns to Ekwendeni.

Haidee Swanby has worked for the past 20 years as a researcher and activist in African food and social movements. Her focus has been on traditional agriculture, indigenous knowledge, and the privatization of African agriculture and corporate control of the food system. Her current research focus connects back to what she loves most – the magic of food, the people that prepare it, and collaboration with nature. This research explores how we draw on the visceral to build political identities and catalyze activism, how food feels in our bodies and connects to intellect, emotions, and histories, and the choices we make as a result.

Jaci van Niekerk completed a master's degree at the University of Cape Town (UCT) in 2009. Her thesis examined the contribution of the international trade in an endemic medicinal plant, *Pelargonium sidoides*, to rural livelihoods in South Africa and Lesotho. Upon graduation she took up a position as researcher in the Department of Environmental and Geographical Science at UCT. She has been closely involved in several projects relating to environmental and social justice and has actively participated in the Seed and Knowledge Initiative since its inception. She is currently part of the Bio-economy Research Chair team and continues to conduct research into themes related to the chair such as agrobiodiversity, farmers' rights, and traditional ecological knowledge. She has a special interest in the ethics of conducting research as well as strengthening university–community relationships.

Jennifer Whittingham is a PhD student in the Bio-economy Research Chair in the Department of Environmental and Geographical Science at the University of Cape Town and holds an MPhil in environment, society, and sustainability. Her master's work focused on the potential contributions of feminist care ethics to the risk assessment practices of genetically modified crops in South Africa. Her current doctoral work explores different ways of knowing the ocean, aiming to find synergies and conflicts between modernist science and Indigenous knowledge.

John Wilson is a Zimbabwean free-range facilitator and activist working with organizations across Africa, from community-based organizations to large regional and continental networks such as the Alliance for Food Sovereignty in Africa and the Seed and Knowledge Initiative. His focus is on strategic and collaborative initiatives that can help grow the agroecology and food sovereignty movement in Africa and globally. He has worked in the agroecology field for over 40 years.

Dr Melaku Worede was born in Addis Ababa in 1936. After obtaining his master's and PhD degrees in genetics and plant breeding from the University of Nebraska in the USA, he returned home and got involved in the planning and establishment of the national Plant Genetic Resources Centre, of which he became director in 1979. He served in that capacity until his retirement in 1993. In 1989, he received the Right Livelihood Award, an international honour conferred on exemplary leaders who offer practical solutions to the urgent and pressing challenges facing the world. Thanks to his efforts and many accomplishments, several initiatives which support biodiversity conservation and utilization in Africa have adopted the Ethiopian experience as their model. Dr Worede passed away in August, 2023.

Rachel Wynberg is a scholar-activist with a special interest in biopolitics, sustainable agricultural futures, and agroecology. As a professor in the Department of Environmental and Geographical Science at the University of Cape Town (UCT), she holds a national Research Chair on Social and Environmental Dimensions of the Bio-economy and is actively involved with environmental policy debates and civil society movements in Southern Africa. Rachel has led UCT in its partnership with the Seed and Knowledge Initiative (SKI) since SKI's inception in 2013, working towards a future where smallholder farmers in Southern Africa are empowered to secure seed and food sovereignty. Her publications include over 220 scientific papers, technical reports, and popular articles, and five co-edited books and monographs. Further information can be found at www.bio-economy.org.za and https://www.researchgate.net/profile/Rachel_Wynberg.

Abbreviations

ACB	African Centre for Biodiversity
ADMARC	Agricultural Development and Marketing Corporation (Malawi)
AFSA	Alliance for Food Sovereignty in Africa
AGRA	Alliance for a Green Revolution in Africa
CAADP	Comprehensive Africa Agriculture Development Programme
CBD	Convention on Biological Diversity
CBI	confidential business information
CoP	community of practice
CSA	climate-smart agriculture
DALRRD	Department of Agriculture, Land Reform and Rural Development (South Africa)
DSI	digital sequence information
ECOWAS	Economic Community of West African States
ESAFF	Eastern and Southern Africa Small-scale Farmers Forum
FAO	Food and Agriculture Organization of the United Nations
FISP	Farm Input Subsidy Programme
GM	genetically modified
GMO	genetically modified organism
GR	Green Revolution
HT	herbicide tolerant
IARC	Internatonal Agency for Research on Cancer
INCA	National Institute of Agricultural Sciences (Cuba)
IP	intellectual property
IPES-Food	International Panel of Experts on Sustainable Food Systems
ITPGRFA	International Treaty on Plant Genetic Resources for Food and Agriculture
KSC	Kenya Seed Company
MELCA	Movement for Ecological Learning and Community Action (Ethiopia)
MRL	maximum residue level
MVP	Millennium Villages Project
NEPAD	New Partnership for Africa's Development
NGO	non-governmental organization
NSCM	National Seed Company of Malawi
OAF	One Acre Fund
OPV	open-pollinated varieties
OSCA	Owen Sitole College of Agriculture

PASS	Program for Africa's Seed Systems
PELUM	Participatory Ecological Land Use Management
PIAL	Local Agricultural Innovation Programme
PROPAC	Plateforme Régionale des Organisations Paysannes d'Afrique Centrale
RDC	rural district council (Zimbabwe)
ReSCOPE	Regional Schools and Colleges Permaculture
ROPPA	Réseau des Organisations Paysannes et de Producteurs de l'Afrique de l'Ouest (Network of Farmers' and Producers' Organizations in West Africa)
RWA	Rural Women's Assembly
SADC	Southern African Development Community
SAGENE	South African Committee for Genetic Experimentation
SFHC	Soils, Food and Healthy Communities
STS	science and technology studies
TSURO	Towards Sustainable Use of Resources Organization
UNAC	União Nacional de Camponeses (Mozambique National Peasants' Union)
UNDROP	UN Declaration on the Rights of Peasants and Other People Working in Rural Areas
UPOV	International Union for the Protection of New Varieties of Plants
USAID	United States Agency for International Development
US EPA	United States Environmental Protection Agency
WTO	World Trade Organization
ZIMSOFF	Zimbabwe Smallholder Organic Farmers Forum

Foreword

People have had a domestic relationship with plants for more than 10,000 years. Through this relationship, based on continuous experimentation and adaptation, farmers have co-evolved and adapted genetic resources, resulting in increased agricultural biodiversity. Relying on reproductive genetic recombination and mutation for novelty, farmers have driven innovation and agricultural biodiversity by selecting which seeds to save, growing the seeds, and distributing them within and among communities through gift, exchange, or sale.

Today, broadly, there are two different types of seed system: farmers' seed systems and commodity seed systems. Farmers' seed systems – defined by the continuous renewal of biodiversity and the free distribution of seeds and knowledge among peoples – make food systems more resilient against climate change, pests, and pathogens. This is because the more diverse a food system and the more dynamic the global ecosystem, the higher the chance that any one species has a particular trait that enables it to adapt to a changing environment (and, in turn, pass that trait along).

This book contextualizes the high-stakes debates around seed and knowledge in sub-Saharan Africa within the context of debates regarding agroecology. Seeds are the source of life but also carry with them knowledge and culture.

Agroecology is essential to fulfilling the right to food, adapting to climate change, and increasing biodiversity. It is a science and a practice, the primary goal of which is to mimic ecological processes and biological interactions as much as possible in order to design production methods so that food producers' systems can generate their own soil fertility and protection from pests, and increase productivity. As an agricultural practice, agroecology is labour-intensive and encompasses a range of production techniques derived from local experience and expertise that draw on immediately available resources. Thus, it also relies heavily on experiential knowledge, more commonly described as traditional knowledge.

As a social movement, producer-based agroecology acts as an important driver for strengthening social cohesion through the gradual reduction of social inequalities, the promotion of local governance and sovereignty, and the empowerment of local communities. Studies continue to confirm that agroecological production can meet the global community's dietary needs and that on-farm biodiversity can lead to dietary diversity at the farm level and beyond.

The growing consensus is that communities all over the world should shift to agroecological practices in order to mitigate and adapt to climate change, while at the same time reversing biodiversity loss. Agroecology also deepens people's relationship with the land and each other.

This book is unique in that it focuses on the role of seeds in relation to agroecology. It brings in a rich array of experiences and knowledge, pointing to practical and theoretical ways forward.

Michael Fakhri
UN Special Rapporteur on the Right to Food

Preface

Beginnings

Sometimes, things are meant to come together.

In 2010, building on two decades of work with small-scale farmers in South Africa, a research and policy initiative on farmers' rights and seed security was launched. Comprising a partnership between my research group at the University of Cape Town (UCT) and the non-governmental organization (NGO) Biowatch South Africa, an organization that had recently emerged victorious from a nine-year battle against Monsanto and the South African state, it involved research with farmers in the province of KwaZulu-Natal on household seed systems, and an investigation into the policies and laws that constrained their ability to save, use, exchange, and sell their traditional seed.

Several research papers emerged from this process, as well as a policy brief, which formed the basis for a dynamic workshop held at Cape Town's Centre for the Book in 2012, with the participation of government departments, parliamentarians, farmers, NGOs, researchers, lawyers, and international guests. A proposal was developed to advance the research, centred on agroecology, reviving agricultural indigenous knowledge systems, and extending the collaboration to include other organizations working on seed in the region.

Around the same time as these ideas were unfolding, I took a call from a man with a strong Swiss accent. 'I'm interested to talk to you about seed and in supporting your work,' he said. 'I've read the policy brief you and your colleagues have written about farmers' rights. These issues are so important.' He introduced himself as Thor Maeder, the Deputy Regional Director of Cooperation of the Swiss Agency for Development and Cooperation (SDC). He and Juliane Ineichen ran what was then the Pretoria office of the SDC. They were at the time funding work on the harmonization of seed laws in Southern Africa, as well as other market-related seed projects, but it was clear that this approach was more helpful to the seed industry than to smallholder farmers. They were interested in supporting more progressive strategies that placed decision-making in the hands of farmers themselves. Fortuitously, we already had a draft proposal in hand!

Following the call I immediately contacted long-standing allies and friends in the agroecology movement: Rose Williams, the Director of Biowatch; Elfrieda Pschorn-Strauss, an activist who had worked in Biowatch, GRAIN, and the Mupo Foundation, all organizations involved in fighting for the rights of small-scale African farmers to maintain their traditional seed systems;

and Liz Hosken, leader of the Gaia Foundation, an organization that had its roots in supporting the rights of Indigenous peoples in Latin America, and had co-founded the Mupo Foundation (now EarthLore) to revive traditional knowledge in Venda, a largely rural area in the north of South Africa. A week later, the four of us met Thor and Juliane in the boardroom of the Department of Environmental and Geographical Science at UCT.

The inception of the Seed and Knowledge Initiative

By the close of this vibrant meeting we had reached an agreement to draft a proposal for a pilot phase aimed at supporting smallholder farmers in protecting and restoring their traditional seed and knowledge systems. If this was successful, the SDC would commit as a long-term funding partner. But there was a condition: while the 18-month pilot period could centre on work in South Africa, in the long term the work had to be regional, extending beyond South Africa's borders. A proposal was drafted, with three founding partners: UCT, Biowatch South Africa, and the Mupo Foundation, and by 2013 the so-called 'pilot phase' of what became the Seed and Knowledge Initiative was launched.

A series of live-in workshops were convened through 2014 with core members of these organizations, and allies, to think and dream together of an agricultural future that would restore the dignity and culture of smallholder farmers, respect plural knowledges, honour diversity, and restore the land, water, and habitats of Southern African agroeco-systems. In addition to the core group they included John Nzira from the Ukuvuna foundation, well known for its restorative work on agroecology; Mphatheleni Makaulule from the Venda-based Mupo Foundation; Method Gundidza, at the time the bookkeeper for Mupo and later the director of EarthLore; and Lawrence Mkhaliphi, the 'hands and heart' of Biowatch. Facilitator Davine Thaw shepherded the process from its outset (and continues to do so), with her gentle but firm mien, helping us to think and do our work in a way that gave meaning and energy. Others involved in these early days include Stephen Heyns, who supported the recording of meetings (and continues to do so), and Jaci van Niekerk, who provided research and writing support for the proposal and its implementation (and continues to do so).

We agreed that Biowatch should be the contracting party due to the development nature of the work, the NGO's ability to act swiftly, and its strong reputation as an organization supporting the interests of smallholder farmers and an agroecological future. The programme was dubbed the 'Seed and Knowledge Initiative', or SKI, pronounced 'sky' – not, we made clear, 'ski', the snow-dependent sport! And thus unfolded the remarkable partnership that has emerged over the past 10 years.

Extending the work as a regional initiative evolved naturally from partner-ships that were already well developed. By then, the African Biodiversity

Network was firmly established. Mupo and Biowatch were active members of the network, and a number of cross-border exchanges of knowledge and seed had commenced between South African and Zimbabwean farmers.

Centring research and policy in the Seed and Knowledge Initiative

As SKI grew, it was clear that in addition to the impactful and practical work on the ground across several Southern African countries, a body of research was emerging that told a different and often disregarded story about seed and the knowledges that journey with it. The research encompassed multiple shapes, ranging from long-term investigations by an active group of postgraduate scholars at UCT and elsewhere, through to policy research conducted by NGOs and allied organizations, and more experiential research carried out by farmers. Through bringing together these different strands, UCT's role in the SKI partnership was to deepen the research and policy discourse on seed and knowledge, with the aim of starting to shift the dominant narratives.

Under the umbrella of SKI, several multi-actor regional seminars were convened by UCT to communicate these experiences. Each taking place over a period of two or three days, the seminars brought together SKI partners, government officials, civil society organizations, seed specialists, postgraduate students, and well-known international activists and scholars. Several contributions in this volume arise from those seminars.

The first seminar was held in September 2014 at Mont Fleur, a conference resort nestled in the Helderberg mountains of Stellenbosch, and well known as a 'future-forging' space. From across the continent, a diverse array of voices interrogated knowledge exchange, collaboration, and innovation between the so-called formal and farmer-led seed systems and explored ways in which small-scale farmers' seed systems could be strengthened.

A follow-up seminar in November 2015, held beside the ocean at the Monkey Valley Resort in Cape Town, turned our focus to the research, education, and training needed to support farmer-led seed and knowledge systems in Southern Africa, again bringing together about 50 people representing a range of disciplines, interests, and actors.

In October 2016 a third seminar was organized, this time at the Salt Rock Hotel in Durban, and at KwaHhohho, a small settlement outside Mtubatuba in KwaZulu-Natal, which provided an opportunity for deeper exchanges with farmers. At the time, the region was experiencing one of the most severe droughts in recent memory and the challenges faced by farmers were accumulating and acute. The seminar thus focused on exploring ways in which farmers could maintain, restore, and strengthen local and diverse seed systems, especially in a context where the loss of seed had been dramatic.

In January 2019 – at the start of the next four-year phase of SKI – a far more ambitious gathering was organized in Cape Town, under the banner of the international Agroecology for the 21st Century Conference, which over three days brought together more than 250 participants from 12 countries, and a wide array of inputs. These ranged from researchers across the humanities and sciences through to food system activists and other members of civil society networks, farmers, practitioners, the private sector, policymakers, and those in the creative and performance arts. The location of the conference in the Company's Garden, alongside a public art exhibition, facilitated greater engagement by a wide range of publics, and focused spaces were created to ensure that the events shifted beyond an academic mode to strongly include the voices of farmers and civil society organizations.

The evolution of this book

It was also at the conference that this book was officially 'born'. After the proceedings, potential authors came together to discuss the book's form and structure, with the intention of designing a contribution that unequivocally conveyed African perspectives on agroecology, and included the voices of African farmers, activists, scientists, scholars, and policymakers. Several young scholars from the region, many linked to SKI through their postgraduate research at UCT, were also invited to contribute, thus gaining a platform to publicize their findings. COVID-19 put paid to plans for an authors' meeting the following year, but once chapters were complete, a final two-day event in November 2021 had authors assembling once again at Monkey Valley, to present their research, discuss their findings, and enable a collective analysis of their work. It was the first in-person event for many of us, after two years of barren Zoom calls, and the discussions were impassioned, inspiring, and vibrant, embracing a crystal-clear imperative for systemic change and a shift to agroecology.

It is hard to convey the richness and inspiration that have infused this process. But it simply would not have been possible without the support of many, many people. I am especially grateful for the assistance of staff within the Biowatch and SKI teams – in particular Rose Williams, Vanessa Black, Lawrence Mkhaliphi, Des Pelser, Blessing Zama, Fanie Nothnagel, Stephanie Aubin, Elfrieda Pschorn-Strauss, Caili Forrest, Lindy Morrison, Ruby Essack, Pumla Mabizela, and the late Nick Molver. The active and enthusiastic participation of SKI partners and others who contributed to the regional seminars and agroecology conference provided both the inspiration and the material for much of the book. Although you are too numerous to mention by name, I thank you all!

At UCT, I am hugely appreciative of Jaci van Niekerk, who helped to support authors, keep the book on track and finesse drafts, and Fahdelah Hartley, who with her usual aplomb and supreme efficiency organized flights, accommodation, and conference logistics. Shirley Pendelbury is gratefully acknowledged for nurturing the writing spaces that helped bring the book to fruition. The delightful UCT bio-economy postgraduate student group that accompanied the process was always a source of respite and motivation, and I am very grateful to Witness Kozanayi, Kudzai Kusena, Bulisani Ncube,

Maya Marshak, Jennifer Whittingham, Morgan Lee, Tsekiso Ranqhai, and Mpho Kganyago for the energy and enthusiasm they have injected into this book. At home in St James, my family – Carl van der Lingen, Art and Mia – indulged my frequent times away and never-ending deadlines (sometimes with exasperation), and I warmly acknowledge their unstinting support.

The copy-editing process has been an easy ride due to the meticulous and painstaking work of 'never miss a beat' Paul Wise, who kept an astonishingly level head, despite the protracted nature of the process. Within Practical Action Publishing, I thank Chloe Callan-Foster, Rosanna Denning, Jutta Mackwell, Clare Tawney, and Jenny Peebles for bearing with me through extended deadlines, and for their enthusiasm in seeing the book to print. Patrick Mulvaney is warmly acknowledged for his detailed review of the book, and for the helpful suggestions he offered for its improvement.

Particular thanks are owed to the SDC, which has played a pivotal role in enabling SKI's success, due in no small part to their commitment to a long-term programme. Within the SDC, warm thanks are due especially to Bulisani Ncube, who has accompanied SKI since its outset, going over and beyond his responsibilities. In addition to SDC, Brot für die Welt and SwedBio are graciously acknowledged for supporting the regional seminars and agroecology conference and for their ongoing and generous support of SKI. The National Research Foundation (NRF) of South Africa (Grant 84429) is also thanked for its generous funding of my Bio-economy Research Chair and associated students.

A final word of thanks is due to the many farming communities across Southern Africa who have journeyed with SKI and its associated research initiatives, and who continue to bring into being the seed, soil, and nourishment that give us life.

Rachel Wynberg, Cape Town, January, 2024

CHAPTER 1
Introduction

Rachel Wynberg

> *The seed is mine. The ploughshares are mine.*
> *The span of oxen is mine. Everything is mine.*
> *Only the land is theirs.*
>
> —Kas Maine
>
> In *The Seed is Mine: The Life of Kas Maine, A South African Sharecropper, 1894–1985* (Charles van Onselen, 1996)

Seed embodies life, power, and culture. From Africa's deserts and drylands to its mighty river systems and tropical forests, from those growing a multiplicity of grains, legumes, and vegetables, to others struggling to produce enough to feed their families, seed provides the mainstay for each one of the continent's 500 million small-scale farmers. Beyond food, seed is at the heart of rich and varied cultures in Africa, accompanying brides on their nuptial journeys, delivering ancestral blessings for good fortune, and enabling the bonds for greater social cohesion and harmony.

But seed is under siege. As the world's food and agricultural systems become increasingly industrialized, homogenized, and privatized, seed has become something of a poster child for the struggles involved. It is also symbolic of the deep injustices that have emerged through years of colonization and exploitation. These range from the policies and laws that have for decades propped up the interests of commercial farmers and multinational seed and agrochemical companies – at severe cost to the environment, climate, and small-scale farmers – through to the new wave of philanthropy that is sweeping the continent, promoting Green Revolution approaches of genetic modification and quick-fix nutritionism as a remedy for the poor, despite their failure elsewhere in the world (Patel et al., 2015). Access to seed and control over its ownership are at the core of the conflict, overlaid by a context where more than 65 per cent of the world's commercial seed is owned by six corporations and some 20 million hectares are under land grabs (Batterbury and Ndi, 2018). While these trends are well described and documented for agriculture around the world, there is a missing narrative for Africa, whose seemingly 'unproductive' lands are now viewed as the last frontier for agribusiness. At the same time, there remains little documented about the resilience of local seed systems, and about the innovative approaches being adopted by small-scale farmers across the continent to retain agrobiodiversity, while pursuing agroecological approaches to farming that not only produce sufficient food but also eliminate harmful inputs. The call to decolonize knowledge systems forms an integral part of the picture. New ways of seeing and doing research are evolving, moving from extractive to facilitative approaches. Western, scientific, and traditional knowledges are beginning to mingle in transformative ways, and inspiring pioneers in the formal structures of government and research institutions are beginning to show that another way is possible. Social movements, long silent in Africa, are emerging as a powerful force for change, alongside the range of NGOs that provide support to farmers at different levels.

This book aims not only to provide critical perspectives about the onslaught on seed and knowledge in sub-Saharan Africa, but also to demonstrate the viability and necessity of agroecological systems that are diverse, nutritious, and environmentally sustainable. Uniquely, the book offers a contribution that is enriched by the collaborative, creative, and critical voices of African farmers, activists, scientists, scholars, and policymakers. Their viewpoints combine in this volume to articulate a shared and dynamic vision of a world where agriculture is productive, diverse, and sustainable; where different ways of seeing and knowing are respected; and where seed and food systems are in the hands of farmers and local communities.

As a state-of-the-art collection, the book consolidates an array of experiences, analyses, and critical perspectives to help inspire and shape the science and practice of agroecology, and the social movements that support it. As a tool for advocacy, it gives impetus to the growing disquiet about existing models of ownership and production for food and seed, and provides suggestions for more sustainable, healthier, and fairer futures. As an approach for working together, it provides an opportunity for reflection about the challenges and opportunities of cross-organizational and cross-sectoral collaboration. Taken as a whole, the importance of the book is underpinned by the urgency of shifting the discourse about agriculture, seed, and knowledge, before our options are completely foreclosed.

The book is divided into four main parts. We first lay the groundwork by describing the remarkable diversity and resilience of Africa's seed heritage, especially in the context of rapid environmental change. After that, a set of contributions explains how African seed and knowledge systems are under threat – driven by the conversion of biodiverse, functioning landscapes to industrial wastelands, the profit-seeking interests of seed, agrochemical, and food companies, and the inadequacies of resource-poor and under-capacitated governments, seduced by capital and the dream of 'development'. Part 3 reflects on the potential of different ways of seeing, knowing, and learning, and finally Part 4 foregrounds the transformational potential of agroecology, and the critical role of civil society movements in enabling such transitions, concluding with a set of recommendations for decision-makers.

Seed, resilience, and diversity

The story of Fakazile Mthethwa, known affectionately to many as 'Gogo Qho', is a fitting opening to the book (Box A). Mvuselelo Ngcoya describes how Gogo Qho, until the last days of her life, 'befriended and tended indigenous plants and trees, frustrated her neighbours with her insistence on ecological stewardship, and inspired thousands of visitors with her commitment to agroecological farming methods'. Living in the hills of KwaBhoboza in northern KwaZulu-Natal, South Africa, Gogo Qho was a passionate advocate of agroecology, and a fearless critic of the industrial food system. 'How can you claim to be free if you are not in control of what is on your plate?' she asks Mvuselelo, exclaiming about both the feasibility and necessity of agroecology as a reparative future.

The undeniable necessity of this future is elaborated in Chapter 2, in which Elfrieda Pschorn-Strauss takes us on an evocative journey across Southern Africa, relating vividly how seed is valued by small-scale African farmers, not only as nourishment, but also in the celebration of culture and the reinforcement of kinship and social relations, and in supporting agrobiodiversity and livelihoods. From the imposing mountains of Chimanimani to the mist-laden forests of Venda, she reveals the reverence in which seed is held, but also the shared stories of loss and fragmentation, and of how industrial agriculture and landscape transformations have not only destroyed ecosystems and habitats but also triggered an epistemicide of the knowledge systems that sustain biocultural diversity. The violence against indigenous knowledge and the natural world has had cascading negative impacts on community cohesion, identity, farmer autonomy, livelihoods, landscapes, and the overall resilience of communities. Even so, farmers have continued to practise multiple strategies to safeguard their seed, ranging from innovative storage methods and relationally complex networks of exchange, through to the maintenance of agrobiodiversity in their fields. Food and seed sovereignty have emerged as a strong counter-paradigm to the ecological damage and deep inequities created by the industrial agricultural system, but, as Elfrieda Pschorn-Strauss explains, fulfilling the 'emancipatory potential' of these paradigms will require active support. She proposes the development of transformative learning opportunities where people develop the agency to ask critical questions and the capability to collectively find creative solutions.

The chapter engages with the question of what it would take to transform agriculture and seed systems to another paradigm and suggests that applying resilience thinking to communities and their seed systems could help to deepen appreciation of the qualities and elements that support their vitality. Drawn from Brown (2016), three themes are proposed to understand the lived experiences of resilience at a community level. *Resistance* places local agency over seed systems at the centre, by challenging the institutions and laws that create enclosures and restrict community seed practices. *Rootedness* connects people with place and meaning, invoking stewardship and connectedness through dialogue, reinvigorating customary systems of governance, and actively restoring sacred sites and traditional crop varieties and farming practices. It is about 'bringing back the meaning of seed from being a "thing", a commodity, to being a set of relationships with people and the earth that are fundamental for creating and sustaining the seed commons for the future'. *Resourcefulness* refers to the way in which resources are accessed, but is also about the capacity to use resources at the right time, and in an appropriate way, and to make the right decisions about when and how to share seed, for example. It is about equitable sharing, stewardship, and inclusive knowledge.

The potential of agroecology for resilience is a theme that carries through to Chapter 3, where Witness Kozanayi and Jaci van Niekerk recount the devastation caused by Tropical Cyclone Idai, which tore through eastern Zimbabwe in March 2019, affecting 80 per cent of the arable land in the Chimanimani district. The authors describe how conventional farming lands were severely damaged by the cyclone, whereas agroecologically managed land fared much better, largely due to the use of practices such as crop cover, mulching, and soil conservation works. Areas managed in this way were less eroded than overstocked grazing areas or those where trees had been removed. A fascinating account is provided of how local people drew on their social networks as a part of the recovery process, sharing resources to rehabilitate damaged landscapes and infrastructure and restore lost seed. Importantly, seed sharing and exchange became even stronger post-Idai, as households developed solidarity networks and strategies to rely on each other rather than on government, the private sector, and external aid, which often delivered inappropriate, expensive, and untimely seed. The chapter underscores the importance of implementing agroecology at a landscape level, especially in the face of unpredictable climate patterns, yet is also candid about the challenges of landscape governance in a context of competing land uses, multiple institutional layers, and political interference.

The histories, contestations, politics, and ironies of maize emerge repeatedly across the book, with its centrality in African agricultural landscapes foregrounded by the late Paul Hebinck and Richard Kiaka in Chapter 4, concerning maize in West Kenya. Although maize is not indigenous to Africa, it has become indigenized over time and today forms an important food staple for many people on the continent. Elfrieda Pschorn-Strauss refers to maize as a crop that both 'feeds and robs' Africa, a portrayal that is given conceptual life in Chapter 4, which refers both to its significance as a food and to its associations with colonization, slavery, and agrobiodiversity loss, as well as to its control, in recent decades, by agribusiness and the subsequent loss of autonomy by farmers.

As in many other African countries, both 'local' (or traditional *Nyaluo* or *Lilimini*) and 'modern' (or hybrid) varieties are grown and consumed in Kenya, often in combination. Paul Hebinck and Richard Kiaka describe how these different varieties are associated with the 'enactment' of two structurally different agrarian 'revolutions' in West Kenya: an indigenous agricultural revolution based on local varieties that inspires a form of agroecology by default, and the Green Revolution, which promotes modern farming with commoditized inputs such as hybrid seed, aiming to replace and displace local varieties. Through exploring the impact of both revolutions on the lived experiences of farmers in Luoland and Luhyaland, they show that farmers continuously shift between autonomous farming and commoditized forms of farming and make multi-dimensional choices based on their interpretations of risk and changing circumstances.

Drawing on longitudinal research beginning in 1996, they argue that the enactment of maize farming defies many of the 'received wisdoms' claiming that, for instance, rural poverty prevents people from enacting modern farming, or that local maize is unproductive and exacerbates poverty and food insecurity. Farmers consistently cited their preference for *Nyaluo* due to its early maturation, drought and pest resistance, and ability to tolerate poor soils and resist weeds. *Nyaluo* was also widely preferred for its taste and nutritional value and because it is more filling than hybrid maize. The claim that modern maize is higher yielding when compared to *Nyaluo* or *Lilimini* was vehemently contested by those who plant and consume these maizes. For them, the nutritious qualities of *Nyaluo* outperformed hybrids and provided more energy. Some farmers appreciated the hybrids because of their higher yield capacity, but cultivating them required much higher monetary inputs than *Nyaluo*. Moreover, hybrid maize was perceived as a *Nyareta* (strange seed) and, unlike local maize, did not easily become family seed for saving and exchange. Echoing findings across different contexts and countries (see, for example, Chapters 3, 5, and 6, which present research from Zimbabwe, Malawi, and Ghana, as well as Box C on Zimbabwe), the authors conclude that maize landscapes in West Kenya are consistently misread by experts from seed companies, donors, state extension services, and NGOs. Such misconceptions arise from pre-conceived notions that farming is or should be only a commercial and entrepreneurial venture, and a persistent broadcasting of the message that *Nyaluo* is unproductive and aggravates poverty.

These fallacies are not unique to Kenya. Across Africa, maize has long been promoted as a driver of modernization to propel development and foreign investment (Bezner Kerr, 2014). In Malawi, Zambia, and elsewhere, the state has essentially sponsored multinational agrochemical and seed companies by introducing subsidy programmes for synthetic fertilizer and hybrid maize seed. However, despite overwhelming policy support for the formal seed sector, many farmers continue to save seed. In Chapter 5, Noelle LaDue, Sidney Madsen, Rachel Bezner Kerr, Esther Lupafya, Laifolo Dakishoni and Lizzie Shumba present findings based on a collaboration between Soils, Food and Healthy Communities (SFHC), Cornell University in the United States, and Western University in Canada. The research seeks to understand the challenges that smallholder farmers in the Northern and Central regions of Malawi face in obtaining seed, the multiple pathways they use to access seed, and the implications of these strategies for food sovereignty.

Drawing on 60 semi-structured interviews they describe five major ways that farmers access seed: seed saving, seed purchasing, seed sharing, *dimba* (dry-season cultivation), and casual day-labour employment, known as *ganyu*. Few of the farmers sourced seeds through only one pathway; most instead utilized a combination of methods. Seed saving was a preferred sourcing method, and along with seed sharing and exchange was also the approach that best maintained crop diversity. Farmers without saved seed often purchased seed through earnings from *ganyu*, although this expense reduced a household's budget for meeting other needs. *Dimba* was not an option for all farmers due to limitations of land and water access. A central finding is that the seed sourcing pathways available to farmers are shaped by their social, economic, and political contexts, and that different pathways reflect varying levels of agency. Certain seed sourcing pathways increased the risk of food insecurity, a finding that supports the Food Sovereignty movement's assertion that farmers' well-being is shaped by the conditions under which they access seeds.

Jaci van Niekerk, writing about the experiences of farmers in the north-west reaches of KwaZulu-Natal, explains how seed saving also fosters preparedness and independence (Box B). 'When the rains come, I take my seeds and plant them,' explains a small-scale woman farmer in Ingwavuma, underscoring the importance of timely crop production, but also autonomy and self-sufficiency, both vital attributes when farming in resource-limited conditions. Women farmers recounted how seed saving instilled knowledge about the provenance of the seed, and afforded them dignity, respect, and status. The incursion of cash crops was rapidly displacing the diverse, local food crops they once grew, also undermining women's roles as seed custodians. Being seed secure translated into enhanced food security, in turn fostering social cohesion.

The relationship between seed security and food security is examined further by Bulisani Ncube in Box C. His research in the Chimanimani district of eastern Zimbabwe reveals how the link between seed and food security is influenced by a range of factors, including the broader socio-economic and political contexts within which farmers are located. He centres his analysis on the classical definition of seed security: when farmers have 'sufficient access to quantities of available good quality seed and planting materials of preferred crop varieties at all times in both good and bad cropping seasons' (FAO, 2016). Having examined elements of availability, access, and quality, he concludes that seed security does not necessarily equate to food security. This finding runs counter to the assumptions of development actors who provide seed to ensure improved food access. It also suggests that the provision of seed aid might be counterproductive if it disrupts farmers' local seed systems, which play a dominant role in these farming communities.

The concept of seed sovereignty is brought centre stage in Chapter 6, in which Hanson Nyantakyi-Frimpong and Audrey Carlson critically examine the politics of building seed systems resilient to climate change. Seed sovereignty, like seed security, focuses on seed availability and ready access to seeds, but with the critical distinction of reclaiming seeds and biodiversity as a common good, and including the rights of farmers to save, breed, sell, and exchange their own seeds. The chapter presents long-term research conducted over seven years in Ghana's Upper West Region, an area in which many farmers have lost their farmlands due to land grabs, mainly for gold mining. Neoliberal reforms in Ghana have reduced state involvement in smallholder agriculture development, leading to the involvement of a range of local and international NGOs, many linked to the Gates and Rockefeller-funded Alliance for a Green Revolution in Africa (AGRA) (see also Chapter 8). A suite of 'climate-smart' technological interventions has followed, aimed at raising the productivity of smallholder farmers.

Three key crops – maize, beans, and groundnuts – are considered in the research, which investigated perceptions of climate variability and changes in seed selection and exchange practices from 2012 to 2019. Findings show a dramatic decrease in farmers' seed selection and exchange practices, due largely to the substitution of local seed varieties with hybrids introduced under the banner of climate-smart agriculture. While local seed systems are often well adapted to climate change, such interventions are, ironically, radically transforming them. The authors point to the exclusion of farmers and associated indigenous knowledge in the design of climate-smart agricultural solutions, and suggest that existing market-based approaches centred on costly external inputs and hybrid seed reproduce and aggravate existing class and gender inequalities. They conclude that the ongoing transformation of local seed systems directly undermines seed sovereignty and contradicts a transformational agenda based on agrobiodiversity, local resources, and indigenous knowledge.

South Africa is a country that bears witness to many of these transformations. It is also a nation that is well known for its highly promotional approach to genetically modified (GM) crops. Not only was it the first African country to commercialize GM crops, but it was also the first in the world to produce a GM subsistence crop and staple food: insect-resistant white maize. GM crops are deeply embedded in the rhetoric that, by analogy with the 'Green Revolution' of the 1960s and 1970s, a new 'Gene Revolution' is needed to save African agriculture. However, despite promises of higher yields, greater economic gains, and improved food security, increasing evidence shows that the first generation of GM crops has failed to meet the needs of African smallholders (Schnurr, 2019). Nonetheless, pressure for African governments to adopt GM crops has been relentless.

In many ways, South African experiences embody the development paradigm forecast for the rest of the continent, and learning from them provides important insights for other African countries where GM crops have not yet been commercially introduced. In Chapter 7, Rachel Wynberg and Angelika Hilbeck consider the dualism and contradictions of maize as both a subsistence and a GM commodity crop and analyse the implications for smallholder farmers of the contamination of their fields and seed systems by transgenic gene flow. The repercussions of contamination are especially profound for farmers who wish to pursue agroecology farming, raising questions about farmers' rights to choose what they plant and eat, the potential legal consequences of contamination, and impacts on agrobiodiversity, food security, and farming practices such as seed saving and seed selection.

Transgenic contamination occurs through multiple pathways, all usually beyond the farmers' knowledge and capacity to detect, and is widely reported in South Africa and other countries. However, few studies explore the social, cultural, and psychological implications for smallholder farmers. In this chapter, the authors describe a process in KwaZulu-Natal that brought together farmers, scientists, gene-bank managers, and NGOs from across the region to travel together to help farmers understand the pathways of contamination so as to manage the impacts and reduce negative effects. They narrate the deep trauma experienced by farmers on learning that their maize had tested positive for GM protein, but also the practical and generative ideas that emerged from farmers to manage contamination. The authors conclude that with support, farmer-led strategies might help not only to mitigate the negative impacts of GM contamination, but also to strengthen the agency of farmers and their communities to secure productive and healthy food systems.

The dramatic loss of agrobiodiversity across African landscapes has profound implications for farming communities. The last two contributions in Part 1 centre on the role of gene banks in supporting smallholder farmers' efforts to reintroduce traditional varieties and strengthen on-farm conservation. In the first seminar convened by the Seed and Knowledge Initiative in 2014, the late Dr Melaku Worede presented a keynote address titled 'Putting farmers first', with reference to his approach: that of a farmer-led system which values farmers' knowledge and practices and uses genetic science to support small-scale farmers, rather than imposing Western science on them (Box D). Together with Dr Regassa Feyisa, Dr Worede was responsible for establishing the first African gene bank, in Addis Ababa, Ethiopia, and is renowned for his contribution to the conservation of agricultural biodiversity and for helping to restore food security in Ethiopia after years of drought-induced famine. Small-scale farmers, he observes, 'are the original "plant breeders" as they employed selective breeding to raise yield, improve quality, and promote diversity long before formal plant breeding became an established discipline'.

In Box E, Kudzai Kusena, the former curator of the Zimbabwean gene bank, and Jaci van Niekerk, set out the history, objectives, and activities of gene banks, providing a useful overview of the different types that exist. Despite the potential for gene banks to link with smallholder farmers, they explain, in practice their design and approach are typically more useful to plant breeders. Challenges that constrain interactions with smallholder farmers include high levels of bureaucracy and cost constraints, the genetic quality of collections, the low quantities of seed available, a focus on elite germplasm, and the 'freezing' of evolutionary pressures that occurs in collections. Nonetheless, as experiences with Cyclone Idai reveal (Chapter 3), gene banks have played an important role in times of disaster in restoring and rebuilding lost crop diversity.

Both contributions make clear the need for a reimagining of gene banks and their role and place in supporting African smallholders and restoring agrobiodiversity. Central to this rethinking is the role of community seed banks and approaches such as farmer field schools and participatory plant breeding, which form vital components of an integrated strategy for agrobio-diversity conservation and use, in addition to their role in supporting food and nutrition security (Andersen et al., 2022).

Privatizing profit, socializing cost

Part 2 of the book, titled 'Privatizing profit, socializing cost', turns towards the drivers of agrobiodiversity loss and epistemicide. At a global level, Stephen Greenberg (Chapter 8) describes how commercial seed markets morphed from a base of small-scale businesses to large-scale, multinational corporations that integrated biotechnology, agrochemicals, and seed. Their shareholders today include global financial firms such as BlackRock which, together with other asset management firms, control the global biotech-seed complex. These trends, combined with the extension of intellectual property rights to living organisms in the 1980s, have set the stage for wealth extraction based on proprietary rights, rather than the resource itself, and the subsequent enclosure of seed and knowledge for private gain (Kloppenburg, 2004). As Jason Moore (2017) describes, the cheapening of land and natural resources, labour, food, and energy (the 'Four Cheaps') has produced a 'Cheap Nature' that systematically denigrates the knowledge and resources of smallholder farmers and local communities as backward and obsolete.

Stephen Greenberg takes us back in history to explain how such trends coalesced with structural adjustment in Africa in the 1980s and 1990s, resulting in the systematic dismantling of public sector support for agriculture and an embracing of market-led approaches to agricultural development. In the seed sector, this included the privatization of formerly state-owned seed companies and the subordination of public plant breeding to commercial imperatives, with a focus on commodity crops such as maize and soya. Such efforts have accelerated with the renewed push to modernize African agriculture, sponsored by private philanthropic institutions such as AGRA (Wise, 2020; Vicedom and Wynberg, 2023). Through introducing improved hybrid seeds, agrochemicals, and linkages to markets, this 'new' African Green Revolution aims to bring to Africa the high crop yields and technological innovations experienced in Asia and Latin America in the 1960s and 1970s (Ignatova, 2017). Such efforts align strongly with the development aid-funded Comprehensive Africa Agriculture Development Programme (CAADP) under the New Partnership for Africa's Development (NEPAD), which recentres agriculture as the main endeavour for African economic development. Here, the focus is on developing strategic partnerships, especially with the private sector; harmonizing policies, including seed; and investment in agricultural research and technology dissemination. A similar model of leveraging private-sector investment to enhance agricultural yields is adopted by the G8 and USAID-supported 'New Alliance for Food Security and Nutrition', which aims to invest over US$10 bn in African agriculture (Schnurr, 2019).

As Greenberg explains, the social, economic, and ecological implications of adopting this approach are stark. Despite the diversity of crops that farmers cultivate, just a handful are targeted for investment, motivated primarily by profit. Despite wide recognition of the need for fundamental change to agricultural production, the business-as-usual model continues to promote large-scale monocultures and the associated package of hybrid and GM seed, synthetic fertilizers, and agrochemicals. Because 'financial power purchases political power', state systems and policies themselves are increasingly captured and privatized, characterized by the seemingly endemic norms of corruption, deceit, and secrecy.

Nowhere on the continent is this matter more relevant than in South Africa, a country that is heavily invested in industrial agriculture and which, as David Fig reveals in Chapter 9, exemplifies corporate capture, or 'state capture' as it has come to be known in the country. He reminds us that corporate capture of the agrifood sphere can be traced back to colonial rule in the 17th century, when a Dutch mercantile corporation took control of the Cape of Good Hope to provide the food needs of passing ships. From the 1860s, food production began to serve emerging mining and other corporate interests more directly. The colonial expropriation of arable land, combined with forced urbanization and the application of new technologies, enabled large family-owned and corporate entities to dominate, and later completely shape, food production and consumption. This was always at the expense of smaller farmers, especially those forced to submit their labour power to the extractive economy.

The chapter describes how pro-corporate laws and policies have continued to prevail, and how they have failed spectacularly to address hunger, health security, and environmental protection in South Africa. A quarter of the nation's population go hungry every day, and half are at risk of hunger, a situation that was exacerbated by the COVID-19 pandemic. The country is witness to epidemics of obesity, heart disease, diabetes, and other non-communicable diseases. Agricultural lands and watercourses are heavily contaminated by pesticides, herbicides, and chemical fertilizers, and are regulated by laws and policies that favour the agrochemical and seed industries. The logical conclusion is that policymakers have succumbed to pressures from the transnational and local corporations that dominate the agrifood value chain, and to the lure of philanthropic bodies such as AGRA which promote an industrial model favouring high inputs and large-scale farming.

A central actor in the agrifood complex is the agrochemical industry. Industrial agriculture is increasingly dependent on agrochemicals, largely due to structural transformations such as decreased innovation, increased regulatory costs, industry consolidation, and a shift to generic chemical products (Shattuck, 2021). The profitability of the industry is astounding. About 3.5 billion kilograms of agrochemicals are sprayed every year across the globe, generating a market value of some $215 bn in 2016. In Chapter 10, Morgan Lee takes a deeper look at the agrochemical industry in South Africa, and the slow and structural violence that it elicits. South Africa accounts for about 2 per cent of global agrochemical use (Handford et al., 2015), with an estimated 4,500 agrochemical products registered for use in the country, several of which are banned elsewhere (Clausing et al., 2020). While many countries of the global North have embarked on processes and policies to reduce agrochemical usage, their use continues to expand in South Africa. Poor capacities for testing and monitoring, combined with fragmented, outdated, and ineffective laws, present a concerning picture, both for South Africa and other African countries which have considerably less capacity and regulation.

Using the herbicide glyphosate as a lens, Morgan Lee recounts how the South African agrochemical regulatory arena is deeply flawed, resulting in harm which is significantly underestimated. Glyphosate is the world's most utilized herbicide and is one of the top five agrochemicals used in Africa, largely by countries producing cotton and GM herbicide-tolerant crops. Through analysing 'chemical geographies' of violence, she explains how human exposure to glyphosate and its persistence in the environment constitute a slow violence: harm that occurs gradually, while largely out of sight (Nixon, 2011). Over time, glyphosate affects the well-being of farmworkers and farming communities and those ingesting glyphosate residues in food. The use of glyphosate also inflicts slow environmental violence through weed resistance, soil and water contamination, and biodiversity loss.

In a similar way, the failure of government to safeguard human and environmental health is seen as a form of structural violence: harm that is concealed, ingrained, and institutionalized beyond recognition. Based on Mbembe's conceptualization of 'necropower' (2003), the 'violent inaction' of regulatory structures may not be 'making people die' but is most certainly 'letting people die'. A 'pesticide culture' of non-interference has emerged in South Africa, with a seemingly complicit relationship between government and industry and a 'wall of silence' that precludes effective public participation. Addressing regulatory failure and acknowledging the different forms of harm agrochemicals cause are clearly critical to safeguard biodiversity and human health.

Ways of seeing, knowing, being, and learning

Cultures of violence extend beyond those epitomized by the agrochemical industry and are embedded in the exclusions and epistemicide that accompanied the colonization of the African continent. Part 3 reflects the impacts of this loss for smallholder farming communities, whose knowledges and skills have been silenced over generations; on cognitive justice; and on the importance of recognizing other ways of seeing, knowing, being, and learning.

In Chapter 11, Vanessa Farr reminds us of how women's caregiving, agricultural, and food gathering practices, knowledge, and activities have been rendered unimportant and unmeasurable, with localized knowledge systems 'routinely sacrificed as a homogeneous agricultural world order is imposed', brought about through settler expansion and the forced relocation of communities to inferior soils. Drawing on a range of visionary African feminists, from Wangari Maathai back to Olive Schreiner, she explains how land and labour mechanisms for leaving women behind were imposed. Agriculture was designed for and about the needs of elite men, requiring, as Whittingham et al. (Chapter 12) describe, the assimilation of 'other ways of knowing and the knowers themselves'. We are taken back to the interview with Gogo Qho (Box A) and how this astonishing woman refused the 'colonially imposed disassociation of intellect from soul and soil'. These stories, and those of Skywoman, who scattered her handful of seeds across Turtle Island and created a garden for the well-being of all life, leave us hopeful about the possible futures we can create.

Using the lens of GM crops in South Africa, Jennifer Whittingham, Maya Marshak, and Haidee Swanby demonstrate in Chapter 12 how the 'political and ideological machinery' that supports these crops has been aggressively promoted and how other ways of knowing have been pushed to the peripheries of decision-making processes which recognize only the 'one-world-world'. GM crops, they argue, cannot be understood independently of their political, economic, and social-cultural context. This chapter suggests that the hegemony of Science,[1] and its historical entwinement with politics and economic growth, has been key in engineering a landscape receptive to GM crops. The authors interrogate the centrality and neutrality of Science-based risk assessment that has accompanied and enabled GM crops in social-agricultural landscapes around the globe, noting that the process privileges its own set of values of objectivity, profitability, and efficiency. While scholars and activists have called for the inclusion of a wider range of concerns in risk assessment, ranging from social-cultural, political, and social-economic to eco-toxicological and social-ecological, the authors articulate the need to move beyond this framing and to challenge the worldviews that inform and legitimize risk assessment and the decision-making processes they permit. In doing this, they suggest the importance of including a more diverse set of knowledge systems and ontologies enabling of more appropriate and equitable approaches to imagining and co-creating agroecological futures.

So many of the challenges we face in our agrifood systems have their roots in the monolithic approaches that have been introduced under the guise of 'development'. Mugove Walter Nyika (Box F) recounts how his grandmother farmed in south-central Zimbabwe and grew an array of local crops and fruit trees. This all changed when the government extension officer arrived and advised farmers to remove all trees from their arable land and plough it uniformly. Twenty years later, his grandmother had become a 'modern farmer' who purchased chemical inputs and mostly grew just maize. He recalls that 'things seemed only to get worse' as the transformation was accompanied by deforestation, soil erosion, siltation, dependency on external inputs, and malnutrition.

A key problem that blocks the adoption of more transformative and sustainable agroecological approaches is the nature of the education and training received by agricultural extension officers. Curriculum change is a critical component of this transformation but is not a simple task. These challenges are laid bare in a case study narrated by Vanessa Black of the South African NGO Biowatch (Box G), who describes a pilot intervention by Biowatch at the Owen Sitole College of Agriculture in KwaZulu-Natal, an institution which trains many of the agricultural extension officers that work in the area. Despite the presence of significant numbers of smallholder farmers in the province, much of the training is directed towards commercial monoculture production, with little emphasis on food and nutrition security. Extension officers thus lack the necessary community development skills and knowledge required to support complex and diverse smallholder systems. A new extension policy recognizes these gaps, but the focus on commodity value chains and Green Revolution inputs continues to dominate.

The pilot course was designed to teach the elements of agroecology and demonstrate its effectiveness. Although the course was carefully designed, with wide consultation, the case study describes the trials of integrating a holistic, interdisciplinary agroecological course into a predetermined curriculum that is structured as discrete courses. Some lecturers and students were alienated by the ideological approach adopted, which advocated for agroecology, with important lessons emerging that emphasize the need to centre the course on experiential learning, dialogue, and self-reflection. Initial scepticism of agroecology was countered to some extent by the establishment of a thriving demonstration plot that produced abundant, chemical-free vegetables and herbs, dispelling the notion that farming could only be undertaken as a large-scale commercial enterprise.

Across the border in Zimbabwe, Shepherd Mudzingwa and Jaci van Niekerk paint a similar picture (Box H). As with the extension system in South Africa, extension staff are equipped with skills to support industrial agriculture, and there is limited capacity in and knowledge of ecologically sensitive methods of production. In response, the Fambidzanai Permaculture Centre, in collaboration with Bindura University of Science Education, set out to develop a Diploma in Agroecology, aiming to train officials in an approach that resonates with the practices and needs of local communities. This successful and inspiring programme has practical, hands-on, and experiential learning at its core. As part of the learning, students must work closely with farming communities to identify key challenges, and together come up with potential solutions. In this process, Fambidzanai and Bindura

University have developed a working model for strengthening extension training which can be scaled up across the region and beyond, and which, through its partnership, also brings quality assurance and compliance with recognized standards.

The closing contribution of Part 3 (Box I) affirms the transformative power of collectively learning, doing, and being together in advancing agroecology and seed sovereignty. Elfrieda Pschorn-Strauss sketches the history of a community of practice (CoP) started by partners in the Seed and Knowledge Initiative to fill the 'knowledge and confidence gap created by years of policies and agricultural extension that devalued community seed systems and farmers' know-how'. A learning-by-doing approach was adopted that included exchange visits, trainings, and practical experiences, also allowing space to think more deeply, cross boundaries, and self-organize as needs arose. The CoP led not only to practical outcomes, such as the uptake of small grains and a strengthened focus on soil health and landscape restoration, but, importantly, also to relationship building, leadership development, and personal growth, providing a heartening example of the possible futures we can effect.

Transitioning towards agroecology: working together and moving the struggle forward

Part 4 of the book is a promising look to an agroecological future where smallholder seed and knowledge systems are supported, strengthened, expanded, and protected, providing the foundation for a food and agricultural revolution that benefits people and planet. In Chapter 13, Mvuselelo Ngcoya takes us on an exciting journey to Cuba, whose farmer-centred and farmer-driven sustainable agriculture model has fundamentally transformed the country's food system. Yet this was not always so. Drawing comparisons with South Africa, he describes how Cuba until recently had an unworkable land model that denied large portions of society ownership of land and security of tenure. The country's agricultural sector was also characterized by conventional, high-input industrial farming methods, dependent on mechanization, agrochemicals, and a formal seed management system run and regulated by the state to benefit large-scale industrial agriculture.

However, the collapse of the socialist bloc in 1990 left Cuba facing unprecedented economic challenges that derailed its agricultural sector. The 'special period', as that epoch was called, decimated the country's seed production capacity for common crops. However, it also opened new spaces for innovation, such as a participatory seed production, improvement, and dissemination programme, the Local Agricultural Innovation Programme (PIAL), which placed greater control over seed production, management, and distribution in the hands of farmers themselves, stimulating the conversion to a low-input, farmer-focused, sustainable farming system. Researchers and farmers began to collaborate: 'it was no longer the research institute extending itself to the farm, but the farmer extending himself to the research institution'. Initiated in 1999 by a small group of radical researchers, by 2012 the network had extended to more than 50,000 farmers working with 12 Cuban institutions. PIAL uses decentralized formal and informal channels including national agrobiodiversity centres, collaborations between farmers and scientists, biodiversity fairs, and farmer experiments. The impressive results of this programme include greater crop diversity, improved food security, and nutritional and health indicators in the population. Although South Africa and Cuba have different historical, political, and economic trajectories, the Cuban experience suggests that with political will, organization, and faith in the ability of farmers, it is possible to design and implement a seed system that enhances farmers' ability to organize themselves and provides wider societal and environmental benefits.

In Chapter 14, Kristof Nordin illustrates how small steps towards agroecology can make a big difference to people's health, centring his analysis on Malawi, where smallholder farmers produce some 80 per cent of the nation's food. With just three crops – rice, wheat, and maize – supplying more than half of the world's plant-derived calories, and only 12 crop and 5 animal species providing 75 per cent of the world's food (Bioversity International, 2017), we have, he submits, forgotten the connection between agriculture, food, nutrition, and health. Reducing the world's agricultural and dietary needs from 200,000 edible species to a mere handful of high-carbohydrate, low-nutrient staple foods promotes poor dietary diversity, contributing to undernutrition, overnutrition, and non-communicable diseases such as diabetes.

The chapter describes how traditional agricultural systems in Malawi once gave farmers access to nutritional diversity and seasonal harvests throughout the year, yet today an estimated 60 per cent of adults in the country have suffered from stunting as children due to undernutrition. There is an increasing disconnect between the causes and effects of human actions about ecology, agriculture, and nutrition, which has led Malawi to divert substantial social, economic, and political resources towards short-sighted and unsustainable interventions such as programmes to fortify products like sugar and cooking oil, which adds to the nutrition crisis, or to subsidize fertilizer use. Part of the problem is due to the obsessive focus on high-input, fertilizer-dependent, and nutritionally poor maize, a crop which displaced more nutritious indigenous crops such as bulrush millet, finger millet, and sorghum, once the predominant staple grains of East and Central Africa. However, encouragement can be found in the wide range of initiatives across the country aiming to address issues of 'earth care, people care and fair share'. These include the Kusamala permaculture training centre, the SCOPE programme, which helps schools redesign their grounds in an ecologically sound manner, and community-based initiatives such as Never Ending Food, which has produced a government-approved training manual on sustainable nutrition. Paraphrasing Margaret Mead, the chapter concludes by reiterating the power of the capacity of a small group of thoughtful, committed citizens to change the world.

Such sentiments find profound expression in Chapter 15, the final contributed chapter of the book, in which Haidee Swanby draws on her experience as an activist to articulate the roots and evolution of African food sovereignty: 'food sovereignty', she reminds us, is the unbroken thread of local knowledge and practices that have nourished the continent, co-creating an astounding richness of agrobiodiversity, whereas 'Food Sovereignty' is the political movement struggling for the rights of producers to shape and control their food systems, and to produce sufficient and healthy food in culturally appropriate and ecologically sustainable ways (Edelman et al., 2014). Food Sovereignty is now part of the international discourse on hunger, nutrition, and agriculture, and is firmly embraced by the 2019 United Nations Declaration on the Rights of Peasants and Other People Working in Rural Areas (UNDROP), which many African countries have now signed.

The African Food Sovereignty movement has emerged from multiple actors and networks of actors 'based in African fields and informal settlements, in kitchens and local seed fairs', extending their reach to affiliate with ever-larger networks in solidarity across issues but with shared perspectives and values. These 'rooted networks' were built in the 1990s during negotiations of the Convention on Biological Diversity and the International Treaty on Plant Genetic Resources for Food and Agriculture (the 'Seed Treaty'), with a pivotal moment in 2004, when the Network of Farmers' and Producers' Organizations in West Africa (ROPPA) and Mozambique's National Peasants' Union (UNAC) joined the Latin American-based La Via Campesina Movement. In 2007, at a historical gathering in Nyéléni, Mali, more than 500 representatives from five continents affirmed the centrality of family producers and women in feeding the world, and of agroecology as an answer to transforming and repairing our food systems and rural worlds (Nyéléni, 2007). At a follow-up meeting held in Sélingué in February 2015, agroecology was validated as a 'key form of resistance to an economic system that puts profit before life' (International Forum for Agroecology, 2015: 4). Since then, multiple organizations, networks, and platforms have emerged across the continent, engaged in Food Sovereignty advocacy. They include umbrella farmers' organizations such as ROPPA, the Central African PROPAC, and the Eastern and Southern Africa Small-scale Farmers Forum (ESAFF); NGO networks such as the African Biodiversity Network and the Alliance for Food Sovereignty in Africa (AFSA), which consolidates 21 African networks; and a range of international partners and allies. The movement has clearly grown from strength to strength, with African voices an integral part of regional and international policy spaces.

Two short, visionary anecdotes conclude the book. The first, a contribution by Million Belay, who heads up AFSA, rejoices in the ability of celebration to connect the youth to elderly knowledge holders within their communities (Box J). 'Culture', he enunciates, 'is like a river. The river has a source, and if the source is kept alive, it keeps on flowing … The river is not flowing because of the gap between elders and youth.' Belay tells two stories to demonstrate the formidable ability of celebrations to heal this rift. The first recounts the revival of barley varieties in Telecho, an agricultural community in Ethiopia. The second reveals how cultural biodiversity celebrations brought together children from a conflict-ridden region in Ethiopia in song and dance. Celebrations, he concludes, have an amazingly galvanizing power, enthusing youth to study the names of seeds and the value of biodiversity, to participate in the arts, to work in teams, and to bring the community together. In doing so they revive nature and culture and let the 'river flow' through celebration.

The second anecdote, related by John Wilson (Box K), a long-standing advocate of agroecology, tells the story of Julius Astiva, a former teacher at an agricultural college, who transformed his lands in Vihiga, western Kenya, from barren maize fields that had once been dense, subtropical forests, to a thriving food forest. Julius terraced his sloping land so that he lost no rain as run-off. He ensured as much ground cover as possible to increase water infiltration. He dug two fishponds, to provide income, and he planted many kinds of trees. Twenty years later, his farm is thriving, and exemplifies the possibilities of a transition to agroecology. At the heart of this transition, explains John Wilson, is a mimicking of nature, transforming our practices to be in line with nature's processes. Such 'nature-based solutions' are now on the policy table as an approach to addressing the interconnected crises of biodiversity loss and climate change, but are entwined with corporate agendas linked to offsetting initiatives by the largest emitters of greenhouse gases (Wynberg et al., 2023). In contrast, the solutions pursued by Julius Astiva and small-scale farmers across the continent promote systemic changes based on the redesign and diversification of agroecosystems through ecologically and relationally based diverse cropping, agroforestry, and agro-sylvo-pastoral systems.

With nearly all landscapes across Africa in decline, a transition to agroecology is no longer negotiable. It will not happen overnight, but it is the beginning of a long journey of transition from degradation to regeneration, and from a monolithic 'one-world world' that recognizes only one way of knowing and being, to multiple worlds that embrace seed and knowledge in all their diversity, wealth, and wisdom. The final chapter of the book provides some pointers for decision-makers as to how this could happen, identifies new challenges on the horizon, and suggests approaches that require further investigation.

Note

1. After Harding (2008), who differentiates 'Science' with a capital 'S' as an instrument in service of politics and capital from 'science' as a tool of enquiry.

References

Andersen, R., Vasquez, V.M. and Wynberg, R. (2022) *Improving Seed and Food Security in Malawi: The Role of Community Seed Banks*, FNI Policy Brief 1/2022, Fridtjof Nansen Institute, Lysaker, Norway <https://www.fni.no/getfile.php/1315836-1659965574/Filer/Publikasjoner/FNI%20Policy%20Brief%202022%2001%20English.pdf>.

Batterbury, S. and Ndi, F. (2018) 'Land-grabbing in Africa', in T. Binns, K. Lynch and E. Nel (eds), *The Routledge Handbook of African Development*, pp. 573–582, Routledge, Abingdon.

Bezner Kerr, R. (2014) 'Lost and found crops: agrobiodiversity, indigenous knowledge, and a feminist political ecology of sorghum and finger millet in northern Malawi', *Annals of the Association of American Geographers* 104(3): 577–593 <https://doi.org/10.1080/00045608.2014.892346>.

Bioversity International (2017) *Mainstreaming Agrobiodiversity in Sustainable Food Systems: Scientific Foundations for an Agrobiodiversity Index*, Bioversity International, Rome <https://hdl.handle.net/10568/89049>.

Brown, K. (2016) *Resilience, Development and Global Change*, Routledge, London.

Clausing, P., Luig, L., Urhahn, J. and Beushausen, W. (2020) *Double Standards and Hazardous Pesticides from Bayer and BASF: A Glimpse behind the Scenes of the International Trade in Pesticide Active Ingredients*, Rosa Luxemburg Stiftung, Berlin, Hamburg, Johannesburg <https://www.rosalux.de/fileadmin/images/publikationen/Studien/Double_Standards_and_Hazardous_Pesticides_ENG_20210422.pdf>.

Edelman, M., Weis, T., Baviskar, A., Borras Jr, S.M., Holt-Giménez, E., Kandiyoti, D. and Wolford, W. (2014) 'Introduction: critical perspectives on food sovereignty', *The Journal of Peasant Studies* 41(6): 911–931 <http://dx.doi.org/10.1080/03066150.2014.963568>.

Food and Agriculture Organization (FAO) (2016). *Seed Security Assessment: A Practitioner's Guide*, FAO, Rome <http://www.fao.org/3/i5548e/i5548e.pdf>.

Handford, C.E., Elliott, C.T. and Campbell, K. (2015) 'A review of the global pesticide legislation and the scale of challenge in reaching the global harmonization of food safety standards', *Integrated Environmental Assessment and Management* 11(4): 525–536 <https://doi.org/10.1002/ieam.1635>.

Harding, S. (2008) *Sciences from Below: Feminisms, Postcolonialities, and Modernities*, Duke University Press, Durham, NC.

Ignatova, J.A. (2017) 'The "philanthropic" gene: bio-capital and the new green revolution in Africa', *Third World Quarterly* 38(10): 2258–2275 <http://doi.org/10.1080/01436597.2017.1322463>.

International Forum for Agroecology (2015) *International Forum for Agroecology. Nyéléni Center, Sélingué, Mali.* 24–27 February 2015. <https://www.ukfg.org.uk/2015/international-forum-agroecology-nyeleni2015/>

Kloppenburg, J. (2004) *First the Seed: The Political Economy of Plant Biotechnology, 1492–2000.* University of Wisconsin Press, Madison, WI.

Mbembe, A. (2003) 'Necropolitics', *Public Culture* 15(1): 11–40 <https://doi.org/10.1215/08992363-15-1-11>.

Moore, J. (2017) 'The Capitalocene, Part I: on the nature and origins of our ecological crisis', *The Journal of Peasant Studies* 44(3): 594–630 <http://dx.doi.org/10.1080/03066150.2016.1235036>.

Nixon, R. (2011) *Slow Violence and the Environmentalism of the Poor*, Harvard University Press, Cambridge, MA.

Nyéléni (2007) *Synthesis Report, Nyéléni 2007 Forum for Food Sovereignty*, Sélingué, Mali <https://nyeleni.org/en/synthesis-report/>.

Patel, R., Bezner Kerr, R., Shumba, L. and Dakishoni, L. (2015) 'Cook, eat, man, woman: understanding the New Alliance for Food Security and Nutrition, nutritionism and its alternatives from Malawi', *The Journal of Peasant Studies* 42(1): 21–44 <https://doi.org/10.1080/03066150.2014.971767>.

Schnurr, M.A. (2019) *Africa's Gene Revolution: Genetically Modified Crops and the Future of African Agriculture*, McGill-Queen's University Press, Montreal.

Shattuck, A. (2021) 'Generic, growing, green? The changing political economy of the global pesticide complex', *The Journal of Peasant Studies* 48(2): 231–253 <https://doi.org/10.1080/03066150.2020.1839053>.

Van Onselen, C. (1996) *The Seed is Mine: The Life of Kas Maine, A South African Sharecropper, 1894–1985*, Jonathan Ball Publishers, Johannesburg.

Vicedom, S. and Wynberg, R. (2023). 'Power and networks in the shaping of the Alliance for a Green Revolution in Africa (AGRA)', Third World Quarterly, https://doi.org/10.1080/01436597.2023.2276820

Wise, T.A. (2020) *Failing Africa's Farmers: An Impact Assessment of the Alliance for a Green Revolution in Africa*, Working Paper No. 20–01, July, Global Development and Environment Institute, Tufts University, Medford, MA <https://sites.tufts.edu/gdae/files/2020/07/20-01_Wise_FailureToYield.pdf>.

Wynberg, R., Pimbert, M., Moeller, N., McAllister, G., Bezner Kerr, R., Singh, J., Belay, M. and Ngcoya, M. (2023) 'Nature-based solutions and agroecology: business as usual or an opportunity for transformative change?' *Environment: Science and Policy for Sustainable Development* 65(1): 15–22 <https://doi.org/10.1080/00139157.2023.2146944>.

Box A 'I'm inferior to nobody'

Fakazile Mthethwa in conversation with
Mvuselelo Ngcoya

Photo A.1 Fakazile Mthethwa (aka Gogo Qho)
Credit: Jarek Plouhar

Background

Affectionately known as Gogo Qho, Fakazile Mthethwa lived an illustrious life in the hills of KwaBhoboza in the town of Mtubatuba in northern KwaZulu-Natal. For over two decades, she befriended and tended indigenous plants and trees, frustrated her neighbours with her insistence on ecological stewardship, and inspired thousands of visitors with her commitment to agroecological farming methods. That was until the last days of her life in December 2020, when she suddenly passed away after being diagnosed with cancer. Her methods, and indeed her life, were entirely engaged with the questions of 'land and bread and freedom'. She worked tirelessly to wean herself off the agroindustrial food system. She was committed to eating mostly what she grew herself. I had the fortune and pleasure of following and being inspired by her for eight years. I am still pondering her key question to me: 'How can you claim to be free if you are not in control of what is on your plate?'

During my first visit to her half-acre garden on a hillside, I was struck by the riotous colours and fragrances of the diverse crops that she cultivated. Various trees canopied a lush garden of indigenous herbs and vegetables such as *unsukumbili*,

(Continued)

Box A Continued

umsuzwane, *iboza*, *umhlonyane*, and *imbuya*, and the more conventional crops such as tomato, spinach, and cabbage. Even more impressive, however, was her encyclopaedic knowledge of the culinary and medical uses of her plants. Hers was not merely a garden, but a testament to her unrelenting struggle to make manifest a prophetic vision for a reparative future.

Against the odds of a brutal economy and an unforgiving sociopolitical landscape, Gogo Qho showed that agroecology was not only possible, but necessary. This recollection of our conversations is my attempt to grasp the contours of her philosophy and practice.

Agroecology as spirit

Mvuselelo Ngcoya (MN): How did you decide that you were going to grow your food this way, Gogo?

Gogo Qho (GQ): I was injured in a car accident, and I could not recover. I went to hospital and saw many health specialists, but my condition didn't improve. My health was deteriorating until one day my grandmother comes to me in a dream and tells me I will not be well – I wouldn't be well if I kept eating the poisonous food I was pumping into my system and using the medicines from the doctors. She showed me that my health and life (*impilo yami*) was in the soil. I had to change. Since I followed that voice, I have not visited the doctor once. Look at me! When I go to *impesheni* (the government pension paypoint) people ask why *itshitshi* (a young maiden) is joining a pension queue. [She laughs.]

The madness of agroecology

MN: Is that why you continue, because farming like this keeps you youthful?

GQ: [Laughing] Yes and no. It's not easy, this work. We have had no water in a long time, so you need to learn how to harvest water, how to mulch to prevent evaporation and protect the soil, you need to manure and build heaps of compost. Yes, it's not easy. You have to continue even though people think you are mad.

MN: Mad, how? You seem the sanest person to me.

GQ: Look throughout history: all the people who have struggled for anything, who have made some difference, were seen as mad. *Kufanele uzitshalele ngendlela oyithandayo, uzitshalele wena.* (You should grow food as you like, to your liking.) When you do that, you look a little off to a lot of people. I grow all kinds of things in this crazy mix that you see here. But everything works together. People don't know this anymore; it looks chaotic and therefore mad. They see the frogs, the worms and piles of dirt, and they think I've lost it. So, in their view, there is a particular kind of madness that has possessed me. *Abantu bazibonela uhlanya lwesalukazi olwenza amakhekhe omsuzwane. Nayo le yonto bayihleka kuvele amabamba.* (Generally, people see a mad old woman who makes cakes with indigenous herbs. And they laugh till their wisdom teeth show.) I don't blame them.

Feasting on poison

MN: But ultimately, you must laugh back at them. It's as if you hold a sacred secret and they have no clue, no?

GQ: Well, the modern way of farming is fast. It looks easy. You buy chicks, fatten them with God knows what, and boom, you are a millionaire because people will buy them. But look at the people who eat *olamthuthu* (industrial battery chickens); is it any wonder that they walk and look like them? Industry pumps these things with poison and people are dying from it without knowing.

(Continued)

Box A Continued

MN: So why does my generation continue eating this poison?

GQ: It's not only your generation. Mine too. It's lack of knowledge and simple laziness as well. We like it easy. I see people buying *imbuya* (amaranth) in plastic bags in the taxi ranks. *Imbuya pho*? (Of all things, *imbuya*?) It grows so easily. When you buy it, it means you don't have a garden because this thing grows like a mad weed. *Noma uthanda ukuthamela isigcaki*. (Or you like basking in the sun.) Then you buy poisoned cabbages. Why don't you grow your own food? You have no idea where the thing came from, what chemicals it was treated with, but there you are spending money on it.

A prophetess is never popular in her village

MN: I've heard people call you *ugogo womsuzwane*. How did the name come about?

GQ: Heeeh, *we* Mvu, people don't take me seriously. They were laughing at my unique muffins of *umsuzwane* [a herb, *Lippia javanica*, or literally 'little fart']. It was given that name because it relieves digestive problems and helps you release gas. So people would not buy my muffins. But one day, I was struggling to sell them in town and a white friend of mine saw me and took my crate and put them on the back of her van. She sold all of them in no time. When I sell them, they are little farts, but when a white woman does, they're seen as fancy food. *Abantu bakini-ke labo!* (That's your people right there!) But I'm inferior to nobody. One day they will wake up, but maybe it will be too late.

Seeding community

MN: One of the things that has always impressed me about your project is your zealous conservation of seeds. What motivates you?

GQ: Yes, for example I have collected and saved 14 varieties of *imbuya*. That is just one crop. I am trying to do the same with all the rare indigenous seeds. You ask why. Well, you remember when we visited MaSibiya during your previous visit? What did she say?

MN: Yes, she complained about how expensive spinach seeds are. If I recall correctly, she said she paid about R55 [US$3] for spinach seeds that covered just one 4-metre bed.

GQ: There! A time will come, and I think it's soon, say about 2040 to 2050, when these commercial seeds won't be available anymore. *Bazoyijova yonke imbewu* (They will doctor all the commercial seed) so that it will disappear or become too expensive. The natural, the indigenous will always be there. We need to cherish the natural so that it will thrive.

MN: When you can't conserve your own seeds?

GQ: I like keeping my own seeds, but I depend on other farmers too. I got yarrow from MaMfekaye the other day. I give you seeds to preserve on your project as well. We give and take among our community of crazy people. There are seed festivals and ceremonies to protect these valuable seeds. We need to do more. Otherwise, what will happen when we run out of seeds?

PART I
Seed, resilience, and diversity

CHAPTER 2

Reviving seed and knowledge towards more resilient communities: the power of transformative learning

Elfrieda Pschorn-Strauss

Introduction

> The stories that define our thinking today describe an eternal battle between good and evil, but these terms are just metaphors for something more difficult to explain, a relatively recent demand that simplicity and order be imposed upon the complexity of creation, a demand sprouting from an ancient seed of narcissism that has flourished due to a new imbalance in human societies. (Yunkaporta, 2019: 3)

The story of seed[1] is braided together with our human story, of our exploration and manipulation of nature, of using seed for both cooperation and domination. Tracing the evolution of seed reminds us of our place in this world, of human genius and generosity, but also of greed. Dr Melaku Worede, celebrated for establishing the first gene bank for seeds and plant materials in Africa (Hosken, 2015), would say: 'A seed is a plant at rest', evoking a magical image of the lifegiving, infinite possibilities of seeds.

Since the domestication of wild plants, farming communities have developed ways to protect, improve, and share seed, in the process creating highly functional seed systems. However, in recent times a range of external factors and actors has converged to erode these ancient seed and knowledge systems, with farmers shifting from being seed savers to becoming seed buyers. In Southern Africa, colonialism and the industrial farming system have had far-reaching cultural, political, and environmental consequences. Traditional farming and seed systems have been demeaned and rendered invisible, resulting in a rapid decline in biocultural diversity and a breakdown of local seed and knowledge systems. This shift has meant that within a short time, whole communities have lost the knowledge, skills, and social and institutional frameworks required to maintain their autonomous, resilient seed systems.

'We now have our own seed and can plant it when it rains' is an expression of farmer autonomy that has layers of meaning (Wilson, 2017: 8). Autonomy is both a condition for success and an outcome of farmer-led seed systems.

There is now mounting recognition from organizations as diverse as La Via Campesina and the Food and Agriculture Organization of the United Nations (FAO) that a transition to a more sustainable farming system is urgent and that agroecology can fulfil that role (Rosset and Altieri, 2017). However, widely different interpretations exist of what this means in practice. Embedded in the more progressive interpretation of agroecology is the sociopolitical framework of food and seed sovereignty. This interpretation suggests that a food system based on diversity, agroecology, and human rights has the greatest prospect of success in addressing the multiple crises of climate change, biodiversity loss, and declining soil fertility (Frison, 2016).

Agroecology is based on the co-creation of knowledge and values diverse ways of knowing and being. It merges the cosmovision and knowledge of Indigenous people with the best of science and develops a farming practice and seed system that is locally contextualized (Pimbert, 2018). Unlike 'modernist' agricultural development projects, agroecology challenges the systemic causes of climate change and socio-economic injustices in the food and seed system.

Vanloqueren and Baret (2018) describe how the prevailing scientific and technological paradigm has aligned with private interests to channel funding towards technologies such as genetic engineering while 'locking out' funding for more holistic agroecological approaches. Over decades this has developed into a systemic bias in research that acts in combination with the agroindustry and political structures to keep industrial agriculture in place and agroecology and local seed systems at the margins (Frison, 2016).

A number of international assessments and studies have challenged the prevailing view that smallholder agriculture and agroecology cannot feed the world and propose a radical change in scientific and technological paradigms, a shift from reductionism to holism, and a movement towards ecologically based science, knowledge systems, and economies (Frison, 2016; Rosset and Altieri, 2017: 68). This requires a collective paradigm shift that comes from transformational learning (Anderson et al., 2019b). Approaches that support such learning can be found in the critical pedagogy pioneered by Paulo Freire ([1970] 2000) and have been widely used by social movements in the South. Critical pedagogy was specifically developed to help people identify and recognize oppressive power dynamics and destructive practices that keep inequalities in place.

Saving seed is a basic right for farmers, but it also expresses a memory of the land, soil, climate, weather, cultural values, stories, and beauty. As Moeller (2018: 205) explains, '[T]he struggle surrounding the protection of traditional knowledge is not only a struggle over access to resources, but also a struggle over meanings and values.' Knowledge needs to not only focus on ways of knowing, but also enliven ways of being. Moeller calls this developing 'the capacity to know' (ibid.). Through valuing people's know-how, stories, and ways of being, we can begin to reverse the epistemicide of knowledge and start reconstructing a new narrative for the future.

Resilience thinking supports the deliberate transformation from one paradigm to another by assessing the merits of change and also 'fostering resilience of the new development trajectory' (Folke et al., 2010: 6). As such, a resilience framework can contribute towards understanding the transformation that is needed to weave back communities and their seed and knowledge systems.

Seed stories about meaning and loss

A story of millet and maps that speak

'Do you have any special rituals for seed?' I asked the *mukadzis*[2] from Chikukwa. Surrounded by the towering mountains of Chimanimani, we were enjoying the afternoon sun, three elderly women shelling cowpeas and me asking questions about seed.

> We have a ritual called Mabota. This ritual is usually done in August or September to prepare for the planting season. Only us old women are allowed to prepare the *rapoko*[3] for the ritual. We collect the *rapoko* seeds, soak them and when they start shooting, we grind them at home. We cook the porridge using no metal utensils, put it in a calabash and early in the morning, before people wake up, we take it to the sacred spring in the forest. The name of the spring is Mandenge. Nobody else is allowed to go with us. Once we have done this ritual then the planting can start.

Rituals like these can only persist if the farmers in Chikukwa continue to grow finger millet and save seed, and if the forests and springs continue to be looked after because they are considered sacred and are experienced as an integral part of the community. I wondered if the survival of finger millet as a crop was dependent on the extent to which people continue to value such rituals. 'How will the loss of *rapoko* seed affect this ritual?' I asked. One of the women looked at me as if I had asked a very daft question. 'We have enough seed,' she answered. 'We look after these seeds very carefully to enable us to keep the connection between the human world and the spiritual world' (Photo 2.1).

I was reminded of another finger millet conversation with Vhavenda seed keepers under a sycamore fig tree in the very north of South Africa. A group of women, most of them *makhadzis* from Tshidzivhe village, were discussing the cultural meaning of *mufhoho* (finger millet) and lamenting the disappearance of this important crop, but they were talking in circles, and could not quite put their finger on the reason why it had disappeared. 'Why don't we do an ecological calendar[4] of *mufhoho*?' one of them suggested.

The women had been making ecological calendars and maps for a while and relished the process of uncovering and sharing layers of knowledge. First, they drew three big concentric circles on paper, depicting the natural order of life in the village. The outer circle represented the seasons and natural cycle,

Photo 2.1 Maria Chikukwa, custodian of seed and stories in Chikukwa, Zimbabwe
Credit: Elfrieda Pschorn-Strauss

the middle one depicted the cropping cycle for finger millet, and the inner circle comprised cultural activities, all cycles taking place in synchronicity. As they were discussing and drawing the life cycle of finger millet, the map started speaking the story of millet and it became clear to everyone why people had stopped growing it. Maize was easier to grow and process, men were away working elsewhere, the children were not at home any more to chase the birds, and the government extension officers strongly promoted maize. Thus, over time, finger millet had been replaced, and now they had forgotten how to grow it. During the first dialogues with this community, they discussed the important link between performing rituals in their sacred forest, finger millet, and the role of women as keepers of both seed and rituals.

> But these days, there is no order in our communities because our sacred sites are not respected and we have lost our seed. Young people do not understand the importance of culture, of our seed and sacred sites in the forest. We can only bring back order in our community if we start doing our rituals, bring back our seed, and protect the forest. Then the healing can begin.

Conversations like these go to the heart of the seed issue, illustrating the meaning and interconnectedness of seed diversity, culture, women's stewardship, and human and ecosystem health (see also Hosken, 2015). They tell stories of loss but also of the persistence of diversity, knowledge, culture, and landscapes. Community seed systems maintain multiple functions, and farmers have supported the vitality and resilience of these systems across time (McGuire and Sperling, 2016). Understanding what these functions and strategies are and how to support their resilient capacity is of vital importance in supporting their future.

Community strategies to safeguard their seed

Wars, droughts, and floods are not new to Africa, but farmer-managed seed systems have endured these and other stresses over millennia, with farmers developing unique strategies to cultivate resilience in the face of such calamities. Some of the important functions inherent in community seed systems include safe storage, growing and selecting for quality and dynamic diversity, improving soil fertility, social networks, seed exchange systems, seed backup systems, and shared knowledge and meaning (Almekinders and Louwaars, 1999; Kusena et al., 2017).

Safe storage and multiple backup systems for seed have always been a crucial part of farmers' strategies for managing risk. Ethiopian farmers describe how they are not allowed to eat from the harvest before the seed has been secured in an underground storage pit, safeguarded by being sealed with a noxious sorghum. They also have a secret place in their fields for storing emergency supplies of seed. In times of hardship they may move away to where there is surplus and return after the rains have come, take out their seed, and plant again (Ethiopian farmers, personal communication, 2007). Farmers protect their seed against insects with strong-smelling plants, wood ash, or a layer of fine finger millet seed.[5] A growing number of local and international initiatives promoting farmer-managed seed systems and agrobiodiversity are organizing on-farm and community seed collections as well as seed fairs (Photo 2.2). These all contribute towards an effort to create layers of backup systems and to facilitate the conservation and distribution of agrobiodiversity (Peschard and Randeria, 2020).

Maintaining crop, varietal, and wild diversity in seed populations is another important way in which farmers spread risk and ensure food and nutrition security. Millets originated in Africa, and their wild–domesticated interface is permeable and dynamic, creating a space for interactions between ecological processes and indigenous knowledge (Gári, 2002) (Photo 2.3). Closely linked with maintaining varietal diversity are farming practices to achieve plasticity, which is the ability of a variety to be grown over a range of different microclimates (Dr Melaku Worede, personal communication, 2007). This intrinsic resilience of seed populations can only evolve *in situ* and is one

Photo 2.2 Finger millet varieties displayed at a seed fair in Harare
Credit: Elfrieda Pschorn-Strauss

Photo 2.3 The importance of diversity in spreading risk: a hairy pearl millet variety is resistant against damage by birds
Credit: Elfrieda Pschorn-Strauss

Photo 2.4 Women selling beans on the main road to Mzuzu play an important role in disseminating varieties, both locally and across Malawi
Credit: Elfrieda Pschorn-Strauss

reason why the global emphasis on the *ex situ* conservation of agrobiodiversity is misdirected, although both strategies are clearly needed.

Seed exchange systems are as old as the first domestication of crops and are recognized as vital arrangements for maintaining crop diversity, knowledge transfer, and agricultural innovation. Kinship, social networks, and local markets have various degrees of openness but jointly they have created seed exchange systems, a crucial safety strategy inherent in most Southern African smallholder farming communities (Coomes et al., 2015) (Photo 2.4).

Knowledge and learning opportunities. From my conversations with the *makhadzis* and other farmers it is clear that it is not only diversity that has been lost but also the knowledge that farmers have gained over centuries. Some farmers have forgotten how to grow finger millet, how to protect seed against insect damage, how to distinguish between seed and grain, how to use antheap soil for soil fertility, and other soil fertility practices. While a wealth of knowledge and practices persists, much has also been lost. Hall and Tandon (2017) write of how Western knowledge and education systems have led to an epistemicide of indigenous knowledge systems. While religion plays a role in demonizing African culture and customs, the education system must also take responsibility for the devaluing of endogenous culture, knowledge, and agricultural practices. Much culturally held knowledge and custom around seed has been replaced by reliance on the advice of extension officers who have learnt only about 'modern cultivars', fertilizers, and pesticides. A study on the role of agricultural extension services in promoting biodiversity in KwaZulu-Natal quotes farmers as saying: 'Extension don't like African farming, they tell you to throw your seeds,' and, 'Extension officers just come with their styles, they don't want to listen' (Abdu-Raheem, 2014: 1025).

> Southern African narratives are permeated by stories of loss and fragmentation – not only of seed systems but also of culture, communities, and landscapes, culminating in an erosion of meaning and identity. In Southern Africa it is maize that most poignantly tells this story (McCann, 2007) (see Box F).

The story of maize: the crop that both feeds and robs Southern Africa

It took me a while to realize that when farmers in Southern Africa discuss seed, they are typically referring to maize and not to the diversity of seed I have in mind. Other seed, and therefore food, is seemingly secondary. I have often heard people say, 'If we have not had a meal with *sadza*[6] for the day, we feel like we have not eaten.' Fischer (2021: 97), in her study of smallholders in South Africa, says: 'Maize is central to the local culture, and farming implicitly means planting maize.' Woven into the story of maize is a history of colonization, slavery, loss of culture, food, health, and land, and, most recently, the rise of seed hegemony.

Maize arrived with Portuguese traders from the Americas. At first, adoption was slow, with maize mostly eaten green during the hunger season, while farmers waited for sorghum and millet to ripen. The first quick-maturing and hard, flinty varieties were well adapted to the climate and long dry seasons, and smallholder farmers soon grew it alongside their other crops (McCann, 2007; Scoones and Thompson, 2011). In the British colonies the rise of maize production by settler farmers was motivated by a lucrative market, and they successfully lobbied for marketing and trade policies that would eliminate competition from African maize farmers (Smale and Jayne, 2003). Once mining took off, there was a growing demand to feed African labourers, and the production of maize grew exponentially, making maize the preferred food and a part of local culture. And so, over time, nutritious small grains were eliminated from people's diet and biocultural diversity declined (Smale and Jayne, 2003; Fischer, 2021).

After independence, 'maize had become the cornerstone of a "social contract" that the [new] governments made ... to redress the neglect of smallholder agriculture during the ... colonial period' and they allocated most of their agricultural budgets to provide cheap white maize to the population (Smale and Jayne, 2003: 17). Millions of smallholder farmers could now benefit from the government-controlled marketing systems and maize breeding programmes established during the colonial era. But fiscal budgets eventually ran into trouble and the notorious structural adjustment programmes were implemented, taking away government support to farmers and radically affecting the profitability of maize, sparking a dependency on donors in a number of countries. Since then, farmers have been at the receiving end of ad hoc programmes to subsidize agricultural inputs and mostly hybrid maize. These input subsidy programmes are a marketing coup for seed and fertilizer multinationals, as otherwise most smallholders would not have the financial means to buy their products (Smale and Jayne, 2003; Scoones and Thompson, 2011).

In Southern Africa, hybrid and genetically modified (GM) maize has been at the forefront of the push for a new Green Revolution, promising the end to hunger and the rise of a new class of African entrepreneurial farmers (McCann, 2007). But for smallholder farmers, 'modern' hybrid and GM maize varieties are risky as they are expensive and vulnerable to drought and disease, and post-harvest crop losses are high.

The rapid expansion of hybrid maize has changed the agricultural landscape. In addition to the rapid disappearance of the more nutritious and drought-resistant crops such as millets and sorghum, the 'Three Sister'[7] planting system of maize, beans, and pumpkin has been sacrificed. This system is not only very productive and good for the soil, but provides a balanced diet to families. The reality is that in sub-Saharan Africa, declining average rainfall, a disastrous decline in soil fertility, and the increasing levels of stunting in children under 5[8] illustrate the urgency of diversifying away from maize monoculture and synthetic inputs to move towards embracing agroecology and agrobiodiversity (Bezner Kerr, 2014). Yet farmers invest their best land and resources in maize in a devotion sometimes called the 'maize mindset'. However, there are also many smallholder farmers that have held on to a level of independence and diversity and have continued planting both their indigenous crops and local maize varieties alongside hybrid maize (Fischer, 2021) (Photo 2.5). Such farmers have more options, more autonomy, and resilient capacity, as they pay heed to the African proverb: 'It's only a fool who tests the depth of the river with both feet.'

Transforming minds: from scarcity to resilience

Instead of farming communities relying on their own experience and knowledge, extension officers and representatives from agrochemical companies have now become the 'experts' on seed and agricultural practices, with farmers dependent on handouts and inputs, and in the process losing a range of important skills and, with them, agency (Stone, 2007; Marshak et al., 2021). Losing agency is losing your adaptive capacity, a quality of resilience. Amartya Sen (in Brown and Westaway, 2011) presents poverty as a deprivation of the capabilities that provide people with the freedom to act in response to changing environmental conditions. Losing autonomy over their own seed is a deprivation that impoverishes farmers and removes their freedom to act.

Photo 2.5 Over time, Southern African smallholder farmers have created their own diversity of maize, as displayed by the Chikukwa Ecological Land Use Community Trust (CELUCT), an organization working in eastern Zimbabwe
Credit: Elfrieda Pschorn-Strauss

The persistent narrative that local seed is inferior and that the industrialization of agriculture is the only way for African farmers warrants a critical analysis, as it can be considered a continuation of the colonizing of people's minds. It does not allow space for a different system to emerge.

Understanding the dominant narrative about seed systems

At the core of the narrative of its superiority is the 'narcissism' of Western society that Yunkaporta (2019) mentions in the quotation that opens this chapter. Vanloqueren and Baret (2018) explain how the values and worldviews of modern science have reinforced this narrative, skewing the focus of research and public policies and generating flawed assumptions about past and future agricultural systems and the nature of innovation.

Scoones et al. (2019) offer an in-depth analysis of the narratives of scarcity that introduce most development- and policy-speak in Africa. They explain how this narrative has been used to frame the rush for land and resources in Africa and argue that 'notions of scarcity are presented as a deliberate political strategy, justifying resource control, appropriation, dispossession, population restrictions and the securing of exclusionary property rights' (ibid.: 234) This scarcity narrative goes something like this: 'Over the next 20 years, we will need to feed another 1.8 billion people' (Syngenta, in Scoones et al., 2019: 235). Or, according to the FAO, 'the world must double food production by 2050' (SIFCA, in Scoones et al., 2019: 235). A recurring theme is the yield gap between African farms and 'modern' farms. Increased production is presented as the answer to poverty and food insecurity, to be achieved through more investment and improved technology.

This gives rise to a *productivist* narrative, where the key challenge for agriculture is seen as producing as much food as possible. It is a very compelling one, given global statistics on malnutrition and poverty, but the irony is that most agricultural land is not used to feed people. Research by Deepak Ray (2022) predicts that in 2030 only 29 per cent of the global harvest of 10 major crops will be consumed as food by people in the country of production.

More specifically, narratives of scarcity in food and seed are constructed to justify restrictive seed laws, the undermining of farmers' rights, the appropriation of public resources for global companies, and the push for 'improved' seeds and fertilizers. It further promotes the belief that smallholder farmers are unproductive and inefficient.

Creating a story for the future

Despite a constant call for more evidence, an established body of evidence demonstrates that agroecology is farming for the future and that smallholders are not only productive, but are also stewards of agrobiodiversity (Gliessman, 2014; Frison, 2016; Ricciardi et al., 2021).

A Zimbabwe seed security assessment found that in spite of the 2008 financial collapse and long drought, farmer seed systems were sufficient (McGuire and Sperling, 2016). Ncube (2021) confirms this, showing that in Zimbabwe farmer-led seed systems have displayed resilience over time despite a lack of policy support and political and environmental challenges. His research confirms that farmer-led seed sources are the most reliable in ensuring that seed is available on time and in closer proximity.

Contrary to commonly held views, Kusena et al. (2017) reveal that local seed systems in Zimbabwe provide and circulate good-quality, fungal-free seed. They conclude that these seed systems are delivering food and nutritional security in sub-Saharan Africa and have the potential to provide solutions that are resilient to changing climates.

Reclaiming seed and knowledge systems is a complex undertaking in that it requires a transformation of attitudes to both seed and food. In Zimbabwe, farmers are shifting to growing more sorghum and finger millet, but it is not so easy to change eating habits from maize *sadza* to finger millet *sadza*. 'Such a transformation is a long-term process and can only be brought about by sustained, consistent and participatory engagement with local communities. It cannot be imposed from outside' (Muchineripi, 2014: 24).

Despite unfavourable policy frameworks, farmer-managed seed systems have persisted and the interest in agroecology is growing. The question is how to create a social dynamic for widespread adoption so that these pockets of resistance and restoration can become catalysts for broader social and environmental transformation.

Learning for transformation

In situations where agroecology is not yet practised, making a shift to agroecology first needs a transformation in the farmer's understanding and management of seed, soil fertility, water, and landscape health. James and Brown (2018: 1) observe: 'Transformation creates space to consider the profound changes necessary for society to pursue just and sustainable social-ecological systems.' It involves deep-seated and complex change that works across the scales of political, practical, and personal spheres of transformation. Simply put, 'convert yourself before you convert your farm' (ibid.).

Such transformations occur when people change their worldviews – about how society works, about what knowledge is, and even about who they are. This can be an endogenous process but can also be externally facilitated if development workers adopt an attitude of valuing and learning from local knowledge. In the words of Caps Msukwa, who facilitated a number of participatory community mapping processes in Zambia and Malawi:

> People know their situation best and are experts of life within their milieu. What was profound and often not highlighted was that we who are going into community spaces should not assume people don't know aspects about their lives. We are not trainers or teachers, leaders or bosses. We are facilitators who enter into the spaces not knowing but trying to understand what the situation is. (Msukwa, 2020)

The decolonization of knowledge is a starting point for seed sovereignty and agroecology and demands a democratic approach to knowledge-generating processes, whether in agricultural education, research, or development programmes. Learning needs to be transformative in both politics and practice (Anderson et al., 2019b). 'There is at the heart of food sovereignty a radical egalitarianism,' Raj Patel reasons in his exploration of what food sovereignty looks like. Egalitarianism 'is a prerequisite to have the democratic conversation about food policy in the first place' (Patel, 2010: 194).

The 'Campesino a Campesino' (farmer-to-farmer) methodology pioneered in Guatemala in the 1970s is a successful methodology for promoting farmer innovation and horizontal sharing towards wider social change and adoption (Holt-Giménez, 2006; Rosset and Altieri, 2017). The pedagogical methodologies used were developed by Paulo Freire in response to oppression and poverty and are an approach to learning that continues to make a vital contribution to the construction of a democratic global society. Social movements such as La Via Campesina have rooted much of their learning approaches in critical pedagogy (Anderson et al., 2019b). Networks such as the African Biodiversity Network and the Seed and Knowledge Initiative draw on these pedagogies for their work with practitioners, farmers, and NGO staff (see Box I).

Critical pedagogy is more than knowledge transfer, in that it addresses social injustice, questions power relations, and is based on the premise that once a person perceives and understands their circumstances and recognizes the possibilities of a response, they will act. In the context of agroecology and food sovereignty, this approach is a way of ensuring that knowledge is produced by farmers and that agroecological learning is 'used as a path towards cognitive justice' (Anderson et al., 2019b: 534). Cognitive justice emphasizes the right for different forms of knowledge and their associated practices, livelihoods, and socio-ecological contexts to coexist.

The point of this tentative exploration of what it would take to transform agriculture and seed systems from one paradigm to another is to come to a better understanding of the capabilities that would provide communities with the freedom to act in response to changing environmental and social conditions. Resilience theory and thinking expand this understanding because they respond to the inherent and necessary complexity in the socio-ecological systems which characterize seed systems.

Transformation requires resilience thinking – and vice versa

Resilience can be a useful approach towards an improved understanding of and support for the dynamism and complexity of seed systems, which cannot be separated from the resilience of a farming community. Community resilience is broadly seen as the collective ability to respond to and influence an environment that is characterized by continuous change, uncertainty, and crisis (Faulkner et al., 2018).

Faulkner et al. (2018) emphasize the importance of understanding resilience as an emergent property of human–environment relationships, which means that it is shaped by the unpredictable interactions within a complex system. Transformation at a local scale, multiplied and spread across scales, can enable resilience at a larger scale, while large-scale resilience supports the capacity of smaller systems to change (Folke et al., 2010).

Using a political ecology perspective, Brown (2016) develops a practicable framework for understanding resilience at community level. The framework uses themes of rootedness, resistance, and resourcefulness as a pragmatic basis for enhancing community resilience that is relational, interactive, and enabling.

These three themes are a useful lens through which to understand and describe the lived experience of communities, institutions, and civil society in Southern Africa that are collaborating in their efforts to regenerate their seed and farming systems.

Resistance places local agency over seed systems at the centre

Resistance addresses issues of power and inequality by disrupting established economic, social, and political practices and creating new ways of doing towards a more equitable and sustainable environment. In the political sense of the word, resistance 'puts agency at the heart of resilience' (Brown, 2016: 196).

A broadened understanding of resistance in community seed systems includes challenging the institutions and laws that create enclosures and restrict community seed practices. Haidee Swanby describes in Chapter 15 how social movements lobby for international and national recognition of African farmers' rights.

Resistance can also include localized but transformative ideas and practices which shift perceptions and conscientize people on a personal or community level. The simple act of reviving seed saving and exchange practices can be a quiet activism but can also be a radical act of resistance in countries where the practice is criminalized (Peschard and Randeria, 2020).

Resistance means paying close attention to the ways in which everyday discussions and social practices reinforce power relations and undermine community cohesion and agency. The push for hybrid maize has impacted most negatively on women, whose agency and wealth lies in their capacity to maintain diversity and to ensure good nutrition for their families. The work of Soils, Food and Healthy Communities, an organization in northern Malawi, powerfully demonstrates the importance of shifting power and equity in gender relationships within the home before societal outcomes such as improved child nutrition can be achieved through crop diversity and agroecology (Bezner Kerr, 2014). This work is a case study for the transformative power of agroecology when combined with shifting gender dynamics.

Given the repressive political contexts of many sub-Saharan African countries, NGOs often follow a more politically neutral route of resistance and focus on changing the minds of government officials. These organizations work hard to build relations and conscientize agriculture extension officers who have influence over farmers and may also influence their superiors. Staff at the NGO TSURO (Towards Sustainable Use of Resources Organization) in the eastern Zimbabwean region of Chimanimani, for example, reported encouragingly: 'Last year Agritex officials announced publicly that famers should use OPV [open-pollinated varieties] and small grains saying these are important because of climate change. This encouraged more farmers to plant OPV seed'.[9]

Sustainability and egalitarianism cannot be separated in our thinking, and consistently making these links requires critical thinking, and a radical form of being (Anderson et al., 2019b). These are important qualities for staying rooted in a transformative and resistant perspective of agroecology and food sovereignty.

Rootedness connects people with place and meaning

Rootedness is situated, experiential resilience: resilience in place. When people have a sense of place, a sense of belonging and identity, the building of resilience is supported. Rootedness can be seen as a precondition for building resilience in communities because of the connectedness and stewardship it invokes. This is what motivates people to restore, revive, and maintain culture, seed, landscape, and regenerative practices and to do it together. Rootedness enriches both resistance and resourcefulness with the qualities of responsibility and appreciation, both of which are important drivers for communities and individuals to recognize their stewardship role.

In resisting the forces that aim to erode seed and knowledge systems, it is important to acknowledge that the struggle is about something more fundamental than the material value of seed and knowledge. It is a struggle for recognition, of values, ways of knowing and being, of sacredness, and of the inherent dignity of the non-human world (Moeller, 2018).

EarthLore Foundation, an organization working in South Africa and Zimbabwe, embodies this respect for nature and the sacred by promoting earth jurisprudence, a systemic approach for transformation towards deepening the relationship to place, belonging, and becoming more ecologically literate in the process. As director Method Gundidza remarks, 'Indigenous people ... understand how ... to belong to a territory ... the way a river does' (Gundidza, 2022). For the communities EarthLore engages with, reviving their seed systems and nurturing diversity flow from a process of restoring their customary governance systems. EarthLore uses community dialogues as a process to reveal the layers of knowledge and socio-environmental relationships that still, often precariously, maintain seed system functions in communities. Rediscovering the value of their roots through dialogue has animated farmers towards actively restoring their sacred sites, rituals, and traditional crop varieties and storage methods and resisting the further encroachment of maize monocultures (Gundidza, 2021).

Narratives about the loss of seed have been kept alive in communities for a long time. When these stories are revived and retold, the connection they make to people's roots can galvanize action. For example, farmers in Machavika village, Zambia, mapped their diversity during a participatory event in 2020. They documented the varieties they had lost as far back as the 1950s: *gankata* maize, good for *nshima*, with a big grain and good taste, lost in 1997; *lubele*, a variety of finger millet, used for *nshima* and traditional drink, lost in 1958; *chigaligari*, a sorghum variety used in brewing and with a good taste, and not easily attacked by birds, lost in 1987 (Msukwa, 2020). After the mapping process farmers resolved to find their lost seeds, to plant and eat more diversely, and to reforest their woodlands. From a discussion about the loss of seeds, this community remembered the qualities and meaning of specific varieties and their relationship to place, water, and forests, which reminded them of the importance of diversity and restoring their ecosystem.

Barthel et al. (2013) write about the importance of identifying key places for safeguarding a diversity of practices for food security and biodiversity. These biocultural refugia are territories where we still find specific social memories related to food security and biodiversity stewardship. As the stories in this chapter illustrate so poignantly, the Green Revolution has been a powerful eraser of stewardship memory, resulting in a widespread 'generational amnesia' (Barthel, 2013: 3). As Method Gundidza (2021) remarks, 'There is still much to do, but bringing back millet has reminded us of our place on this planet and our responsibility to protect and restore Mother Earth.'

Resourcefulness resides in know-how and sharing this in social networks

Resourcefulness is not only about access to resources but also about the capacity to use resources at the right time, in the right place, and in an appropriate way (Brown, 2016). Resourcefulness has implications for how decisions are made, and for the use of knowledge and innovation, capability, and agency, and supports the capacity for change – of communities and individuals. It is not enough to have resources; it is also important to have the imagination and ability to make good decisions and the right choices, and to make things happen, 'even toward radically altering or revolutionizing the system itself' (Ungar, 2020: 779).

Most smallholder farmers keep some level of crop diversity alive, but certain farmers stand out because of their special commitment to growing remarkable levels of crop diversity. These are custodian farmers, also called seed keepers, seed experts, or nodal farmers, and they are crucial to resourcefulness in communities, as they are a source not only of diversity but also of knowledge. Their value as seed custodians depends as much on the diversity they maintain as on their willingness to share their knowledge and seed with others in the community (Coomes et al., 2015).

Closely linked to seed custodianship is the concept of seed stewardship, emphasizing custodianship on behalf of future generations. Stewardship is a mindset, a cosmovision that requires a transformation in human perceptions of our place in nature and an understanding that systemic change is necessary (Folke et al., 2010). Strengthening practices of sharing seed, organizing seed fairs, and establishing community seed banks as backup systems and to improve access for current and future generations are all acts of stewardship and of *ubuntu*, or reciprocity. Engaging in seed stewardship and seed sharing through social networks strengthens and expands these networks and enhances community resilience (Chapin et al., 2010).

Joint activities of strengthening and governing seed systems at a community level bring a number of paradoxical and ethical dilemmas to the surface, specifically around the commodification of seed versus seed as a commons. For example, how do we reconcile a focus on promoting high-yielding crops or varieties to increase food production with the importance of conserving agrobiodiversity? Putting the focus on a rights approach demands that governments and states recognize and protect those rights, but in the process this can neglect the important link between rights and responsibilities. Therein also lies a paradox, as the discussion on farmers' rights over seed takes place within the framework of intellectual property rights. Recognizing farmers' varieties through formal registration implies an ownership approach that is contrary to the notion of seed as a commons for humankind. Peschard and Randeria (2020) explain how seed activists have moved away from the original concept of farmers' rights, to campaign for the collective rights of local, peasant, and Indigenous communities over seeds and agrobiodiversity. These are all complex issues that are important to clarify and for farmers to debate until they appreciate them in the context of communal principles such as reciprocity, and intergenerational and intragenerational stewardship – all important values in many African societies (Hosken, 2015).

Resourcefulness is most evident when people or communities have the imagination and agency to ask the right questions and to use their critical thinking to either adopt and innovate with new practices or resist potentially destructive interventions from outside. That is when they can make the most of their intimate local knowledge when confronted with rapid socio-ecological change so that they are able to adapt or even transform.

Conclusion

Farmer-managed seed systems are complex systems, embedded in the intricate fabric of social and environment relations, economics, local politics, and institutions, therefore eluding a simplistic analysis or an easy response.

In Southern Africa, people's livelihoods and futures are being held captive by a seductive politics of scarcity that justifies the acquisition and enclosure of seed through comprehensive policy reforms in the name of development, while avoiding the systemic problems of dispossession and social and environmental justice (Scoones et al., 2019). Development interventions, funding, and investment, as well as education and the media, feed this narrative, making generations of farmers aspire to a type of farming and seed system that is not suited to context or culture. This to some extent explains the 'maize mindset', because it is this paradigm that is funded, valued, researched, and supported by the 'agriculture establishment'.

For farming communities there are three levels of responding to environmental and other stressors: coping, adaptation, and transformation (Brown and Westaway, 2011). Coping is not an option any more, as change is confronting farmers at too rapid a pace. They face two significant, interlinked challenges: equity and sustainability. This chapter argues that these must be addressed through agency and capacity. The importance of emphasizing agency is that it shifts the view that people are powerless against environmental and sociopolitical change and recognizes that people are never passive (Brown and Westaway, 2011). Agency also has an aspect of meaning that is embedded in the concepts of care, hope, and possibility.

Using participatory and critical learning and research methodologies, a transformative education can be facilitated that bolsters both agency and capacity (Anderson et al., 2019b). In the context of agroecology and seed systems, this approach to social learning is a way of ensuring that knowledge is produced by farmers and that learning is 'used as a path towards cognitive justice' (Anderson et al., 2019b: 534). It deeply challenges both development and research institutions to shift from being project implementers and trainers to new roles of co-creation. 'Learning for transformation ... knits together the human-ecological relations within territories that have often been stripped of their cultures, people, resources, and autonomy through centuries of capitalism and colonialism'(Anderson et al., 2019a: 526).

Resistance, rootedness, and resourcefulness emerge as three key integrating features towards building resilience capacity for local communities. When these key features of resilience capacity are strengthened together, they become bigger than the sum of their parts, and transformative.

Resistance is the first step towards resilience as the recognition that radical systemic change is needed animates the search for a different way of doing and being. Resilience is constructed at various scales, and recognizing the impact of unequal power dynamics at all levels, from household to international fora, is a first step in building resistance and agency.

A rootedness in place and culture brings back the meaning of seed from being a 'thing', a commodity, to being a set of relationships with people and the earth that are fundamental for creating and sustaining the seed commons for the future.

Resourcefulness that enables the equitable sharing, stewardship, and inclusive governance of seed, knowledge, and territory towards amplifying community cohesion is vital for resilient capacity.

Transformational change in seed and agriculture systems requires getting to the heart of what really matters to people and giving voice to marginalized people. Embracing indigenous ways of seeing the world while opening new possibilities for the future creates space for creativity and hope to emerge towards an increasingly sustainable and equitable world.

> Open yourself to that politics, to those voices, to cosmologies in which private property is abusive, and life begins before – and ends long after – you. (Patel, 2017: 44)

Notes

1. A note on terminology: in this chapter, the term 'seed' is employed broadly to include the plant structure (i.e. seeds, seedlings, cuttings) used in the propagation of crops.
2. *Mukadzi* (chiShona) or *makhadzi* (Tshivenda) is a wise or spiritual woman who, in the traditional system, has special responsibilities, such as leading important ceremonies in sacred sites.
3. *Rapoko*: 'finger millet' in chiShona (*mufhoho* in Tshivenda).
4. An ecological calendar is a tool to document, through a process of dialogue, the interconnectedness of the annual cycle of activities in a community, including the agricultural, cultural, and natural cycles. In the process the community members affirm and transfer local knowledge and revive useful practices and culture.

5. The hard endosperm of the tiny finger millet seed is impervious to insect attack, and this is one way of protecting other seed.
6. *Sadza*: a chiShona term for a stiff maize porridge eaten with vegetables or meat.
7. In most Southern African countries, the Three Sisters are maize, beans, and pumpkins, intercropped in a symbiotic, productive relationship. In Mexico this farming system is called *milpa*: maize in polyculture with other crops, often squash and beans.
8. Over the past 25 years, the number of stunted children in Eastern and Southern Africa has risen from 23.6 million to 26.8 million (UNICEF, n.d.).
9. This information was taken from a 2018 unpublished SKI report.

References

Abdu-Raheem, K.A. (2014) 'Exploring the role of agricultural extension in promoting biodiversity conservation in KwaZulu-Natal province, South Africa', *Agroecology and Sustainable Food Systems* 38(9):1015–1032 <https://doi.org/10.1080/21683565.2014.899283>.

Almekinders, C. and Louwaars, N. (1999) *Farmers' Seed Production: New Approaches and Practices*, IT Publications, London.

Anderson, C.R., Maughan, C. and Pimbert, M.P. (2019a) 'Introduction to the symposium on critical adult education in food movements: learning for transformation in and beyond food movements – the why, where, how and the what next?', *Agriculture and Human Values* 36: 521–529 <https://doi.org/10.1007/s10460-019-09941-2>.

Anderson, C.R., Maughan, C. and Pimbert, M.P. (2019b) 'Transformative agroecology learning in Europe: building consciousness, skills and collective capacity for food sovereignty', *Agriculture and Human Values* 36: 531–547 <https://doi.org/10.1007/s10460-018-9894-0>.

Barthel, S., Crumley, C. and Svedin, U. (2013) 'Bio-cultural refugia: safeguarding diversity of practices for food security and biodiversity', *Global Environmental Change* 23(5): 1142–1152 <https://doi.org/10.1016/j.gloenvcha.2013.05.001>.

Bezner Kerr, R. (2014) 'Lost and found crops: agrobiodiversity, indigenous knowledge, and a feminist political ecology of sorghum and finger millet in northern Malawi', *Annals of the Association of American Geographers* 104(3): 577–593 <https://doi.org/10.1080/00045608.2014.892346>.

Brown, K. (2016) *Resilience, Development and Global Change*, Routledge, London.

Brown, K. and Westaway, E. (2011) 'Agency, capacity and resilience to environmental change: lessons from human development, wellbeing and disasters', *Annual Review of Environment and Resources* 36(1): 321–342 <https://doi.org/10.1146/annurev-environ-052610-092905>.

Chapin, F.S. III, Carpenter, S.R., Kofinas, G.P., Folke, C., Abel, N., Clark, W.C., Olsson, P., Smith, D.M.S., Walker, B., Young, O.R., Berkes, F., Biggs, R., Grove, J.M., Naylor, R.L., Pinkerton, E., Steffen, W. and Swanson, F.J. (2010) 'Ecosystem stewardship: sustainability strategies for a rapidly changing planet', *Trends in Ecology and Evolution* 25(4): 241–249 <https://doi.org/10.1016/j.tree.2009.10.008>.

Coomes, O., McGuire, S., Garine, E., Caillon, S., McKey, D., Demeulenaere, E., Jarvis, D., Aistara, G., Barnaud, A., Clouvel, P., Emperaire, L., Louafi, S., Martin, P., Massol, F., Pautasso, M., Violon, C. and Wencélius, J. (2015) 'Farmer seed networks make a limited contribution to agriculture? Four common misconceptions', *Food Policy* 56: 41–50 <https://doi.org/10.1016/j.foodpol.2015.07.008>.

Faulkner, L., Brown, K. and Quinn, T. (2018) 'Analyzing community resilience as an emergent property of dynamic social-ecological systems', *Ecology and Society* 23(1): 24 <https://doi.org/10.5751/ES-09784-230124>.

Fischer, K. (2021) 'Why Africa's new Green Revolution is failing: maize as a commodity and anti-commodity in South Africa', *Geoforum* 130: 96–104 <https://doi.org/10.1016/j.geoforum.2021.08.001>.

Folke, C., Carpenter, S.R., Walker, B., Scheffer, M., Chapin, T. and Rockström, J. (2010) 'Resilience thinking: integrating resilience, adaptability and trans-formability', *Ecology and Society* 15(4): 20 <http://dx.doi.org/10.5751/ES-03610-150420>.

Freire, P. ([1970] 2000) *Pedagogy of the Oppressed*, 30th anniversary edn, Continuum International Publishing Group, New York.

Frison, E.A. (2016) *From Uniformity to Diversity: A Paradigm Shift from Industrial Agriculture to Diversified Agroecological Systems*, International Panel of Experts on Sustainable Food Systems, Louvain-la-Neuve, Belgium <https://cgspace.cgiar.org/handle/10568/75659>.

Gári, J.A. (2002) 'Review of the African millet diversity', paper for the *International Workshop on Fonio, Food Security and Livelihood among the Rural Poor in West Africa*, IPGRI/IFAD, Bamako, Mali, 19–22 November 2001 <https://www.fao.org/fileadmin/templates/esw/esw_new/documents/Links/publications_other/6_millets.pdf>.

Gliessman, S.R. (2014) *Agroecology: The Ecology of Sustainable Food Systems*, 3rd edn, CRC Press, Boca Raton, FL.

Gundidza, M. (2021) 'Grains of hope', *Resurgence and Ecologist* 324 <https://www.resurgence.org/magazine/article5663-grains-of-hope.html>.

Gundidza, M. (2022) 'Respecting the rights of nature is the only way out of climate chaos and biodiversity collapse', *Daily Maverick*, 7 March <https://www.dailymaverick.co.za/article/2022-03-07-respecting-the-rights-of-nature-is-the-only-way-out-of-climate-chaos-and-biodiversity-collapse/>

Hall, B.L. and Tandon, R. (2017) 'Decolonization of knowledge, epistemicide, participatory research and higher education', *Research for All* 1(1): 6–19 <https://doi.org/10.18546/RFA.01.1.02>.

Holt-Giménez, E. (2006) *Campesino a Campesino: Voices from Latin America's Farmer to Farmer Movement for Sustainable Agriculture*, Food First Books, Oakland, CA.

Hosken, L. (ed.) (2015) *Celebrating African Rural Women: Custodians of Seed, Food and Traditional Knowledge for Climate Change Resilience*, African Biodiversity Network, The Gaia Foundation, African Women's Development Fund <https://gaiafoundation.org/wp-content/uploads/2018/03/Celebrating-African-Rural-Women_largefile.pdf>.

James, T. and Brown, K. (2018) 'Muck and magic: a resilience lens on organic conversions as transformation', *Society & Natural Resources* 32(2): 133–149 <https://doi.org/10.1080/08941920.2018.1506069>

Kusena, K., Wynberg, R. and Mujaju, C. (2017) 'Do smallholder farmer-led seed systems have the capacity to supply good-quality, fungal-free sorghum seed?' *Agriculture & Food Security* 6: 52 <https://doi.org/10.1186/s40066-017-0131-7>.

Marshak, M., Wickson, F., Herrero, A. and Wynberg, R. (2021) 'Losing practices, relationships and agency: ecological deskilling as a consequence of the uptake of modern seed varieties among South African small-holders', *Agroecology and Sustainable Food Systems* 45(8): 1189–1212 <https://doi.org/10.1080/21683565.2021.1888841>.

McCann, J.C. (2007) *Maize and Grace: Africa's Encounter with a New World Crop, 1500–2000*, Harvard University Press, Cambridge, MA.

McGuire, S. and Sperling, L. (2016) 'Seed systems smallholder farmers use', *Food Security* 8: 179–195 <https://doi.org/10.1007/s12571-015-0528-8>.

Moeller, N.I. (2018) 'Plants that speak and institutions that don't listen', in M.P. Pimbert (ed.) *Food Sovereignty, Agroecology and Biocultural Diversity: Constructing and Contesting Knowledge*, pp. 202–233, Routledge, Abingdon <https://doi.org/10.4324/9781315666396>.

Msukwa, C. (2020) *Zambia Participatory Community Mapping Report*, unpublished report, Seed and Knowledge Initiative.

Muchineripi, C. (2014) *Grain revolution: finger millet and livelihood transfor-mation in rural Zimbabwe*, Policy Voices Series, Africa Research Institute, London <https://www.africaresearchinstitute.org/newsite/wp-content/uploads/2014/10/ARI-Policy-Voice-Grain-Revolution.pdf>.

Ncube, B.L. (2021) 'Exploring the relationship between seed security and food security in eastern Zimbabwe: research feedback', Seed and Knowledge Initiative, Durban <https://bio-economy.org.za/wp-content/uploads/2021/11/Ncube-2021-Research-Feedback_FINAL_print-ready.pdf>.

Patel, R. (2010) 'What does food sovereignty look like?' in H. Wittman, A.A. Desmarais and N Wiebe (eds), *Food Sovereignty: Reconnecting Food, Nature & Community*, pp. 186–196, Pambazuka Press, Cape Town, Dakar, Nairobi and Oxford.

Patel, R. (2017) 'Raj Patel', in M. Hodgkins (ed.), *Letters to a Young Farmer: On Food, Farming, and Our Future*, pp. 43–45, Princeton Architectural Press, New York.

Peschard, K. and Randeria, S. (2020) '"Keeping seeds in our hands": the rise of seed activism', *The Journal of Peasant Studies* 47(4): 613–647 <https://doi.org/10.1080/03066150.2020.1753705>.

Pimbert, M.P. (ed.) (2018) *Food Sovereignty, Agroecology and Biocultural Diversity: Constructing and Contesting Knowledge*, Routledge, Abingdon.

Ray, D. (2022) 'A shrinking fraction of the world's major crops goes to feed the hungry, with more used for nonfood purposes', *The Conversation*, 13 May <https://theconversation.com/a-shrinking-fraction-of-the-worlds-major-crops-goes-to-feed-the-hungry-with-more-used-for-nonfood-purposes-181819>

Ricciardi, V., Mehrabi, Z., Wittman, H., James, D. and Ramankutty, N. (2021) 'Higher yields and more biodiversity on smaller farms', *Nature Sustainability* 4: 651–657 <https://doi.org/10.1038/s41893-021-00699-2>.

Rosset, P.M. and Altieri, M.A. (2017) *Agroecology: Science and Politics*, Practical Action Publishing, Rugby.

Scoones, I. and Thompson, J. (2011) 'The politics of seed in Africa's Green Revolution: alternative narratives and competing pathways', *IDS Bulletin* 42(4):1–23 <https://doi.org/10.1111/j.1759-5436.2011.00232.x>.

Scoones, I., Smalley, R., Hall, R. and Tsikata, D. (2019). 'Narratives of scarcity: framing the global land rush', *Geoforum* 101: 231–241 <https://doi.org/10.1016/j.geoforum.2018.06.006>.

Smale, M. and Jayne, T. (2003) *Maize in eastern and southern Africa: 'Seeds' of success in retrospect*, EPTD Discussion Paper No. 97, Environment and Production Technology Division, International Food Policy Research Institute, Washington, DC <https://ebrary.ifpri.org/utils/getfile/collection/p15738coll2/id/77285/filename/77286.pdf>.

Stone, G.D. (2007) 'Agricultural deskilling and the spread of genetically modified cotton in Warangal', *Current Anthropology* 48(1): 67–103 <https://doi.org/10.1086/508689>.

Ungar, M. (ed.) (2020) *Multisystemic Resilience: Adaptation and Transformation in Contexts of Change*, Oxford University Press, New York <https://resiliencere-search.org/wp-content/uploads/2021/02/Multisystemic-Resilience.pdf>.

UNICEF (no date) *Programme: Reduce Stunting*, UNICEF Eastern and Southern Africa <https://www.unicef.org/esa/reduce-stunting>.

Vanloqueren, G. and Baret, P.V. (2018) 'How agricultural research systems shape a technological regime that develops genetic engineering but locks out agroecological innovations', in M.P. Pimbert (ed.), *Food Sovereignty, Agroecology and Biocultural Diversity: Constructing and Contesting Knowledge*, pp. 57–92, Routledge, Abingdon.

Wilson, P. (2017) *Mid-Term Review for 1 February 2015 to 31 January 2017*, Seed and Knowledge Initiative, Durban.

Yunkaporta, T. (2019) *Sand Talk: How Indigenous Thinking can Save the World*, The Text Publishing Company, Melbourne.

CHAPTER 3

In the wake of Cyclone Idai: a holistic look at its impacts and an exploration of the resilience-enhancing potential of landscape agroecology

Witness Kozanayi and Jaci van Niekerk

> *We dedicate this chapter to the memory of those who lost their lives in Cyclone Idai, and express our deepest condolences and sympathy to the bereaved and injured.*

Introduction

On the afternoon of 14 March 2019 Mrs Moyana[1] looked up from tending her millet field with a sigh; rainfall in her native Chimanimani in the Eastern Highlands of Zimbabwe had become increasingly erratic, either arriving at odd times or staying away mid-season, leaving her rain-fed crops to shrivel in the sun. This uncertainty made the cultivation of some crop varieties, for example long-season sorghum such as tsveta, *impossible. She maintained hope, however, that the variety of crops remaining in the field, supplemented by the seed she had saved and exchanged with her neighbours and family members, as had been her custom since she first started farming, would see her and her family through to the next season. She also had a steady supply of fruit from the papaya and other fruit trees she had interplanted with her crops, she kept chickens at her homestead, and her grandson tended her small flock of goats. She looked up at the sky: it was ominously dark, with a blustery wind starting to pick up strength. Just then she got a text message from her son in Harare, warning her of an impending cyclone, predicted to be much bigger than any that had passed through before, approaching from the east. She rushed home.*

The cyclone which Mrs Moyana encountered that night was named Idai. First making landfall in Mozambique, Idai reached wind speeds of up to 195 km/h, making it a Category 3 storm, bringing with it heavy rains, and leaving flooding, landslides, infrastructure damage, and crop losses in its wake (Ninteretse, 2019). On record as the most destructive natural disaster in Southern Africa in living memory, Cyclone Idai affected millions of people in the region and, in Zimbabwe alone, took the lives of hundreds and uprooted thousands (RINA, 2019). Livelihoods, the majority based on small-scale agriculture, were severely disrupted, causing major food shortages as not only crops, but also seed, stored grain, and livestock, were swept away. Although the international aid community sprang into action to avert a food security disaster, efforts were hampered by flood-damaged roads and bridges.

In this chapter we focus on Chimanimani, a district in the far east of Zimbabwe, one of two districts severely affected by the cyclone. We explore the impacts of the cyclone at farm, field, and homestead level, also looking at the landscape more broadly, as we attempt to understand whether employing agroecology at a landscape level could build the resilience of agrarian communities exposed to climate-induced disasters. We believe that this enquiry is salient in the face of an increasingly unpredictable and erratic climate in the region, which may precipitate tropical cyclones that are as intense as Cyclone Idai, or even more so (RINA, 2019).

In order to address our aim, we draw on experiences from the ground, relayed by one of the authors (WK), who not only travelled to the area in 2019 and 2020 as part of a transdisciplinary research team, but had grown up there. Data informing this chapter were collected through key informant interviews, observations of affected forests, plantations, fields, homesteads, and communities, and attendance at multi-stakeholder workshops on Cyclone Idai, as well as reviews of donor and government reports on the impact of the cyclone. This was supplemented with the results from a household survey carried out as part of a research programme on agroecology and landscape management in Chimanimani in the aftermath of the cyclone (see TSURO, 2020).

Chimanimani: a place of contrasts

Chimanimani District: a brief profile

Chimanimani District is 3,353 km^2 in area, with a largely rural population of 133,810 who predominantly practise agriculture-based livelihoods (CSO, 2012) (see Figure 3.1a and 3.1b). The district comprises 23 wards, categorized as communal (18), commercial (3), resettlement area (1), and peri-urban (1) (CRDC, 2017).

Two distinct geographical areas can be discerned in Chimanimani: the uplands and lowlands. The mountainous eastern areas comprise the uplands, with the highest peak, Mount Binga, rising to 2,440 m above sea level. The uplands typically experience high rainfall of up to 1,400 mm per year (CRDC, 2017). On the other hand, the lowlands, situated in the western part of the district, are at an altitude of around 350 m above sea level and receive 300–800 mm of rain in an average wet season (CRDC, 2017). These clearly distinct conditions result in different agricultural production systems. Farmers in the uplands grow maize, finger millet, taro, and a range of citrus fruits, integrated with large livestock. In the low-lying areas, in addition to the staple maize, farmers grow a diversity of small grain crops such as sorghums and millets, as well as melons, cucumbers, groundnuts and roundnuts, and cowpeas. Farmers in the valley keep both small and large livestock.

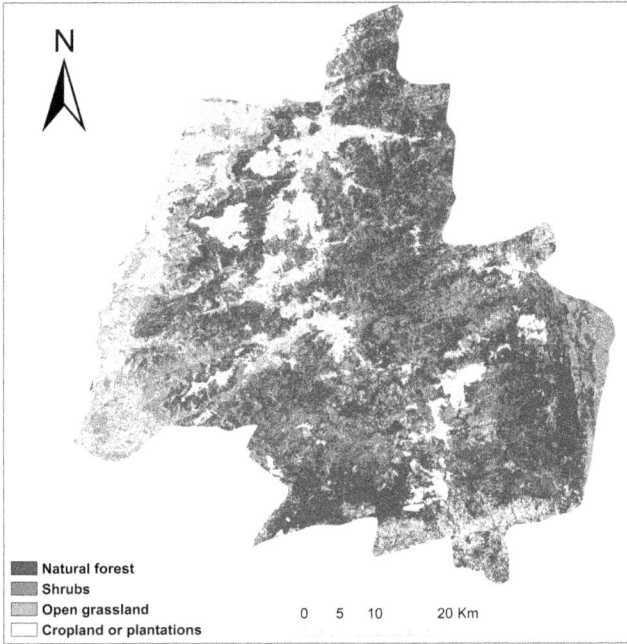

(a)

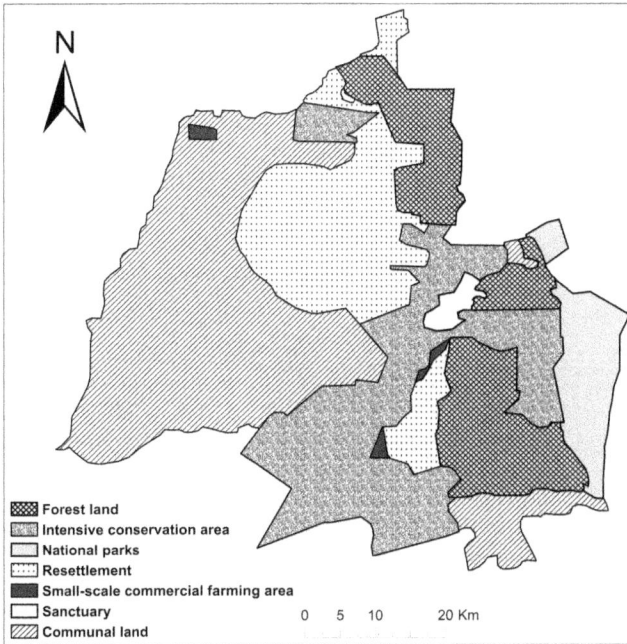

(b)

Figure 3.1 Map of Chimanimani indicating (a) land cover and (b) land use

The land comprises an intricate and constantly changing mosaic of uses nested within each other: for example, crop fields may be located within grazing areas or illegal crop farming might take place in timber plantations. The latter situation arises where peasants encroach on timber plantations to establish crop fields, often, as we will describe, with the support of politicians who seek to garner votes from those who settle in the timber plantations.

At farm level, two types of fields can be distinguished: irrigation schemes (*maeka*) and dryland fields (*munda wekunze*). Cash crops dominate the irrigation schemes while diverse traditional food crops are grown in the dryland fields (Kozanayi, 2019). The majority of farmers in Chimanimani have adopted conventional agricultural production methods, largely owing to land use pressures and cash incentives offered by the Grain Marketing Board (a parastatal that buys cereals from farmers) for mainly monocropping hybrid maize with the use of synthetic fertilizer. Agroecology is practised by some, but not extensively (TSURO, 2020).

As observed in the Towards Sustainable Use of Resources Organization (TSURO) (2020) report, annual rainfall has been decreasing (see Figure 3.2), and the frequency of years with negative rainfall anomalies per 10-year interval is increasing, from 3.5 years per decade between 1898 and 1908 to more than 6.5 years per decade between 2008 and 2018. The latter is evidenced by the years preceding Cyclone Idai, when the district experienced episodes of erratic rainfall distribution, characterized by droughts in the middle of the rainy season which made the cultivation of certain crop varieties impossible (TSURO, 2020).

Chimanimani has witnessed devastating droughts, the worst on record being in 1992, which resulted in a complete loss of crops and livestock in most parts of the district, affecting the lowlands particularly severely (Scoones

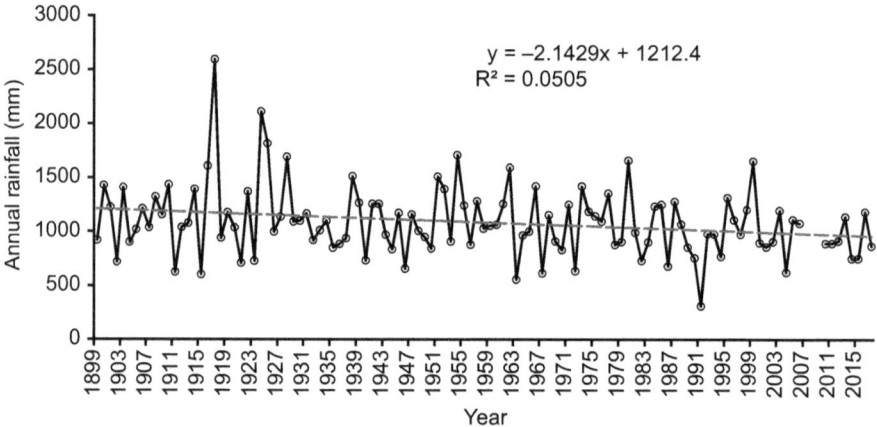

Plot annotation: $y = -2.1429x + 1212.4$
$R^2 = 0.0505$

Figure 3.2 Long-term changes in annual rainfall in Chimanimani District
Source: TSURO, 2020

et al., 1996). More recently, the El Niño phenomenon induced a drought in the 2015–16 cropping season which resulted in harvest failure. This was compounded by an outbreak of the fall armyworm, which devastated maize fields, also affecting cowpea and sorghum crops.

Tropical cyclones are not unknown in the Chimanimani District. In the past two decades a number of cyclones have made landfall, notably Cyclone Eline (February 2000), Cyclone Japhet (March 2003), and Cyclone Dineo (February 2017). No loss of human life was recorded from these cyclones, which, in the view of local residents, rendered their impact benign in comparison to Cyclone Idai.

Farmers in Chimanimani may all be affected by the same disaster, but because of differences in topography, land use, and soil cover, they experience it to different degrees. For example, the uplands have greater topographic relief and slope than the lowlands, making them more susceptible to landslides. Other factors which might contribute to landslides are timber plantations, deforestation, veld fires, cultivation on marginal lands, and the construction of settlements on steep slopes (TSURO, 2020).

The highlands are planted to commercial timber (pine and gum trees) interspersed with fireguards (firebreaks). These plantations are owned by conglomerates such as Border Timbers. Since the 2000s, there have been conflicts between timber companies and local landless people who invaded the plantations during the implementation of the land reform programme. Though the timber plantations were not targeted for land redistribution, land-hungry peasants took advantage of the land reform programme to settle there. The government has struggled to resolve the impasse between the timber companies and the land invaders, who chop down timber to establish crop fields, some of which are situated on hilly ground.[2]

Impacts of Cyclone Idai

Cyclone Idai had a devastating impact on Chimanimani: the 13 wards directly in its path bore the brunt of the high winds and heavy rains that culminated in floods and landslides. Human lives were lost, children were orphaned, many were injured, and some of those who survived the ordeal found themselves internally displaced as the land where their homesteads and fields had once stood was swept away. Strong winds were closely followed by heavy rains which continued for five days, some reports estimating that a year's worth of rain fell during that period (TSURO, 2020). A resident of Chimanimani described the cyclone's advance as follows: 'First came the trees, big trees, and then the mud, then the boulders, some of them huge, and lastly the water' (see Photo 3.1).

The landscape

In April 2019, the World Bank initiated a joint exercise with the Zimbabwean government to assess the losses and damages arising from the cyclone and

Photo 3.1 Crop field washed away by flash floods, Ngorima, Chimanimani
Credit: Witness Kozanayi

to develop a strategy for immediate recovery as well as longer-term resilience building under the Zimbabwe Rapid Impact and Needs Assessment (RINA) initiative (RINA, 2019). The Tudor Trust (United Kingdom) and Bread for the World (Germany) later supported a research project commissioned by TSURO and conducted by several universities and local institutions to understand the causes and impacts of Cyclone Idai as well as to inform recovery and development options. The TSURO (2020) report indicated that 80 per cent of arable land had been affected, of which 57 per cent was conventionally managed land, compared to only 29 per cent of agroecologically managed land. As indicated in Table 3.1, undisturbed forested areas and timber plantations fared equally well, with only 3 per cent of the area covered by each impacted (TSURO, 2020).

Land use patterns determined impacts on the landscape. For instance, veld fires deliberately set to clear land for cultivation destroyed vegetation, leaving the soil vulnerable to agents of soil erosion (WK, personal observation, September 2020). The fireguards around timber plantations acted as waterways, with disastrous consequences for the areas below as gullies and mudslides formed along them. Patchy ground cover led to increased surface flow which in turn induced the formation of gullies in grazing lands.

Cases of unsustainable farming practices such as monocropping along riverbanks, and the opening up of crop fields on hilly areas without the requisite soil and water conservation measures, were reported (TSURO, 2020). In some areas, local leaders settled people in fragile areas such as lands formerly designated for grazing, or waterways. Across much of the district farmers had stopped constructing or maintaining soil conservation structures such as contour ridges, ostensibly as a result of government officials reducing the frequency of

Table 3.1 The level of destruction experienced by different land use types in Chimanimani District (n = 817)

Area (ward)	Land use system				
	Conventional farmlands (%)	Agroecology farmlands (%)	Grazing lands (%)	Forest plantations (%)	Undisturbed areas (%)
Chakohwa	29	57	9	4	1
Chayamiti	49	41	2	6	2
Chikukwa	82	15	1	0	2
Martin	45	6	47	1	1
Nyahode	64	17	4	4	11
Biriiri	71	20	1	2	6
Ngorima A	64	28	4	2	2
Ngorima B	52	47	0	1	0
Average	**57**	**29**	**8**	**3**	**3**

Source: TSURO, 2020

their monitoring of these structures. Without these in place in crop fields, soil erosion had become rampant. There had been considerable loss of trees prior to the cyclone, mostly as land was opened up for farming and energy needs. This again led to bare soils that were susceptible to erosion (TSURO, 2020).

In the low-lying areas, plant cover had been reduced due to the recurrent droughts which preceded the cyclone. From December 2018 to February 2019 very little rain fell, resulting in extremely dry soils, which prevented the torrential rainfall from being absorbed, in turn worsening flash floods (RINA, 2019). Other soils were compacted due to tillage and livestock movements, which also reduced infiltration rates (TSURO, 2020). In some areas, crop fields were either washed away or flooded and silted up. Even fields located more than 30 m away from riverbanks, as recommended by Zimbabwe's Environmental Management Agency, were washed away as rivers burst their banks and changed course (RINA, 2019).

Areas managed agroecologically, and thus exposed to practices such as crop cover, mulching, and soil conservation works such as swales and terraces, were less eroded than areas where trees had been removed and grazing lands overstocked (RINA, 2019). Post-Idai soil analysis showed more nitrates in the topsoil, higher rates of infiltration, and more organic matter in agroecologically farmed areas than those managed conventionally (Madanzi et al., forthcoming).

In the aftermath of the cyclone farmers practising agroecology drew on their social capital to galvanize people to rehabilitate damaged landscapes and infrastructure and restore lost seeds. Informants gave vivid accounts of how farmers who were organized around agroecology easily came together and worked as a team after the disaster. Group solidarity was key in the rendering of support to affected families before and after external assistance from the government and NGOs arrived.

Table 3.2 A summary of the destruction caused by Cyclone Idai in Chimanimani District

Type of loss	Extent of loss
Arable land	18,244 ha[1]
Maize	7,100 ha
Bananas	1,626 ha
Pineapples	131 ha
Grazing lands	105 ha
Plantations: mangoes, oranges, macadamia nuts	85 ha
Seeds lost as crops in the field (finger millet, sorghum, maize)	52%
Irrigation schemes with damaged weirs	7
Critical bridges	10
Households displaced	8,000[2] (24.6% of district total)

[1] Representing 28.3 per cent of the total of 64,457 ha of arable land in the district (RINA, 2019).

[2] The total number of households in Chimanimani as at the last population census (CSO, 2012) was 32,578.

Sources: Chikukwa, 2019; RINA, 2019

Table 3.2 summarizes some of the destruction caused by Cyclone Idai.

Livelihoods

Since Chimanimani is a largely agrarian district, it comes as no surprise that the most severe livelihood impacts were felt by those who practise farming. Beyond the immediate loss of arable land and crops in the fields, grain and seeds in storage, beehives, and livestock were also lost.

Many livelihood streams were affected by the cyclone, as summed up by one farmer: 'The major sources of income and food in the district have been long-term affected by the cyclone: small livestock production and sales, crop production, value addition, agricultural casual work, petty trading, beekeeping, barter trading, community-based seed systems, and brick moulding.' Collectively, these losses, combined with knock-on effects such as higher prices for food and other commodities, would bring hardship to many households.

In response to the dire state of affairs they found themselves in, local residents started getting involved in various enterprises which were not directly linked to their farmlands. While some engaged in gold and diamond panning in the streams, others tapped into the bounty held within the remaining indigenous forests which survived the cyclone largely intact, collecting wild products such as honey, baobab fruit, and various edible insects (TSURO, 2020). These products of the 'hidden harvest' (Campbell and Luckert, 2002) were either consumed, bartered, or, in some cases, sold.

Local NGO TSURO has for a number of years supported agroecology-based enterprises such as livestock keeping, peanut butter manufacturing, fruit processing, and the production of local seed. These enterprises served as a source of either food or income soon after Cyclone Idai passed through the district.

Food security

Food security was significantly impacted by the cyclone, with around 77 per cent of Chimanimani residents needing food assistance (TSURO, 2020). Residents remarked: 'The nutritional status of many households is likely to worsen, and this is especially so for vulnerable people,' and 'The effect of the cyclone is that many households in the district now survive on food handouts.' Support from NGOs and the government started flowing in soon after the president of Zimbabwe declared a state of disaster. The government sought the support of the World Bank, other donors, and NGOs, spending the money collected through these channels chiefly on large infrastructure projects. A portion of the donations was earmarked for 'enhanced agriculture resilience', which, in effect, turned out to be mainly irrigation infrastructure (RINA, 2019).

The provision of food was aimed at vulnerable groups, partly through a miniscule budget (US$40,000, or 0.008 per cent of a total of $494 m), which was allocated to establish and support 'nutrition gardens'.[3] A much bigger role was played by seed and fertilizer companies, which were mobilized by the government to prioritize selling their products either directly to affected communities or to companies which were supplying goods to the affected communities. Heeding the government's plea, seed companies and agro-dealers set up outlets in Chimanimani. To buy their products, however, farmers needed cash, which was in short supply after the disaster. Any savings farmers had had been used for more pressing needs such as food, clothes, and shelter. The success of this government-led intervention was also thwarted by delays. By the time the seed companies were able to move their stock to the affected areas (after damaged roads had been repaired) it was too late for farmers to plant rain-fed crops. Therefore, not only could farmers not utilize those seeds, but other factors made farming difficult, if not impossible, as soils had been washed away from the uplands and stream banks, irrigation schemes had been flooded, and fields in the valley had silted up (TSURO, 2020).

In order to stem growing food insecurity, households adopted various strategies to acquire food, some used historically in times of food scarcity, others new. Many households relied on multilayered strategies that buttress local food systems, including harvesting, consuming, or trading forest products, exchanging labour and livestock, inter-household food sharing, and drawing on social relations, some of which transcended national borders. The state and NGOs did offer assistance, but this only happened long after the affected communities had triggered local food security strategies.

A number of enterprising residents turned to barter and trade, exchanging their excess grain, acquired in times of excess long before the cyclone hit, for grocery items or transportation. Though households had always traded food at the local level, there was evidence of more traders acting opportunistically and selling food in the affected communities after Cyclone Idai. Some cross-border traders and bus crews served as couriers of food, mainly maize meal, from South Africa.

Soon after the cyclone hit the area, residents' first line of defence was reliance on social networks to acquire, exchange, and buy food. As remarked by an elderly man in Ward 17, 'Food is to be shared, regardless. The notion of segregated food distribution is alien to us, it came with NGOs and government'. What little food households had, they shared with those less fortunate. And when food aid parcels arrived and some residents were denied them because they did not meet the donor criteria, recipient households reportedly shared their food with them.

Seed

Many farmers in Chimanimani rely on saving and exchanging seed in order to have enough locally adapted planting material available throughout the year (TSURO, 2020). Although a lot of seed was lost in the cyclone, or damaged by floodwater, a number of farmers managed to save some of their stocks. Post-Idai, seed sharing and exchange became stronger, with many local seeds, such as cowpeas, roundnuts, sorghum varieties, millets, cocoyam, sweet potatoes, and pumpkins, shared. 'We were sharing local seed we had remaining – with neighbours, church members, farmers from far-off areas, everyone. TSURO also organized a look-and-learn tour for us to the Valley (Nyanyadzi Ward) and we used that trip to borrow seed from other farmers,' remarked one farmer.

Once the road network was restored, some NGOs and the government started distributing seed to affected communities. Different organizations used different criteria in the selection of beneficiaries, depending on donor preferences: for example, some organizations targeted widows specifically. Neither the government nor NGOs (with the exception of TSURO and Jekesa Pfungwa) supported local farmers with local seeds; yet these constitute the bulk of the food basket. Where other NGOs distributed what they labelled as 'local' seed, it was not the type the locals were familiar with and therefore not the farmers' preference. For example, farmers in the dry valley reportedly received a late maturing type of cowpea, meaning it would not withstand the hot, dry valley conditions. Some locals shared the hybrid maize seed which they got from government and NGOs, something that would not normally happen, as those donors usually make follow-up visits to check if the seed they have distributed has indeed been planted as allocated, effectively discouraging beneficiaries from sharing the seed.

Zimbabwe's national gene bank provided sorghum seed to farmers, also carrying out research to establish the ex-ante and ex-post trends and status of local seed systems. Fortuitously, the gene bank had collected accessions from Chimanimani before Cyclone Idai hit the district and was able to return these to farmers to bulk up.

Psychosocial needs and social cohesion

Less direct, but equally devastating, were secondary impacts such as the psychological toll experienced by those who witnessed the destruction and lost loved ones. As remarked by a farmer, 'Everyone was lacking in some respect ... some families needed psychological support after the loss of relatives or infrastructure, others needed land, some food.' An unintended positive outcome was that the disaster brought people closer together than before: in the words of a local resident, 'Your neighbour was the first line of support before the government and donors came.'

According to the TSURO (2020) report, farmers who practised agroecology demonstrated a stronger inclination to share labour and seed, as well as to organize mutual aid for early response and recovery. These farmers mobilized around collective actions for landscape restoration to mitigate the impact of future events and reported a renewed appreciation of the importance of lost social-ecological connections.

Government response to Cyclone Idai

When Cyclone Idai struck, the government of Zimbabwe responded through the Department of Civil Protection, its disaster management coordination arm. The response exposed multiple weaknesses, however, including those related to risk management, policy, institutions, and capacity (ACB, 2020). With close to $10 bn in debt and repayments mounting, the government simply did not have the finances to operationalize the resources needed for recovery efforts, obliging it to depend on donor support (ACB, 2020).

Relying on donor and NGO support, the government first established the Zimbabwe Idai Recovery Project and later the Post Cyclone Idai Emergency Recovery and Resilience Project. The latter, worth $24.5 m, aimed to restore essential services including transport, electricity, water, and sanitation to the most severely affected communities (UNDP, 2020).

Decision-making post-Idai was highly politicized, which resulted in unclear plans and disorganized implementation (TSURO, 2020).

Discussion

Three central findings emerge from this study, all of them linked to resilience, whether related to ecological functioning, climate, or human–environment systems. First, farms under agroecological management fared better after the cyclone than those managed conventionally; in other words, agroecological farms were more resilient than conventional farms. Second, the response from government and aid agencies was deemed to be uneven and inappropriate, at times undermining community resilience. Third, elements of landscape agroecology practised before the disaster pointed the way towards a more resilient agrarian future in Chimanimani.

In this discussion we will analyse the impact of Cyclone Idai in Chimanimani in terms of resilience, probing the elements which contributed to residents' ability to resist shocks, and examining which factors influenced their capacity to recover from shocks (TWN and SOCLA, 2015).

With evidence pointing to links between Cyclone Idai's destructiveness and climate change, it is imperative to look at ways in which agrarian societies can be more resilient in the face of a changing climate, and one pathway which is emerging strongly is agroecology (Knutson et al., 2010; Hernandez, 2019). Agroecology is viewed as crucial to building resilience to climate change through presenting a framework for a range of sustainable practices and approaches which, importantly, also incorporates social and political dimensions (SKI, 2020). In the context of climate change, resilience is defined as 'the capacity of social, economic, and environmental systems to cope with a hazardous event or trend or disturbance, responding or reorganizing in ways that maintain their essential function, identity, and structure, while also maintaining the capacity for adaptation, learning, and transformation' (IPCC, 2014: 5).

Agroecology as a vehicle for resilience

Proof of the resilient nature of agroecologically managed farming systems can be found in the observation that millions of small-scale farmers around the world have been successfully practising resource-conserving agriculture for thousands of years while routinely facing environmental and economic change (Altieri and Koohafkan, 2008). The fact that the farms in Chimanimani under agroecological management had less damage after the cyclone than those managed conventionally accords with findings observed elsewhere in the world. For instance, when Hurricane Mitch devastated parts of Central America in 1998, farms utilizing a wide range of crops and soil conservation measures suffered less damage than those planted to monocrops without any soil conservation measures (TWN and SOCLA, 2015). Similarly, when Hurricane Ike struck Cuba in 2008, diversified farms utilizing hedgerows, inter alia, suffered 50 per cent losses, compared to 90–100 per cent losses on conventional farms (Altieri and Toledo, 2011; TWN and SOCLA, 2015).

In Chimanimani, farms under agroecological management contain more agrobiodiversity (TSURO, 2020), a component of agroecological farming which, according to Altieri et al. (2015), increases resilience to climate-induced disasters. Agroecological management of agrobiodiversity takes many forms, conferring resilience in different ways: intercropping, for example, minimizes the risk of losing the entire crop during a natural disaster whereas polycultures survive shocks such as droughts better than monocultures (Altieri et al., 2015).

Areas managed agroecologically were able to absorb and sink more excess rainwater than those managed conventionally, thanks to high levels of organic matter in the soil and low levels of compaction (TSURO, 2020).

According to SKI (2020), the following factors encourage rainwater run-off by disrupting the hydrological cycle: deforestation, growing monocultures, and poor grazing practices. Avoiding these is vital for continued farming, as is capturing rainwater in the soil to recharge the water table and reduce the run-off that removes nutrient-rich topsoil (Richmond, 2017; SKI, 2020).

Maintaining intact natural areas such as forests is an important facet of landscape agroecology. As indicated in Table 3.1, both intact natural forests and alien timber plantations weathered the cyclone with minimal damage. The reason for this may well be that the plant cover prevented soil loss. Mendez (2016), however, posits that while timber plantations are grown as even-aged, mostly young, monocultures, natural forests are more resilient as they contain varied growth, soil, and wildlife. The timber plantations in Chimanimani may have withstood the effects of the cyclone, but the fireguards surrounding them acted as pathways for landslides, leading to soil loss and gulley formation.

Disaster responses at multiple levels: enhancing or undermining community resilience?

Small-scale farmers frequently rely on socially mediated networks to cope with extreme climate-induced events (Altieri and Koohafkan, 2008). For example, immediately after the disaster struck Chimanimani, residents reached out to support their neighbours and in turn relied on social networks and relationships for psychosocial support, a core element of community resilience (Patel et al., 2017). The positive aspects of social networks and relationships, such as cohesion and connectedness, contribute greatly towards helping people cope with the uncertainty which so often accompanies a disaster (Patel et al., 2017). Research informing the TSURO (2020) report found that farmers practising agroecology worked together to restore the landscape in anticipation of future climatic disasters, taking actions that are important for decreasing vulnerability.

Reliance on social networks happened long before official assistance in the form of government support programmes and international aid could reach Chimanimani, and, unfortunately, when official support did arrive, it was uneven and, in some cases, unsuitable, thus undermining local recovery attempts. The distribution of seed is a good example. In Chimanimani seed saving and exchange are at the centre of traditional agricultural systems. These activities have been shown to increase the resilience and autonomy of smallholder farmers, also enhancing social cohesion through stronger community and family ties (Van Niekerk and Wynberg, 2017). In Chimanimani, the farmers who managed to hold on to precious reserves of traditional seed demonstrated their solidarity by sharing this seed with those in need, whether relatives, neighbours, or other smallholders. The seeds they shared were familiar to them as they were locally adapted varieties of the crops they would normally consume. This contrasted with the seed distributed by international aid agencies, some NGOs, and government, which were unfamiliar to farmers and, in some cases, unsuited to the local climate. This corresponds with research conducted by Joshi and Gauchan (2017) in Nepal after the earthquakes of 2015. The authors found that the Nepalese government's response indicated a lack of understanding of the value of local agrobiodiversity in rebuilding local seed systems and improving the food security and livelihoods of people in marginal areas.

The Zimbabwean government's intervention of inviting seed houses to sell hybrid seed and fertilizer to the affected communities was disingenuous, as the communities did not have disposable income after their main source of revenue, farming, had been disrupted. As Sinclair et al. (2019) observe, the perverse subsidization of artificial fertilizers and pesticides encourages farmers to adopt industrial agriculture rather than agroecology, even though agroecology is known to confer climate resilience.

In terms of the future recovery of local seed systems, the national gene bank of Zimbabwe has embarked on a promising programme. Through the Foundations for Rebuilding Seed Systems Post Cyclone Idai project, the gene bank has started the process of understanding the extent of local seed loss with a view to reintroducing the missing seed by using local farmers to bulk the seed from genetic material collected before the cyclone struck. This intervention bodes well for the potential recovery and strengthening of local seed systems.

Besides the dissatisfaction with certain segments of the government's seed programmes, the state was also criticized for being unable to restore disrupted livelihoods after the disaster (TSURO, 2020). Largely due to a lack of funds, communities, local NGOs, and individuals were left to rebuild their own livelihoods, meaning that it would take longer to build resilience. Some tapped into their local knowledge of natural resources to generate an income, an essential component of community resilience, according to Patel et al. (2017). One woman farmer, for example, crafted a new livelihood and boosted her household's food security by selling baobab fruit. This accords with research undertaken by Kozanayi (2018), who contends that the contribution of forest products is often overlooked in interventions that seek to address food insecurity, even though the role of these products goes beyond their default recognition as safety nets.

Landscape governance: key to a more resilient future?

In the context of agroecology, social resilience pertains to the management of the landscape by social networks (TWN and SOCLA, 2015). These networks form part of the natural, social, and human assets emphasized by the sustainable and resilient agroecological approach to farming, in contrast to the industrial model, which emphasizes financial and physical assets (Lengnick, 2015). According to the Seed and Knowledge Initiative (SKI, 2020), agroecology is able to support community resilience at the landscape level since it offers sociopolitical values alongside proven agricultural practices.

For a landscape approach to succeed in Chimanimani, whether as part of the restoration efforts after Cyclone Idai or as the foundation for a long-term resilient and sustainable agrarian future, communities and their organizations should participate effectively in the decisions which influence their landscape, and the institutions and processes within that landscape need to be collaborative and flexible (SKI, 2020).

Landscape governance is complex, particularly in Chimanimani, where formal and informal actors' responsibilities for land and land use allocation overlap. The Chimanimani Rural District Council (RDC) is the *de jure* land authority but is supposed to allocate land in consultation with traditional authorities (Kozanayi, 2018). The *de facto* situation is that the RDC is active in the allocation of land for business centres and social amenities, while traditional authorities are responsible for the day-to-day allocation of land in all rural areas. The multiplicity of actors involved in land allocation results in mosaics of land uses nested within other pre-planned land uses. For example, traditional authorities sometimes allocate plots to land seekers in areas designated as grazing areas by the RDC. On the other hand, the RDC, using its powers as the land authority, can override traditional land use plans and reallocate land for uses other than those earmarked by the traditional authorities.

To complicate matters further, politicians frequently get involved in allocating land acquired through the land reform programme, which illustrates how politics can shape landscapes and create territories (SKI, 2020). This situation is likely to complicate the implementation of a landscape agroecology approach in Chimanimani, as it is not clear whether there will be cooperation among the various stakeholders involved in land use (including the powerful business interests vested in the timber plantations), among governmental departments and hierarchies, between the government and communities, or within and among communities (Minang et al., 2015). With frequent and unpredictable changes in land use pattern, agroecological practices such as mulching and agroforestry, which require some degree of permanence, are impossible to implement.

Conclusion

In Chimanimani, Idai was the most devastating tropical cyclone experienced in living memory, and climate experts predict that future cyclones in the area are likely to be as intense, if not more so. In order to plan for a more sustainable and resilient agrarian future in Chimanimani and elsewhere, a holistic appreciation needs to be realized to better inform the design and implementation of land use and agricultural production plans. A piecemeal approach to understanding impacts would overlook important and effective local-level mechanisms for dealing with such disasters.

Resilience to disasters should be viewed from both biophysical and social perspectives. In this chapter we have found sound evidence proving that agricultural land under agroecological management is more resilient to disasters than land under conventional management. Furthermore, undisturbed forests – a feature of landscape agroecology – were shown to constitute the most resilient land use, a fact which traditional authorities and the RDC should take on board when allocating land for cultivation, grazing, and the establishment of settlements.

We also found that local people relied strongly on social networks and solidarity to support each other and cope with the after-effects of the cyclone, highlighting these as the first defence mechanism. In order to design locally appropriate and effective disaster management plans, time should be devoted to integrating local experiences and understanding local livelihood-rebuilding mechanisms.

Post-disaster seed restoration plans would do well to follow in the footsteps of the national gene bank's Foundations for Rebuilding Seed Systems Post Cyclone Idai project, with its close focus on understanding local seed systems and involving farmers in bulking up lost seeds for future use. Returning to the theme of 'local is the first defence', it is also advisable for government departments and NGOs to support household seed saving as this is often the first port of call for farmers to replenish their lost seeds.

For Chimanimani to transition towards a sustainable, resilient agrarian future, many aspects of landscape agroecology need to be taken into consideration. Although there is no blueprint for a landscape approach, it is important for it to be responsive to its social, cultural, and political contexts. For example, strengthening social networks will enhance social and community resilience, as well as support landscape management. Collaborative and flexible institutions and processes will ensure more effective decision-making; and complex land tenure arrangements could be more successful if approached with a view to fostering cooperation between stakeholders.

Acknowledgements

We gratefully acknowledge the input from Elfrieda Pschorn-Strauss on an earlier draft of this chapter, and we thank Rachel Wynberg for her valuable comments, which strengthened our analysis and discussion.

Notes

1. Not her real name.
2. According to the Timber Producers Federation, 17,544 ha of plantation timber have been occupied by illegal settlers (Chifamba, 2017).
3. These gardens are mainly supported by NGOs to promote the production of a diversity of vegetables and herbs. The aim is to meet the dietary requirements of members, who are usually vulnerable groups in society (Muzawazi et al., 2017).

References

African Centre for Biodiversity (ACB) (2020) *Neo-colonial economies and ecologies, smallholder farmers and multiple shocks: The case of cyclones Idai and Kenneth in Mozambique and Zimbabwe*, Discussion Paper, African Centre for Biodiversity, Johannesburg.

Altieri, M.A. and Koohafkan, P. (2008) *Enduring Farms: Climate Change, Smallholders and Traditional Farming Communities*, Third World Network, Penang.

Altieri, M.A. and Toledo, V.M. (2011) 'The agroecological revolution in Latin America: Rescuing nature, ensuring food sovereignty and empowering peasants', *The Journal of Peasant Studies* 38(3): 587–612 <https://doi.org/10.1080/03066150.2011.582947>.

Altieri, M.A., Nicholls, C.I., Henao, A. and Lana, M.A. (2015) 'Agroecology and the design of climate change-resilient farming systems', *Agronomy for Sustainable Development* 35: 869–890 <https://doi.org/10.1007/s13593-015-0285-2>.

Campbell, B.M. and Luckert, M.K. (2002) 'Towards understanding the role of forests in rural livelihoods', in B.M. Campbell and M. Luckert (eds), *Uncovering the Hidden Harvest: Valuation of Woodland and Forest Resources*, pp. 1–16, Earthscan, London.

Central Statistical Office (CSO) (2012) *National Census Profile*, Central Statistical Office, Government of Zimbabwe, Harare.

Chifamba, O. (2017) 'Illegal settlers in timber estates face eviction', *The Herald*, 27 September <https://www.herald.co.zw/illegal-settlers-in-timber-estates-face-eviction/>.

Chikukwa, T. (2019) 'Effects of Cyclone Idai on agriculture in Chimanimani District', presentation at the Cyclone Idai planning workshop, 3 July, Bronte Hotel, Harare.

Chimanimani Rural District Council (CRDC) (2017) *Chimanimani District Climate Change Response and Watershed Management Policy*, adopted 11 September, Resolution No. C3331.

Hernandez, A. (2019) 'Tropical Cyclone Idai, climate change and the North-South divide', *Inside Over*, 14 April <https://www.insideover.com/environment/tropical-cyclone-idai-climate-change-and-the-north-south-divide.html>.

International Panel on Climate Change (IPCC) (2014) 'Summary for policymakers', in C.B. Field, V.R. Barros, D.J. Dokken, K.J. Mach, M.D. Mastrandrea, T.E. Bilir, M. Chatterjee, K.L. Ebi, Y.O. Estrada, R.C. Genova, B. Girma, E.S. Kissel, A.N. Levy, S. MacCracken, P.R. Mastrandrea, and L.L. White (eds), *Climate Change 2014: Impacts, Adaptation, and Vulnerability. Part A: Global and Sectoral Aspects. Contribution of Working Group II to the Fifth Assessment Report of the Intergovernmental Panel on Climate Change*, pp. 1–32, Cambridge University Press, Cambridge and New York <https://www.ipcc.ch/site/assets/uploads/2018/02/ar5_wgII_spm_en.pdf>.

Joshi, B.K. and Gauchan, D. (eds) (2017) *Rebuilding local seed system of native crops in earthquake affected areas of Nepal: Proceedings of a National Sharingshop*, 18 December, Kathmandu, NAGRC, Bioversity International and Crop Trust, Kathmandu, Nepal <https://cgspace.cgiar.org/bitstream/handle/10568/90412/Rebuilding_Gauchan_2017.pdf?sequence=1&isAllowed=y>.

Knutson, T.R., McBride, J.L., Chan, J., Emanuel, K., Holland, G., Landsea, C., Held, I., Kossin, J.P., Srivastava, A.K. and Masato, S. (2010) 'Tropical cyclones and climate change', *Nature Geoscience* 3: 157–163 <https://doi.org/10.1038/ngeo779>.

Kozanayi, W. (2018) *Influences of customary and statutory governance on sustainable use and livelihoods: The case of baobab, Chimanimani District, Zimbabwe.* PhD thesis. University of Cape Town.

Kozanayi, W. (2019) 'Community seed mapping for Nyanyadzi, Chimanimani District, 11–15 November', unpublished report submitted to the TSURO Trust.

Lengnick, L. (2015) *Resilient Agriculture: Cultivating Food Systems for a Changing Climate*, New Society Publishers, Gabriola Island.

Madanzi, T., McAllister, G., Kozanayi, W., Goss, M., Gadzirayi, C. and Chikukwa, T. 'Farming in disaster-prone landscapes: Making the case for skilling-up and scaling-out agroecology across Zimbabwe's Eastern Highlands after Cyclone Idai' (unpublished manuscript).

Mendez, A. (2016) 'Tree harvesting: Old growth forests vs. tree plantations' [presentation] <https://prezi.com/rga30c41fanv/tree-harvesting-old-growth-forests-vs-tree-plantations/>.

Minang, P.A., Duguma, L.A., Alemagi, D. and van Noordwijk, M. (2015) 'Scale considerations in landscape approaches', in P.A. Minang, M. van Noordwijk, O.E. Freeman, C. Mbow, J. de Leeuw, and D. Catacutan (eds), *Climate Smart Landscapes: Multifunctionality in Practice*, pp. 121–134, World Agroforestry Centre, Nairobi.

Muzawazi, H.D., Terblanché, S.E. and Madakadze, C. (2017) 'Community gardens as a strategy for coping with climate shocks in Bikita District, Masvingo, Zimbabwe', *South African Journal of Agricultural Extension* (1)45: 102–117 <https://www.scielo.org.za/scielo.php?script=sci_arttext&pid=S0301-603X2017000100010>.

Ninteretse, L. (2019) 'Cyclone Idai shows the deadly reality of climate change in Africa', *The Guardian*, 21 March <https://www.theguardian.com/commentisfree/2019/mar/21/cyclone-idai-climate-change-africa-fossil-fuels>.

Patel, S.S., Rogers, M.B., Amlôt, R. and Rubin, G.J. (2017) 'What do we mean by "community resilience"? A systematic literature review of how it is defined in the literature', *PLoS Currents* 9 <https://www.ncbi.nlm.nih.gov/pmc/articles/PMC5693357/>.

Rapid Impact and Needs Assessment (RINA) (2019) *Zimbabwe Rapid Impact and Needs Assessment*, May, Government of Zimbabwe, Harare <https://documents1.worldbank.org/curated/en/714891568893029852/pdf/Zimbabwe-Rapid-Impact-and-Needs-Assessment-RINA.pdf>.

Richmond, D. (2017) 'Water retention landscapes: The solution to the water challenge we are facing', *Quinta das Abelhas* [blog].

Scoones, I., Chibudu, C., Chikura, S., Jeranyama, P., Machaka, D., Machanja, W., Mavedzenge, B., Mombeshora, B., Mudhara, M., Madziwo, C., Murimbarimba, F. and Zirereza, B. (1996) *Hazards and Opportunities: Farming Livelihoods in Dryland Africa: Lessons from Zimbabwe*, Zed Books, London.

Seed and Knowledge Initiative (SKI) (2020) *Working Together to Build Resilient Landscapes Using a Regenerative, Agroecological Approach*, Unpublished background paper, The Seed and Knowledge Initiative, Durban.

Sinclair, F., Wezel, A., Mbow, C., Chomba, S., Robiglio, V. and Harrison, R. (2019) *The Contribution of Agroecological Approaches to Realizing Climate-Resilient Agriculture*, background paper, Global Commission on Adaptation, Rotterdam and Washington, DC <https://gca.org/wp-content/uploads/2020/12/The ContributionsOfAgroecologicalApproaches.pdf>.

Third World Network (TWN) and Latin American Society of Agroecology (SOCLA) (2015) *Agroecology: Key Concepts, Principles and Practices*, Third World Network, Penang, and Latin American Society of Agroecology, Berkeley, CA <https://www.twn.my/title2/books/pdf/Agroecologycomplete1.pdf>.

Towards Sustainable Use of Resources Organisation (TSURO) (2020) *Building Resilience to Natural Disasters in Populated African Mountain Ecosystems: The Case of Tropical Cyclone Idai in Chimanimani, Zimbabwe. A Report on Environmental Impact and Climate Resilience Building Strategies*, TSURO Trust, Chimanimani <http://www.tsurotrust.org/wp-content/uploads/2022/11/ Final-Research-Report-_-Cyclone-Idai-as-at-26.11.2020-TSURO-Trust-Chimanimani.pdf>.

United Nations Development Programme (UNDP) (2020) 'Govt of China donates USD 2 million to rebuild houses, schools and clinics destroyed by Cyclone Idai', *ReliefWeb*, 3 July <https://reliefweb.int/report/zimbabwe/ govt-china-donates-usd2-million-rebuild-houses-schools-and-clinics-destroyed-cyclone>.

Van Niekerk, J. and Wynberg, R. (2017) 'Traditional seed and exchange systems cement social relations and provide a safety net: a case study from KwaZulu-Natal, South Africa', *Agroecology and Sustainable Food Systems* 41(9–10): 1099–1123 <https://doi.org/10.1080/21683565.2017.1359738>.

CHAPTER 4

Enacting indigenous and green revolutions in maize in West Kenya

Paul Hebinck and Richard Dimba Kiaka

Introduction

This chapter examines the enactment of two maize revolutions in West Kenya that have transformed and restructured the socio-ecological, institutional, and technological (pre)conditions for (maize) farming. We distinguish between the Green Revolution (GR), which is a part of the global effort to modernize agriculture (Otsuka and Larson, 2013), and an indigenous, peasant type of agricultural revolution (Richards, 1985; Van der Ploeg, 2013) linked to the broad principles of agroecology (Altieri and Nicholls, 2012). The GR dominates the agronomic and scientific food security literature (Onyutha, 2018), while agroecology is well covered in the rural sociology and anthropology literature (Val et al., 2019). Whereas the GR has attracted much of the funding from the global donor community, agroecology has been supported by a coalition of farmers and consumers, as well as farmer organizations, that form the core of the associated agroecology and food sovereignty movement (Holt-Giménez and Altieri, 2013; Peschard and Randeria, 2020). Although there is agreement about improving food security and livelihoods and tackling climate change, each body of literature constructs different trajectories to enrich land, labour, knowledge, seeds, and soils for (maize) farming. The GR targets yield increases through transforming breeding and commoditizing seeds and is widely praised by donors and experts for its success in this regard (Otsuka and Larson, 2013). Agroecology favours using locally available resources and is lauded for contributing to a just and sustainable food system that secures food security and food sovereignty (Holt-Giménez and Altieri, 2013). Both trajectories have also received criticism: the GR for 'betting on the strong' and not fully addressing food poverty and sustainability (Altieri and Nicholls, 2012; Holt-Giménez et al., 2021); and agroecology for not being able to produce enough food for the world population – or only being able to do so in very specific circumstances (Onyutha, 2018) – and for failing to accommodate climate change (Holt-Giménez et al., 2021).

The questions we address are: which of the revolutions in maize matters in West Kenya, for whom, how, and why? What is the nature of the revolutions and how does this manifest in the style of breeding and promoting? We argue, based on our longitudinal research in the region, that the enactment of maize farming defies many of the 'received wisdoms' claiming that, for instance, rural poverty prevents people from enacting modern farming or that local maize is unproductive and enhances poverty and food insecurity. We are critical of such wisdoms as they do not resonate with everyday reality in West Kenya. We have found that the way maize is farmed is much more complex, blurring boundaries that are strategically drawn between a GR-inspired agriculture and one encouraged by agroecology. To forestall an analysis that presumes an 'either or' kind of reasoning, we engage the issues through a grounded approach that incorporates the 'lived experiences' of farmers. Maize growers do not frame their identity and modernity in terms of *either* modern *or* local maize. They regularly make choices that take cognizance of the opportunities offered to them, but they also, perhaps pragmatically, balance their interpretations of risk while taking account of rapidly changing circumstances. They continuously consider shifting their maize farming between autonomous forms of farming and commoditized forms regulated by market forces and NGOs.

We begin by specifying our theoretical and methodological point of entry. We then zoom in on the West Kenyan setting to explore the history of maize varieties and evolving maize practices. We specifically pay attention to the mechanisms by which the GR and the indigenous agricultural revolution have transformed local food economies and how local people accommodate and respond to the challenges both revolutions pose.

Theoretical orientation and context

Our approach revolves around the idea that both agrarian revolutions and the growing of maize are *enacted*. 'Enactment' has entered the social science vocabulary and involves 'doing' or 'performing' (Law and Singleton, 2000). Human actors (e.g. farmers, breeders, traders, policymakers, donor agencies, and philanthropists) enact realities through practising maize (by selecting, breeding, planting, caring, and consuming) and facilitating (maize) research, extension, and marketing; but non-human actors are also involved (e.g. the different maize varieties, soils, rainfall patterns, markets, agrarian policies). The interplay of humans and non-humans generates agential capacities in both (Latour, 2005).

Using the case of maize, we account for a series of enactments that form the foundations for the GR and indigenous agricultural revolution. Enacting maize involves breeding and selecting maize genes, and distributing and sharing seeds, but also planting disciplines and routines. Given that these practices differ, we argue that maize is plural and that different 'maizes' are being enacted (Law and Lien, 2012). Breeders and maize growers enact maize by adhering to their respective scripts, but are steered by different interpretations of modernity, one of which we can associate with 'modern farming', discursively framed in terms of productivity, yield, commoditization, and entrepreneurial values, while another mirrors the principles of agroecology, of which taste, nutritious values, colour, local exchange, and culture are central and strategic elements. Both 'modern' and agroecological farming have built-in requirements for use that users need to comply with for the maize variety to do what it promises. This implies for the enactment that technical instructions and cultural procedures need to be followed. Such prescriptions are often poured into rituals, but also make giving gifts and attributing prestige enticing. This underscores the fact that maize varieties are enacted and embody 'structurally' different realities, traits, meanings, and relationships.

Historically, both revolutions have built on appropriating the accumulated experiences of peasant farmers, who, throughout the history of farming, have actively enriched their resources by improving their knowledge and, through observation and experimentation, the objects (e.g. seeds, soils) and instruments of their labour (e.g. plough, hoe). A strategic trait of agroecology is that it forms a crucial element in the struggle to prevent markets and agribusiness from increasing their control over farming (Van der Ploeg et al., 2019). Agroecology often unfolds in what Van der Ploeg (2013) has described as 'farming economically', which translates into low degrees of commoditization, tapping primarily but not exclusively from the immediate locality, and applying skill-oriented technologies that resonate with the local ecology. This kind of farming is not necessarily ideologically motivated; hence we may refer to it as 'agroecology-by-default'. With regard to maize, agroecology-by-default revolves around 'local' varieties which in the literature are termed 'landraces'. These are cultivars that have a long history of being cultivated in a region. Landraces are genetically heterogeneous and in a constant state of evolution; as a result of domestication by farmers and natural selection, they are well adapted to the local ecologies (Casañas et al., 2017). Maize is not indigenous to Africa (Miracle, 1966; McCann, 2005), which implies that the landraces in Africa originate from elsewhere via different trade-based arrangements involving spontaneous and planned actions by the (colonial) state.

The GR, in contrast, builds on a scientific regrounding of the foundation for agricultural production to increase yields in staple food crops. With regrounding, farmers' fields were superseded by laboratories and experimental stations, and farmer experimentation overshadowed by on-station trials. New plant breeding methods and chemical fertilizers took the place of soil biology, manure, and peasants' knowledge. Plant breeding, however, evolved over the years in significantly different directions (Kloppenburg, 1988). While originally breeding took place in adaptive on-station research conditions, with a focus on improved open-pollinated varieties (OPVs), maize breeding also, and increasingly, shifted to hybridization (Kutka, 2011). The embracing of hybrids occurred under the influence of Nobel Prize winner Norman Borlaug and the advances in molecular biology, and marks the foundation of what we now know as the GR. A major explanation, Kloppenburg (1988) suggests, is that plant breeding methods co-evolved with the privatization of the seed sector, contributing in Africa to a further disconnection of agriculture from its genetic resources, traditions, and belief systems.

Further distinctions in 'modern' maize can be made between OPVs and hybrids. Both are commodities and bred to be sold as improved cultivars and are produced by seed companies that operate differently. Whereas OPVs are bred and introduced with the option to be creolized by farmers within certain limits, hybrids are produced and marketed to be purchased for every planting season afresh. In Kenya, OPVs are nowadays released by the Kenya Agricultural Research Institute and NGOs and independent agronomists (KEPHIS, 2020). OPV-seed producers are generally not profit-minded but motivated to provide food security (Mango and Hebinck, 2004). They also receive funding from donor organizations. Hybrid maize is produced by seed companies that, with the support of the state's extension services, advise farmers to purchase fresh seed every year to generate potentially higher yields. Hybrids are produced to generate profit for shareholders of local (e.g. Kenya Seed Company [KSC], Western Seed) and global seed companies (e.g. Monsanto/Bayer, Pioneer, SeedCo). Hybridizing maize thus not only connects maize gene pools across the globe (i.e. the central Americas with sites in Africa) but also strategically connects donors, think-tanks (e.g. the World Bank), philanthropists like Bill and Melinda Gates, globalizing seed companies, and (biological and economic) scientists (Magdoff et al., 2000) to channel substantial development funding into agronomic research on developing new and high(er)-yielding cultivars, seed testing, and certification. Central to enacting the GR has been the progressive commoditization of knowledge and the objects, as well as instruments, of labour. Making the GR work entailed condensing the relations between producers and extension agencies, banks, consultants, seed and fertilizer companies, and markets.

Notwithstanding both structurally different styles of breeding, the motivation to facilitate the transformation from a 'traditional' peasant into a modern agriculture is common. Peasant agriculture was and still is perceived to lack the productive factors of production and adequate skills. Infusing 'traditional' agriculture with the products and principles of modern scientific agriculture would help close the yield gap. New (grain-fertilizer) technologies were designed as highly divisible and introduced as scale neutral, which would allow them to be integrated (or 'adopted') into existing systems of small-scale agriculture like those in West Kenya.

Data sources

Methodologically, enacting entails following the actors involved (Latour, 2005). This implied making a distinction between 'local' and 'hybrids', something that is actively done in the villages and in the literature. This is not our construct per se. For us, 'local-modern' or 'local-hybrid' serves to convey that maize is plural, representing multiple realities. We also tried to trace the origins of the varieties and examine the different arrangements and connections through which these arrived in West Kenya.

We started our studies in Luoland from 1996 onwards (Mango, 2002; Mango and Hebinck, 2004; Hebinck et al., 2015; Kimanthi and Hebinck, 2018; Kimanthi, 2019), which allowed us to trace the maize varieties and to follow several maize growers and their families through a series of detailed case studies and life histories over an extended period. We consulted the colonial archives in Nairobi to trace the origin of the maize varieties. We followed the actors that actively promote modern maize farming: state-organized research and extension, the Millennium Villages Project (MVP), and the One Acre Fund (OAF). The MVP (operational in Sauri between 2004 and 2015), which is part of the global Millennium and Sustainable Development Goals initiative, and the OAF (operational from 2006 onwards) are organizations implementing programmes with donor funding.

Two surveys were held. The first, in 2000, was limited to three villages in Yala location in Luoland (see Figure 4.1), with a small, purposeful sample of 40 maize-growing homesteads (Mango and Hebinck, 2004). The second, the Maize Variety Survey, was held in 2018–2019, covering nine sites in Luoland and Luhyaland in West Kenya, with a sample of 558 homesteads that together planted various maize varieties in 1,200 fields (Almekinders et al., 2021).

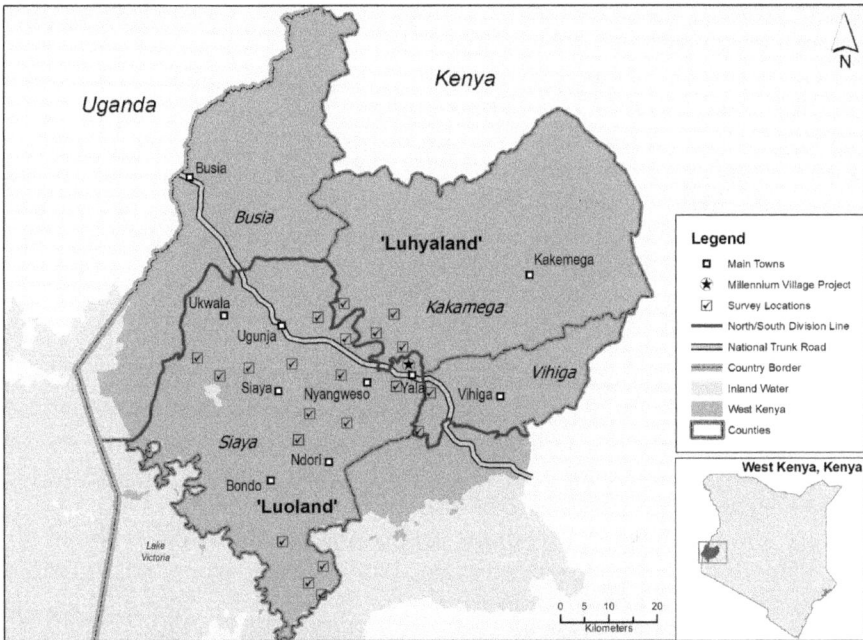

Figure 4.1 West Kenya and the location of Luoland and Luhyaland, Millennium Villages projects and the survey sites

Context

Luoland is primarily inhabited by Dholuo-speaking people of Nilotic descent. Luhyaland is populated by Luhya-speaking people, who are of Bantu descent. Culturally they differ from the Luo and do not share a seniority aspect in their social organization, including agriculture and settlement pattern. The seniority aspect in Luo agriculture, which entails performing rituals such as first planting (*golo kodhi*) and first harvesting (*dwoko cham*), requires the eldest in the homestead to initiate planting and harvesting, after which members of the younger generations follow (Mango, 2002). Apart from these rituals, both Luo and Luhya people plant maize in similar ways: that is, on small fields with local means to feed their families and sell locally. We extend our understanding of Luo agriculture to the way in which the Luhya farm and choose between the different varieties. Their names are different but are phenotypically comparable and identified and selected in similar ways.

We do not categorize those that plant maize as 'farmers' but rather as 'producers of maize' or 'maize growers'. This is because not everybody in villages in West Kenya makes a living through farming. The fields are generally small, sometimes too small to secure a living from farming only. Livelihoods generally revolve around combining several activities (Place et al., 2007). Some have regular and secure cash incomes (pensions, shops), but for the majority, cash flow is insecure and less regular. They trade small quantities of maize and engage in casual labour for cash or in exchange for seed and food. Being cash-strapped is common. However, having cash does not automatically translate to purchasing inputs to grow maize. Even 'rich' people plant local maize, even though they could afford to purchase modern inputs. Money is needed for a variety of purposes. One youngster commented that he also needed money for airtime and sometimes went to a sports café to watch his favourite English soccer team. Yet they rally all their resources to grow maize for food, partly for cash for groceries, airtime, and school fees; and growing maize is also a cultural obligation (Cohen and Atieno Odhiambo, 1989).

Maize in West Kenya

The Luo call their local maize *Nyaluo*, which includes the dent multicoloured and white varieties. The Luhya's call it *Lilimini*. We use the term *Nyaluo* in this chapter to refer collectively to the landraces of maize grown in the region. The traits we attribute to *Nyaluo* also include *Lilimini*. Maize growers identify *Nyaluo* phenotypically, and not genotypically as GR breeders do. Maize growing occurs in West Kenya in a bimodal rainfall context. There is a period of long rains between mid-February and mid-March during which the first planting (*chwiri*) occurs. Fields and kitchen gardens are usually prepared between November and January. If rainfall is sufficient and labour available – and cash to pay for land preparation, seeds, and fertilizer – planting hybrids during *chwiri* potentially yields a bumper harvest. This is, however, not always the case, as rainfall has become erratic. Some do not wish to take risks and rather plant *Nyaluo*, some varieties of which do extremely well during both seasons. Then there is a second short rainy season which begins in mid-August when the second planting (*opon*) takes place. From mid-October, quick-maturing *Nyaluo* is planted and occasionally hybrids. We return to this later. Most maize growers demonstrate, when asked, detailed knowledge about the nature of these varieties, when and where to plant them, and why. They are generally well aware of the soil qualities and arrange their planting accordingly (Mango, 2002).

We pinpoint the arrival of maize in West Kenya to the late 1800s (Mango, 2002; Mango and Hebinck, 2004). Over the years, maize replaced sorghum as the staple crop. Traders, migrants, returning soldiers from the First World War, white settlers, famine relief programmes, African farmers, local and international plant breeders, and projects and programmes introduced and secured the spread and planting of a wide range of maize varieties. Some of these operated in so-called informal arrangements and traded in maize, connecting West Kenya with gene pools in Uganda and the coastal areas (Miracle, 1966). Others are embedded in formal research and development programmes dedicated to spreading modern maize varieties that originate from exogenous genetic sources in Mexico and Peru and are (cross)bred for their higher-yielding capacity or better suitability for Kenya's ecological conditions.

Rachar and *Radier*, both white varieties, were already being cultivated at the end of the 19th century but were not widespread. These were known as *mzungu* (white man's) seeds because of their association with settler farming and famine relief during the colonial period (Mango, 2002). *Ababari*, named *Panadol* in Luhyaland, is also among the early varieties which are still being planted today (see Table 4.3). Both varieties were introduced by the Department of Agriculture of the colonial government. *Ababari* means a 'great thing' and came as part of famine relief. *Oking* was introduced during the great famine of 1906–1907. *Nyamula*, named *Sipindi* in Luhyaland, is a yellow variety that is today widely grown. *Nyamula* can be traced to two different sources. *Nyamula* first arrived in the 1890s and possibly earlier during the pre-colonial period through trading routes. The yellow maize that was introduced during the famines of 1928, 1936, and 1982 does not bear the full characteristics of *Nyamula* as currently known, but is nonetheless still referred to as *Nyamula* because of its yellow colour. Maize growers today confirm that these later varieties are not their '*Nyamula*' (Kimanthi, 2019).

Some of the modern maize varieties cultivated in the early 20th century were brought by settlers migrating from South Africa to Kenya. These included the white dent-type OPVs such as Hickory King and Ladysmith White (Harrison, 1970). Primarily introduced for the white settlers, these slipped through to the 'native areas' such as West Kenya via African farm labourers and traders. On-station research which began in the early 1950s primarily produced improved varieties that were multiplied by the KSC and promulgated by the Department of Agriculture to increase the maize output per unit of land. Michael Harrison, Kenya's chief maize breeder in the 1950s, returned in 1958 with exotic breeding material from a trip to Central America, Mexico, and Colombia that was funded by the Rockefeller Foundation. After screening and crossbreeding with Kitale Synthetic II, Kenya's first varietal hybrid, H611, was released in 1964. H611 has since been the basis of all hybrids developed by the national programme and multiplied by the KSC (Harrison, 1970). Until the 1990s the KSC enjoyed a monopoly in the market for improved and hybrid maize seed. Structural adjustment ended that position in 1990, however, by liberalizing the seed market, which allowed new corporate players such as Monsanto, Pioneer Hybrid, PANNAR, and SeedCo to market their hybrids.

The current pattern of maize

Both modern and local varieties are grown and combined, both in and between seasons. Results from the Maize Variety Survey suggest that some 45 per cent of the maize growers combine modern and local varieties, 28 per cent plant only modern varieties, and about 27 per cent plant only *Nyaluo* (see Table 4.1). Table 4.2 shows this at the level of households and plots, as well as indicating when the varieties are planted. Both tables are a snapshot of the current use of modern maize and *Nyaluo*. We know from our earlier work that the 'adoption' of hybrids, from the first days they were introduced, did not proceed linearly: farmers simultaneously embraced modern varieties and refrained from planting them, and this still varies from year to year (Hebinck et al., 2015).

Table 4.3 summarizes the current selection of hybrid maize and *Nyaluo* that is planted in West Kenya. Monsanto's DK8031, SeedCo's Duma43 and Punda Milia, KSC's H614, and various varieties from Western Seed are popular

Table 4.1 Use of local and modern maize in West Kenya, 2018–2019, *chwiri* and *opon* combined (n = 558 households)

Category	Number	Percentage
Did not plant	2	0.4%
Only local	148	26.5%
More local than hybrids	46	8.2%
Half local, half modern	90	16.1%
More hybrids than local	117	21.0%
Only hybrids	155	27.8%
Total	558	100.0%

Note: Maize growers plant more than one variety within the categories. The distinction between 'more local' and 'more hybrids' is also based on the difference in size of the areas planted, for instance between the main field and the kitchen garden. 'More local' means planted on a bigger plot.
Source: Maize Variety Survey West Kenya, 2018–2019

Table 4.2 Number of plots with local and hybrid varieties in long and short season (2018–2019)

Site	No. of house-holds	Plots in long season (chwiri)			Plots in short season (opon)			Total plots
		Hybrid	Local	Total	Hybrid	Local	Total	
1	58	43	29	72	28	30	58	130
2	50	44	11	55	21	30	51	106
3	70	46	49	95	11	53	64	159
4	54	44	28	72	15	41	56	128
5	50	49	8	57	18	34	52	109
6	70	25	71	96	13	33	46	142
7	96	88	44	132	26	51	77	209
8	49	44	9	53	13	36	49	102
9	60	41	49	90	9	19	28	118
Total	**557**	**424**	**298**	**722**	**154**	**327**	**481**	**1203**

Source: Maize Variety Survey West Kenya, 2018–2019. Courtesy of Conny Almekinders

Table 4.3 Frequency of maize varieties planted in West Kenya in 2018–2019 during *chwiri* and *opon* (n = 558 households)

	'Local'				'Modern'			
Variety	Chwiri	Opon	Total	Variety	Chwiri	Opon	Total	Company
Ababari	7	3	10	DH04	10	2	12	KSC
Anzika	7	7	14	H614	24	0	1	KSC
Nyamula	140	68	208	Other KSC	14	6	43	KSC
Panadol	22	14	36	Katumani	4	4	8	KSC
Rachar	98	87	185	WH505	37	7	44	Western Seed
Radier	4	11	15	Other WS	17	14	31	Western Seed
Wepesi	17	11	28	DK8031	107	53	160	Monsanto
Mwezi Mbili	3	3	6	DK8053	0	1	1	Monsanto
Msamaria	4	14	18	Punda Milia	29	7	36	SeedCo
Sipindi	25	60	85	Duma	103	35	138	SeedCo
'Local' and 'other'	10	14	24	Pioneer30G19	17	1	18	Pioneer
				PH4	2	1	3	Pioneer
				'Hybrid'	32	6	38	Unknown
Total	**354**	**303**	**657**	Total	**396**	**137**	**533**	

Note: Other varieties mentioned are Zaire (1), Sungura (2), Achupa (1), Akasimba (1), Nyalewe (1), Lewa Kwar (1), 'mixture' (1), Oking (1), Rachich (1), Imodi (1), 'local' (12), and 'unknown' (1).
Source: Maize Variety Survey West Kenya, 2018–2019

hybrids. These were introduced by MVP and OAF. Except for DK8031, the hybrids are primarily grown during *chwiri*. The local varieties *Nyamula/Sipindi*, *Rachar*, and *Ababari/Panadol* are the most popular and are planted in both the short and long seasons.

Enacting modern maize

Figure 4.2 shows when the surveyed households first planted hybrid maize. The time span covered coincides with the activities of the OAF, which intensified by 2014; the MVP, which was operational between 2004 and 2015; and the state-led research and extension programme, which ran from the 1970s. They have a common goal – to accelerate the enactment of modern maize – and both promised to solve the pertinent issues of poverty and the lack of markets and skills (Kimanthi and Hebinck, 2018), but each chooses which hybrids to introduce and which seed company to cooperate with (see Table 4.3). They also have in common the offering of incentives to encourage modern hybrid maize farming. The OAF and MVP operate from the premise that small-scale farmers cannot afford inputs because of a lack of cash. They spread the gospel of achieving (at least potentially) bigger yields and enticing their 'clients' with gifts and preferential treatment; they also extend credit, offer advice, and provide training.

The MVP strategically chose to 'give' participants free inputs (hybrid seeds and inorganic fertilizers) for the first year and promoted Monsanto's DK8031 and Duma43 from SeedCo. In the second year, half was for free and the other half on credit. In the third year, the farmers were expected to pay back the loans in kind with maize. To encourage the planting of modern maize,

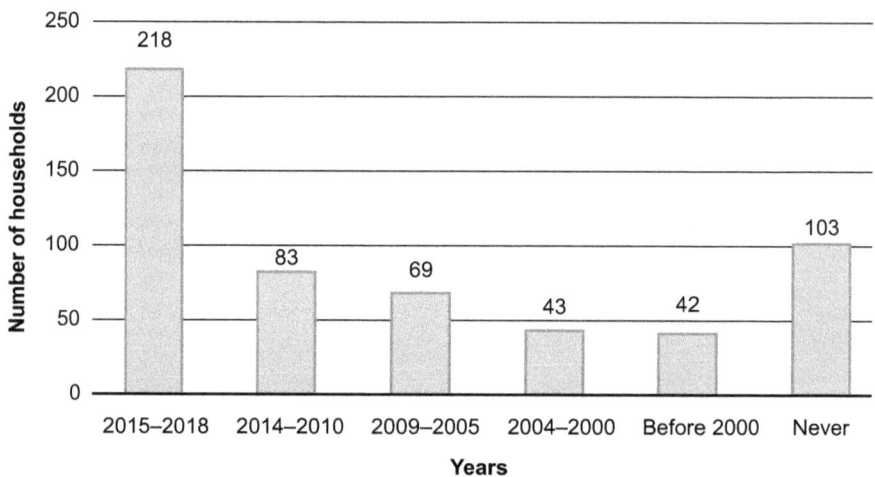

Figure 4.2 Number of households and the time they first planted hybrids in West Kenya
Source: Maize Variety Survey West Kenya, 2018–2019

the MVP's market centre in Yala did not purchase local maize. The majority of farmers tried the inputs, even though most later resumed planting what they were used to. Others stopped using these inputs and started to plant *Nyaluo* with compost and *boma* manure.[1] Kimanthi (2019) documented that some farmers sold the inputs they received from the MVP. The inputs-for-free approach of the MVP only temporarily increased the planting of hybrid maize. 'Never refuse a gift!' was a common answer when farmers were asked why they planted hybrids. The number of farmers planting hybrids dropped to nearly pre-MVP levels after the MVP was phased out in 2015 (Kimanthi and Hebinck, 2018), but increased again with the entry of the OAF (see Figure 4.2).

The OAF intensified its operations in Yala and surrounds in 2015, targeting those cultivating 2 acres (0.81 ha) of land or less. Like MVP, the OAF facilitates access to the inputs through microloans which are paid back over a year in small amounts. It promotes the planting of Punda Milia from SeedCo next to other hybrids. The inputs members receive depend on the size of their fields and individual needs and come as a package, implying that one cannot apply separately for fertilizers and hybrid seeds. OAF members are not allowed to apply the fertilizers to their local maize, and are required to employ the package as instructed, which is confusing, as the instructions are different from the MVP's and from how they plant their *Nyaluo*. An OAF officer made it very clear to us that he did not want to hear about local maize doing better under certain conditions. He commented that 'local maize is not productive and does not represent progress'.

A strategic element of the OAF's strategy is to offer training and other benefits on credit or at lower than market prices. Solar panels, iron roofing sheets, pots and pans, and other cooking equipment can be purchased through the OAF once one is a member of an OAF group. Enrolments and payments for inputs to enable planting for the next long rains start in October; the inputs are distributed in February and the participants are expected to complete repaying their loans by September of the following year after the harvest.

The OAF has designed several disciplinary measures to ensure that members enact maize as instructed. For example, it puts people living close together in groups, assuming that they must be closely related and will therefore work smoothly together. Every group member is required to be present when the inputs are issued. Although repaying the loans is an individual concern, the entire group is disqualified from participating for a whole year if one member defaults. OAF officials monitor planting closely to make sure that members plant according to the OAF's specifications of one seed per hole to avoid competition for nutrients, as this would reduce the yield. If members are found planting more than one seed in each hole, they are instructed to remove the seeds and replant as required. Having members plant as a group encourages them to remind one another of the best planting practices, and also means that members' fields are planted consecutively over a period of time, not all simultaneously.

The reality of the 'farming by instruction' approach of the OAF, as Kimanthi (2019) has documented, is often quite different. Conflicts between members have arisen as they do not necessarily share a similar cultivation schedule. Some members do not like working with others, one reason being that some may have small main fields while others have larger fields, and yet they are all expected to work together, in rotation, on each member's field. For several members, this is a reason to disconnect. They send someone else (e.g. their son or a casual worker) to the compulsory training sessions so that they can use their time more productively in their shop or other businesses. A few added that they had learnt what they wanted to learn and quit the OAF.

The 2018–2019 survey shows that more than half of the purchases of hybrid maize in the surveyed sites occurred not through the OAF but through village-based farm inputs shops known as agrovets and at open-air markets. Agrovets are run by entrepreneurs and are located in small towns like Yala. They purchase their merchandise from registered seed and fertilizer companies and offer a variety of hybrid seeds and fertilizer in any quantity (Almekinders et al., 2021). One of their strong points is that they provide free advice without compulsory training and group formation processes. Their business is strictly on a cash basis. That many people purchase their seeds and other inputs from the local agrovet may be an indication of the waning popularity of the OAF.

Enacting Nyaluo

Nyamula/Sipindi, Radier, Rachar, and *Ababari/Panadol* are planted because they tolerate poor soils and drought, resist weeds such as *Striga* and are not easily attacked by birds and termites. *Nyaluo* matures early, which makes it a good variety to grow during *opon*. Widely mentioned in interviews is that *Nyaluo* reduces and contains the risk of crop failure due to unreliable rainfall and pests and diseases; *Nyaluo* is drought-resistant and secures a harvest. *Nyaluo* requires little or no fertilizer and does well on *boma* manure. Accessing it does not require much cash as one gets the seeds cheap from local sources such as neighbours or friends, or selected from the previous harvest. Some maize growers argue that there is little difference between the yield capacities of local and modern seeds. They claim that with good treatment of the soil and seeds, one gets good yields and hence there is no need to spend lots of cash on fertilizers and seeds from the shops. For many, in contrast to 'farming by instruction', growing local maize revolves around 'farming economically'.

The enactment of *Nyaluo* resonates well with culturally embedded notions of how and why to grow it. Supported by strong and frequently voiced claims, *Nyaluo* is said to have unique qualities which hybrid maize does not have. Women claim that *ugali* (porridge) from hybrids requires twice as much maize as *ugali* from local maize. Hybrids are less filling. This claim is important because the assumed yield superiority of hybrids is outweighed by the much higher nutritious quality of local maize. Women brewing beer claimed that local maize produces higher-quality and sweeter beer than that brewed from hybrids. Extension agents confirmed this, attributing it to the heavier starch content in local maize. Hybrids are also said to be less tasty than *Nyaluo*. Certain local varieties taste good when boiled, while others are good for roasting. Most women mentioned that the seeds from *Nyaluo* have a harder coat than hybrids. They store well when dusted with ash from burnt-bean husks. Hybrid grains are softer and are highly vulnerable to attack by weevils, even after dusting with the same ash. If farmers grow hybrid maize, it is usually only for sale. Some said that they would quickly sell their hybrids in order to buy local maize at the markets. This is not a common practice, however.

Another crucial fact about *Nyaluo* is that *golo kodhi* and *dwoko cham* are incompatible with hybrids. Hybrid maize is perceived as a *Nyareta (*strange seed) and, unlike local maize, does not easily become family seed. *Golo kodhi* may also slow down planting in situations where local and hybrids are combined, causing the late planting of hybrids which need to be planted at the start of the rainy season to realize a good yield. This occurs in situations of polygamy, with the first wife using her power to slow down her first planting to prevent the second wife from planting hybrids on time (Kimanthi et al., 2022). But if the relationship between relatives in the homestead is good, a solution can easily be found to at least some of the problems generated by *golo kodhi*. We observed many situations where the elderly were delayed in land preparation activities and therefore could not sow in time. By sowing just a few square metres of maize, they signalled to the younger generations that they could begin to start sowing their plots. These kinds of accommodations are quite common. One informant said that to circumvent *golo kodhi*, seed could be purchased at the market and planted immediately, without being brought into the homestead.

Discussion

The long history of efforts to extend maize enactment in West Kenya holds fascinating but contradictory experiences. The way maize is enacted not only triggers our sociological imaginations but also defies many of the received wisdoms agencies and experts hold about farming. Whereas OPVs that were introduced a long time ago still render promising results as creolized landraces, such as *Nyaluo*, we tend to be critical of the promotion of hybrid maize as supported by agencies and experts in West Kenya and elsewhere on the continent.

We began by showing that maize growers do not enact maize exclusively according to *either* a commoditized GR *or* an agroecology-by-default script. They frequently rather tap from both and form alliances, at least temporarily, with both sets of supporting institutions. This supports a preliminary conclusion that maize producers in West Kenya have benefited from both the indigenous agricultural revolution and GR in maize. Both have made a lasting impact through significantly transforming the maize landscape in West Kenya by adding maize genes – admittedly in structurally different ways, but ultimately providing a foundation for multiple routes to regional and homestead food security.

The indigenous agricultural revolution builds on past efforts by the colonial state and the coastal trade routes that brought a range of varieties that creolized into today's *Nyaluo*. This indicates the significance of an indigenous agricultural revolution. *Rachar*, *Radier*, and *Nyamula* were introduced to relieve famine and became known as *mzungu* varieties. Hickory King meant for white settlers spread and was adapted as *Oking*. The spreading of maize intensified, so that through years of creolization, planting with local means and knowledge, and trade in local markets, varieties of maize once considered 'modern' are now widely considered local and *Nyaluo*. *Nyaluo* is appreciated because of its taste, colour, drought resistance, and nutritious qualities. The way *Nyaluo* is selected – whether organized or haphazard – does not hinge on seed as a commodity per se. *Nyaluo* – or *Lilimini*, for that matter – works for those who are cash-strapped and want to 'farm economically'. *Nyaluo* provides good seed that grows well during *opon* and allows connectedness with the locality, culturally and economically, providing food security for the family. The claim that modern maize is higher yielding when compared to *Nyaluo* or *Lilimini* is vehemently contested by those who plant and consume these maizes. For them, the nutritious qualities of *Nyaluo* outperform hybrids and provide more energy.

The GR, in contrast, materialized as a result of the shift from breeding OPVs to hybrids, the increasing role of private seed companies in the breeding and marketing of hybrids, the willingness of donors to fund the promotion of hybrids, and related efforts to intentionally replace *Nyaluo*. Maize growers in West Kenya appreciate the hybrids because of their higher yield capacity, even though cultivating them requires much higher monetary inputs than *Nyaluo*. Hybrid maize works for those with regular access to cash for the monetary inputs and who see a future for commoditized maize farming. They liaise with agrovets and NGOs such as OAF and, being willing to take the risks of farming on credit, experience farming as per OAF instructions. The higher yield capacity of hybrids is contested, and not only when nutritional values are included. Ongoing experiments in Sauri with a youth group to compare hybrids and *Nyaluo* show that *Rachar* planted under local conditions (e.g. local *boma* manure, one-time weeding) yielded 3.5 bags (90 kg) on half an acre (0.2 ha), while DK777, a hybrid from Monsanto, yielded 3 bags. Other trials (see Almekinders et al. (2021: 414) for a summary) indicate that yields do not differ substantially.

That maize growers continue to look for hybrid maize varieties and appreciate them, with or without being enticed by the attractions offered, could imply a re-evaluation of the critique on the GR. However, while *Rachar* and *Radier* are examples of early modern maizes, which were all bred as OPVs, the hybrids currently being promoted are not bred to be creolized but programmatically designed to replace the landraces and to be purchased afresh for every planting season. The current set of hybrids will, unlike the older modern seeds, not so easily creolize and cross the boundary between 'modern' and 'local'. The choice to hybridize maize, in other words, has generated conditions that severely limit its impact. The GR did not cater for varieties that maize growers continue to look for: those that can be planted during *opon*, align with local cultural repertoires, are drought- and disease-resistant, taste good, and yield well in local conditions.

Our findings and conclusions defy many of the received wisdoms agencies and experts hold about maize farming. The maize landscapes in regions like West Kenya and elsewhere are consistently misread by experts from seed companies, think-tanks, and NGOs like the OAF and the MVP, and the state rural extension service. The misreading arises from the misconception that farming is or should be a commercial and entrepreneurial venture and the persistent broadcasting of the message that *Nyaluo* is unproductive and aggravates poverty. This has cultivated a discursive promotion of the advance of the GR that not only ignores the fact that growing maize is a cultural obligation but also disregards the socio-cultural-ecological significance of *Nyaluo* for food security. The attendant choice of NGOs and donors to solely promote hybrids is problematic. Relying on global private-sector companies to enrich key genetic resources for food security is not only typical of a failure to appreciate that the boundaries between local and modern are hazy, but also risky in times of climate change.

The misreading also arises from a misconception of the choices maize growers make with regard to which variety to plant. These choices are multi-dimensional and reflect a pattern of maize growers positioning themselves at various points along the continuum from 'farming economically' to 'commoditized farming'. This positioning is not to be understood as a linear progression from 'traditional' to 'modern'. Maize growers shift between these positions and continuously balance, some pragmatically and some strategically, between protecting their autonomy as it interlocks with the local social economy, markets, and needs, and engaging more deeply with the commodity economy stretching beyond the village and region. They do not grow modern maize because they lack the monetary resources to purchase seeds and accompanying inputs. We found 'large farmers' that are well respected in their community because they plant and sell *Nyaluo* and hire casual labourers locally for land preparation and harvesting. Even 'rich' maize growers do not automatically enact modern maize. The variety choices are manifestations of ontologically differently constructed economies that hinge on contrasting meanings of what constitutes food and are structured by the various types of markets that operate in the region to sell and exchange food and seed. Decisions to plant one variety or another are taken pragmatically, bounded by past relationships, obligations, and rapidly changing conditions. This is why no two planting seasons unfold in the same way.

Note

1. *Boma* manure refers to a composted mixture of crop residues, cow dung or goat/sheep droppings, and urine.

References

Almekinders, C., Hebinck, P., Marinus, W., Kiaka, R.D. and Waswa, W.W. (2021) 'Why farmers use so many different maize varieties in West Kenya', *Outlook on Agriculture* 50: 406–418 <https:doi.org/10.1177/0030727021 1054211>.

Altieri, M. and Nicholls, C.I. (2012) 'Agroecology scaling up for food sovereignty and resiliency', in E. Lichtfouse (ed.), *Sustainable Agriculture Reviews*, pp. 1–30, Springer, Dordrecht <http://dx.doi.org/10.1007/978-94-007-5449-2_1>.

Casañas, F., Simó, J., Casals, J. and Prohens, J. (2017) 'Toward an evolved concept of landrace', *Frontiers in Plant Science* 8 <https://doi.org/10.3389/fpls.2017.00145>.

Cohen, D. and Atieno Odhiambo, E. (1989) *Siaya: The Historical Anthropology of an African Landscape*, James Currey Publishers, London.

Harrison, M.N. (1970) 'Maize improvement in East Africa', in A. Leaky (ed.), *Crop Improvement in East Africa*, pp. 21–60, Commonwealth Agricultural Bureau, Farnham Royal.

Hebinck, P., Mango, N. and Kimanthi, H. (2015) 'Local maize practices and the cultures of seed in Luoland, West Kenya', in J. Dessein, E. Battaglini, and L. Horlings (eds), *Cultural Sustainability and Regional Development: Theories and Practices of Territorialisation*, pp. 206–219, Routledge, London.

Holt-Giménez, E. and Altieri, M. (2013) 'Agroecology, food sovereignty and the new Green Revolution', *Agroecology and Sustainable Food Systems* 37: 90–102 <http://doi.org/10.1080/10440046.2012.716388>.

Holt-Giménez, E., Shattuck, A. and van Lammeren, I. (2021) 'Thresholds of resistance: agroecology, resilience and the agrarian question', *The Journal of Peasant Studies* 48: 1–19 <https://doi.org/10.1080/03066150.2020.184 7090>.

KEPHIS (2020) *National Crop Variety List: Kenya*, Kenya Plant Health Inspectorate Service, Nairobi.

Kimanthi, H. (2019) *Peasant Maize Cultivation as an Assemblage: An Analysis of Socio-Cultural Dynamics of Maize Cultivation in Western Kenya*, PhD thesis, Wageningen University, Wageningen, the Netherlands <https://doi.org/10.18174/478306>.

Kimanthi, H. and Hebinck, P. (2018) '"Castle in the sky": the anomaly of the millennium villages project fixing food and markets in Sauri, western Kenya', *Journal of Rural Studies* 57: 157–170 <https://doi.org/10.1016/j.jrurstud.2017.12.019>.

Kimanthi, H., Hebinck, P., and Sato, C. (2022) 'Exploring gender and intersectionality from an assemblage perspective in food crop cultivation: a case of the Millennium Villages Project implementation site in western Kenya', *World Development* 159: 106052 <https://doi.org/10.1016/j.worlddev.2022.106052>.

Kloppenburg, J. (1988) *First the Seed. The Political Economy of Plant Biotechnology, 1492–2000*, Cambridge University Press, Cambridge.

Kutka, F. (2011) 'Open-pollinated vs. hybrid maize cultivars', *Sustainability* 3: 1531–1554 <https://doi.org/10.3390/su3091531>.

Latour, B. (2005) *Reassembling the Social: An Introduction to Actor-Network-Theory*, Oxford University Press, Oxford.

Law, J. and Lien, M.E. (2012) 'Slippery: field notes in empirical ontology', *Social Studies of Science* 43: 363–378 <https://doi.org/10.1177%2F030631 2712456947>.

Law, J. and Singleton, V. (2000) 'Performing technology's stories: on social constructivism, performance, and performativity', *Technology and Culture* 41: 765–775 <http://dx.doi.org/10.1353/tech.2000.0167>.

Magdoff, F., Foster, J. and Buttel, F. (2000) *Hungry for Profit: The Agribusiness Threat to Farmers, Food, and the Environment*, NYU Press, New York.

Mango, N. (2002) *Husbanding the Land: Agrarian Development and Socio-technical Change in Luoland, Kenya*, PhD thesis, Wageningen University, Wageningen, the Netherlands <https://library.wur.nl/WebQuery/wurpubs/fulltext/139846>.

Mango, N. and Hebinck, P. (2004) 'Cultural repertoires and socio-techno-logical regimes: maize in Luoland', in H. Wiskerke and J.D. van der Ploeg (eds), *Seeds of Transition: Essays on Novelty Production, Niches and Regimes in Agriculture*, pp. 285–319, Royal Van Gorcum, Assen.

McCann, J. (2005) *Maize and Grace: Africa's Encounter with a New World Crop, 1500–2000*, Harvard University Press, Cambridge, MA.

Miracle, M. (1966) *Maize in Tropical Africa*, University of Wisconsin Press, Madison, WI.

Onyutha, C. (2018) 'African crop production trends are insufficient to guarantee food security in the sub-Saharan region by 2050 owing to persistent poverty', *Food Security* 10: 1203–1219 <http://doi.org/10.1007/s12571-018-0839-7>.

Otsuka, K. and Larson, D. (eds) (2013) *An African Green Revolution: Finding Ways to Boost Productivity on Small Farms*, Springer, Dordrecht <https://doi.org/10.1007/978-94-007-5760-8>.

Peschard, K. and Randeria, S. (2020) '"Keeping seeds in our hands": the rise of seed activism', *Journal of Peasant Studies* 47: 1–33 <https://doi.org/10.1080/03066150.2020.1753705>.

Place, F., Adato, M. and Hebinck, P. (2007) 'Understanding rural poverty and investment in agriculture: an assessment of integrated quantitative and qualitative research in western Kenya', *World Development* 35: 312–325 <https://doi.org/10.1016/j.worlddev.2006.10.005>.

Richards, P. (1985) *Indigenous Agricultural Revolution: Ecology and Food Production in West Africa*, Unwin Hyman, London.

Val, V., Rosset, P.M., Zamora Lomelí, C., Giraldo, O.F. and Rocheleau, D. (2019) 'Agroecology and La Via Campesina I. The symbolic and material construction of agroecology through the dispositive of "peasant-to-peasant" processes', *Agroecology and Sustainable Food Systems* 43: 872–894 <https://doi.org/10.1080/21683565.2019.1600099>.

Van der Ploeg, J.D. (2013) *Peasants and the Art of Farming: A Chayanovian Manifesto*, Fernwood Publishers, Nova Scotia.

Van der Ploeg, J.D., Barjolle, D., Bruil, J., Brunori, G., Costa Madureira, L.M., Dessein, J., Drag, Z., Fink-Kessler, A., Gasselin, P., Gonzalez de Molina, M., Gorlach, K., Jürgens, K., Kinsella, J., Kirwan, J., Knickel, K., Lucas, V., Marsden, T., Maye, D., Migliorini, P., Milone, P., Noe, E., Nowak, P., Parrott, N., Peeters, A., Rossi, A., Schermer, M., Ventura, F., Visser, M. and Wezel, A. (2019) 'The economic potential of agroecology: empirical evidence from Europe', *Journal of Rural Studies* 71:46–61 <https://doi.org/10.1016/j.jrurstud.2019.09.003>.

CHAPTER 5

Seed matters: understanding smallholder seed sourcing in Malawi

Noelle LaDue, Sidney Madsen, Rachel Bezner Kerr, Esther Lupafya, Laifolo Dakishoni, and Lizzie Shumba

Introduction

Seeds operate at the nexus of biological and social relations in agrarian communities. As Bezner Kerr (2013: 4) writes, 'The seed has become a commodity, while still maintaining other forms, such as a gift, exchange item, cultural icon or source of agrobiodiversity, and as such is a contested site of state, market, community and aid relations.'

A farm cannot begin a season without seeds to plant, which makes seeds vital to subsistence and farming livelihoods. Malawian smallholder farmers face several challenges to seed provisioning, which include limited access to quality seeds, a low or non-existent supply of local or indigenous seeds, and insufficient national seed production and distribution systems (Guei et al., 2011). While some scholars assess farmers' access to seed in terms of seed security, La Via Campesina argues that 'seed sovereignty' – that is, farmer input and autonomy over both how they access seeds and how they were produced (Wittman et al., 2010) – is a vital component of a just food system.

This study provides insight into how seed sourcing is linked to household food security by describing the different seed sourcing pathways used by smallholder farmers in Malawi and exploring the trade-offs and dynamics shaping their sourcing options. Seed sourcing pathways are different methods of obtaining seed, which have varied economic, agronomic, and social implications. In this study, 'pathway' refers to the ensuing implications for the individual farmer based on their seed source (Lentz, 2018). We present findings from in-depth interviews with smallholder farmers carried out in 2018 as part of a broader participatory agroecology project.

Seed systems

The conditions of seed sourcing are greatly determined by the variety of channels through which farmers acquire seed, or their seed systems. Seed systems are commonly categorized as formal or informal. The formal sector comprises sourcing channels that sell certified seed, produced by public or private companies and research institutes, which farmers generally acquire through market transactions. The informal seed sector includes all the channels through which farmers acquire non-certified seed, which commonly is sourced from farmers' own fields through seed saving. While informal and formal seed systems coexist – indeed, many farmers rely on both sectors to access seeds – there has been a recent policy push to limit the informal seed sector. An increasing number of countries in the Global South are implementing certification laws, which require all seeds sold at market to be approved by the national certification board (Mayet, 2015; Wattnem, 2016). These laws enforce a bureaucratic process that is more accessible to large corporate seed producers than independent ones and effectively excludes or prohibits the channels through which most smallholder farmers source seeds (Wattnem, 2016). The enclosure of traits and varieties through intellectual property laws has further limited smallholder farmers' control over their seed systems (Kloppenburg, 2010).

Agriculture and seed systems in Malawi

Malawi's agricultural and political history offers insight into current seed sources and the role of the state in shaping them. During the colonial period, the British government implemented an estate agricultural system which focused on cash crops, mainly cotton, tea and tobacco, for large-scale producers (Kydd and Christiansen, 1982). After Malawi's independence in 1964, maize became the focus of the national policy to address food insecurity, and several support programmes were implemented for smallholders, including subsidies for seeds and fertilizer. The government established the Agricultural Development and Marketing Corporation (ADMARC) and the National Seed Company of Malawi (NSCM). The NSCM and ADMARC served smallholder farmers by breeding maize varieties and other crops, providing rural depots for inputs and seeds, and setting a base price for the main smallholder crops (Kydd and Christiansen, 1982). These state institutions and regulation were dismantled by structural adjustment programmes implemented in the 1980s in response to a debt crisis. A critical moment of liberalization in relation to seed systems was the sale in 1996 of the NSCM to Cargill, which then sold it to Monsanto (Bezner Kerr, 2013).

Maize has been promoted as a driver of modernization to propel development and foreign investment (Bezner Kerr, 2014), a promotion that took the form of subsidy programmes for synthetic fertilizer and maize seed. The national subsidy programme and multinational companies are now the main providers in the formal seed sector (Chinsinga, 2010). Ninety per cent of the formal seed market has been captured by international companies, half of which are controlled by Bayer (formerly Monsanto). Formal seed providers focus primarily on the production and distribution of improved or hybrid varieties of commercial crops; in 2010, hybrid maize made up an estimated 40 per cent of all maize planted in Malawi (Chinsinga, 2010). The influx of subsidized hybrids and highly promoted corporate varieties means that fewer farmer-produced, open-pollinated seeds are circulated and planted, suggesting that these recent seed sector trends have contributed to agrobiodiversity decline in Malawi (Chibwana et al., 2012).

Despite overwhelming policy support for the formal sector, many farmers continue to save seed in Malawi (Mngoli et al., 2015). Although seed saving can give low-income farmers access to seed (Mngoli et al., 2015), previous studies in Malawi have documented challenges associated with both seed saving and commercial seed availability, which farmers have to navigate to sustain their livelihoods (Bezner Kerr, 2013). This study seeks to further understand Malawian farmers' seed sourcing practices. Through farmer narratives, we draw out the difficult decisions underlying farmers' struggles to access seed. These stories highlight the prominent role that seeds play in the successful pursuit of farming, not just technically, but as a way of life.

Methods

This research draws on analysis of in-depth interviews conducted as part of a participatory research project carried out by Soils, Food and Healthy Communities (SFHC). SFHC has been using agroecological and participatory approaches to support sustainable agriculture and improved household nutrition for over 20 years in the Northern and Central regions of Malawi. This research was part of a project called 'Building Sustainable and Equitable Food Systems using Participatory Communication and Agroecology in Malawi' led by SFHC with partners Cornell University in the United States and Western University in Canada. Over a four-year period the project engaged with over 537 households in 10 villages located in Mzimba district in the Northern Region of Malawi and Dedza district in the Central Region (Figure 5.1). Farmers were trained in agroecological farming methods, gender equity, and child nutrition using participatory communication methods involving drama and small-group discussion. Participating farmers also received seeds in the first year of the project, including cowpea, bean, groundnuts, finger millet, and sorghum. A team of local farmer leaders, called farmer promoters, facilitated the intervention using a farmer-to-farmer approach to knowledge dissemination.

The study is based on a total of 60 translated and transcribed interviews, half of which were conducted by the farmer promoters, following training on qualitative participatory research methods and semi-structured interviewing. In alignment with the participatory methods inherent to an agroecological framework (Méndez et al., 2013), farmer promoters played a critical role in this study through their involvement in interviews and the preliminary analysis of results (Lentz, 2018). The interview guide covered farmers' seed preferences (including varietal preferences), strategies to acquire seeds, and

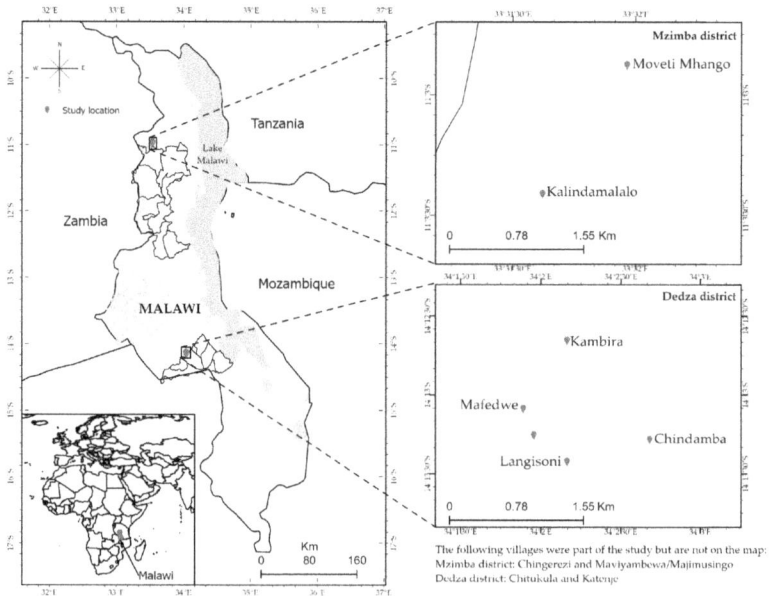

Figure 5.1 Map of study village sites in Malawi
Source: Authors.

barriers to seed access. Participants for each interview were chosen to ensure a range of income levels and household types. We interviewed only women in the households, in order to capture the unique knowledge that they have of household food provisioning due to the gendered division of labour. The research team of students and farmer promoters conducted participatory data analysis in both regions following completion of the interviews. Further qualitative coding of the translated and transcribed interviews was done using Atlas.ti software. The researchers drew on a critical grounded theory approach to analyse and interpret coded data and code group relationships; the resulting analysis revealed the pathways and conclusions presented in this chapter (Silverman, 2005).

Results

The farmers' seed sourcing strategies fell into five major categories: seed saving, seed purchasing, seed sharing, *dimba* (dry-season cultivation), and casual day-labour employment, known as *ganyu* (Figure 5.2). The farmers often utilized multiple pathways simultaneously to obtain seed, and thus we depict the seed sourcing pathways as interconnected. Each pathway could serve as a direct strategy to immediately acquire seed, but pathways also mediated each other; in particular, *dimba*, *ganyu*, and social support pathways could indirectly lead to seed saving or seed purchasing.

Seed saving

Farmer preferences and socio-
economic, environmental and
political conditions

Ganyu

Dimba

Social support

Seed purchasing

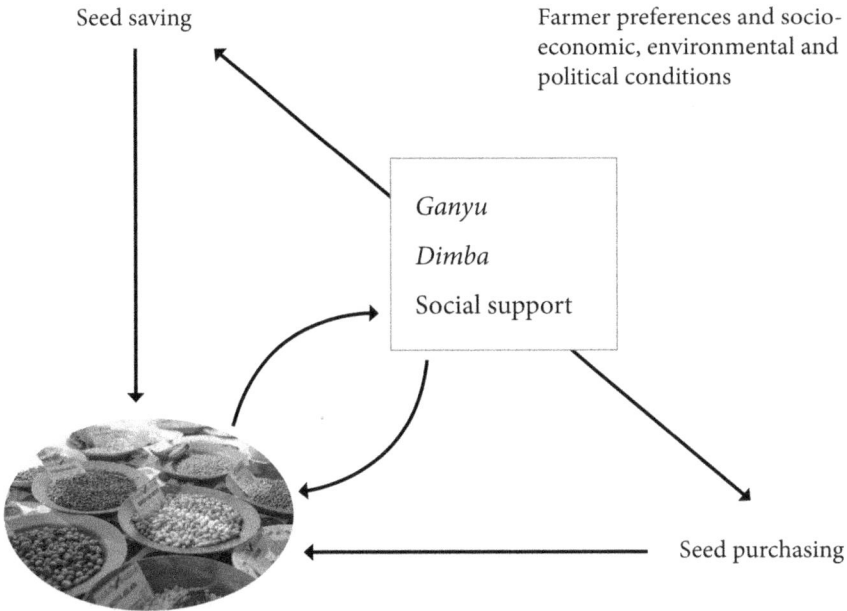

Figure 5.2 Seed sourcing pathways conditioned by a farmer's context

Varietal choice

Seed sourcing was shaped by the farmers' preferences for biophysical traits, which they attributed to certain varieties. Of the crops cultivated, varietal preferences were most strongly voiced for maize, with farmers differentiating between 'local' and 'hybrid' varieties. While there was consensus on some of the respective attributes of hybrid versus local varieties (e.g. local varieties being more suitable for storage, hybrid maize being higher yielding under proper conditions), interviews revealed conflicting viewpoints on other attributes (e.g. which was more drought-tolerant, higher yielding), and in particular contradictory views on the efficacy of recycled hybrid seed.

When purchased, local varieties might be obtained via the informal seed sector, from a neighbour, community member, or vendor, while hybrid seed was more often acquired from small agro-dealers. Seed saving was commonly reported for local varieties, though a surprisingly large number of respondents also saved hybrid seeds. Recycling hybrid seeds goes against advice from agricultural extension agents, who recommend purchasing new seeds every year, since it is generally known that hybrid seeds decline in quality when recycled. Many interviewed farmers accepted this as a fact, and therefore their financial resources shaped the type of maize they cultivated, with poorer farmers preferring open-pollinated varieties (OPVs) that could be reliably recycled without losing valuable traits, and wealthier farmers more likely to consistently access hybrid maize seed. In some farmers' experience, the advantages of hybrid maize compared to local maize – namely its drought tolerance due to quick germination and growth – were preserved even in recycled seed, and another benefit was that it grew faster, so late-season drought did not affect yield as much. Other farmers who consistently saved hybrid maize found that recycled seed generally maintained the same qualities for which it had been selected by the breeders, but only for one or two seasons.

Seed saving

Most of the farmers expressed a strong preference for sourcing their seeds from their own grain stores. The strength of this preference was evidenced by the prioritization of seed saving over other pressing household needs. Tiwine,[1] a single 25-year-old mother in the Central Region, had harvested five 50 kg bags of soya beans, but sold four of them to pay for her children's school fees. Until the next harvest, Tiwine planned to purchase soya from the market to use as food and save her remaining bag of soya to use as seed. Her decision-making demonstrates the value of saving seed for a crop like soya, which provides important nutrition for her children in the form of porridge for breakfast. In this case, however, she exchanged the immediate benefit of this dietary addition for future seed security. The reasons for seed saving were evident to most of the farmers: this sourcing method reduced the cost of production, eliminating the need to purchase seed. It also reduced the grave risk that uncertain and limited income sources posed for their access to this fundamental input. One farmer explained, quite simply, what the alternatives to seed saving implied for her: 'To avoid suffering, after harvesting we save and wait for the rainy season to plant.'[2]

Pride and the desire for market autonomy led Alice, a 58-year-old married woman who farmed 2 hectares of land in the Northern Region, to save seed. Alice[3] remarked that she saved seed in part because 'I don't want to be begging from other people. It is better for me to save, plus sometimes it may happen that we don't have money to buy the seeds, so it is better to save.' She did not save seed from all of her crops but preferred to save at least some so that she could be self-reliant and budget to purchase her maize seed. Many seed savers prioritized maize seed, primarily of local and OPV varieties, above other crop seeds. Other crops considered good for saving were beans, groundnuts, soya beans, and potatoes, both sweet (*Ipomoea batatas*) and 'Irish' (*Solanum tuberosum*). Farmers less commonly reported saving seeds of 'relish' crops, or foods supplementary to the dietary staple (most commonly maize porridge, or *nsima*), including greens, mustard, and other vegetables. Reluctance to save seeds often stemmed from perceptions of lower germination rates from recycled seed and reduced productivity due to inadequate storage conditions.

Inadequate seed storage further limited the farmers' ability to source seeds from their own harvested crop. A number of farmers expressed their preference for OPV or local maize varieties because of their superior storage qualities. The cost of chemical seed treatments to reduce pest damage could be prohibitively high. When asked which seeds were best for saving, Dorothy,[4] a 46-year-old unmarried farmer from the Northern Region with no children, responded: 'Soya and groundnuts, because they don't need medicine, while maize needs medicine [pesticides for storage], and now the maize has already started getting weevils, so I want to sell maize to find money so I can buy seed treatment.' Without alternative sources of income, she would have to resort to selling maize that she had wanted to use for subsistence in order to purchase chemical treatments. Even if the treatment prevented pest or rot damage, selling a significant amount of her maize to purchase these inputs would likely mean that her stored seed would not last to the next planting season, as she would be left without enough to meet both her subsistence and seed needs. Other methods of protecting seed during storage, such as ash, were mentioned as well, but many farmers referenced their inability to access pesticides as an indirect yet notable barrier to saving seed. This problem also demonstrates the value of cultivating a diversity of crops, as legume crops of soya and groundnuts are less susceptible to the storage issues that affect maize.

Alternatives to seed saving

The capacity to prioritize grain stores for seed use was linked to socio-environmental factors that influenced the farmers' resources and their allocation. Households with more income earners and sources were better equipped to meet household needs without selling all of their harvest, especially in years with adequate rainfall and no unexpected expenses. However, if the household incurred expenses from an illness, or the farmer was a single woman who needed to provide for her family on her own, she might resort to selling seed she had wished to plant the next year. Seed saving was also contingent upon a certain level of crop productivity; many farmers reported that dry spells had led to insufficient quantities of grain to source as seeds, while a few had suffered total crop loss of some species. In other cases, illness and migration might have prevented farmers from cultivating their own crops. After losing a seed store, a farmer could rebuild seed stocks through purchasing, seed sharing, *dimba*, and *ganyu*. As a dry-season garden grown in marshy, lowland or riverbed areas, the *dimba* can be used to plant and multiply seed stocks that will be used in future growing seasons. Farmers could also access seed as a form of payment for informal wage labour, locally known as *ganyu*.

Purchasing

If saving seed implied the prioritization of harvested grain for seed rather than food or income, purchasing seed meant giving primacy to this expense over other uses of farmers' scarce income and labour. Without saved seed, the seed for next year's cultivation was often acquired through earnings from *ganyu*, and this expense reduced a household's budget to meet needs for school supplies, purchased foods, and other items. Farmers also reported purchasing seed in years when they had sufficient funds and wished to diversify their crops with a species or variety only available on the market. Local and regional marketplaces provided access to a wider diversity of crops than a local trader or neighbour might, although several farmers recounted stories of when they had wished to purchase certain crops and the seed was unavailable.

Farmers purchased seed from a number of actors: agro-dealers, the parastatal ADMARC, vendors, and family or kin. They typically preferred to purchase seed from small, local agro-dealers, since those dealers were often willing to negotiate prices, unlike the larger, regional corporate dealers. Fewer farmers reported purchasing seeds from ADMARC, which sells seed at a price regulated by the government, but this related less to farmers' preference than their inability to obtain seeds from the state-run depots. Indeed, a few respondents explained that it was common for ADMARC, rather than selling to individual smallholders, to sell seeds in bulk to vendors, who then added a 50 per cent premium when they sold those seeds in the villages. The price of a 50 kg bag of maize from ADMARC was 7,500 kwacha (approximately US$10), and in the course of the season the vendors' price was liable to increase to 11,500 kwacha ($11). Farmers found this market behaviour exploitative, and transacting with vendors was considered by many the least satisfactory way to purchase seeds. Despite their mistrust of vendors, more remotely located farmers, who could not readily engage with the range of sellers available at a local or regional market, often depended on travelling vendors to acquire seeds. Another option for purchasing was through exchange with friends, family, and other village residents. Such arrangements were mediated through social relationships, and therefore seeds acquired this way were most likely provided at a discount to those in need.

Seed sharing

While seeds could be purchased from neighbours or friends, community members also shared seeds without financial compensation as a social practice of reciprocity. Not only did these exchanges reinforce social ties between friends and family, but they also often ensured seed resources for the future. Loveness[5] (aged 57 and not married) from the Northern Region described the concept of reciprocity: 'Sharing seeds is like keeping them; you can ask for [the seeds] later.' While sharing seed freely sacrificed possible income from selling to vendors, it strengthened kinship relationships that seed sharers considered more valuable than ties to transient, often fickle, traders. Not all interviewees practised seed sharing, however. When asked about exchanging seeds in order to have a variety of crops, Melita, an unmarried 29-year-old farmer from the Central Region, responded, 'Ah no, that also doesn't happen, maybe only to the family members.' Seed sharing was infrequently reported in the Central

Photo 5.1 Farmers display crop and varietal diversity at a seed fair
Credit: SFHC

Region, as was food sharing, indicating different social norms, as well as material conditions; farmers in the Central Region had smaller farm sizes, relied more heavily on *ganyu*, and experienced more severe food insecurity.

Seed sharing and exchange were methods for increasing crop diversity, especially where market channels to seed diversification were limited. Faless[6] from the Northern Region was 36 years old and married, with a four-person household. She explained that she had recently chosen to share seed with her neighbours 'so that they should also grow after they have seen the benefit'. The desire to share the benefits of diverse crops was a reason farmers frequently gave for sharing seeds. At a seed fair organized by SFHC, farmers demonstrated a level of crop and varietal diversity that would be difficult to find in local marketplaces (Photo 5.1); seed exchanges gave fellow farmers access to such seed resources. These seed exchanges often implied an experimental use, and thus were not likely to yield sufficient seeds to sow an entire field with the crop, but might provide a friend with starter seed stock that could be further propagated. In both regions, when farmers spoke of the social support practices of sharing, it was more frequently in reference to food than seed. This finding suggests that farmers are more likely to ask for food and share it with each other to address an immediate need, and recipients might expect to find another way to obtain the seeds in time for planting.

Dimba

Most farmers interviewed cultivated a dry-season garden, or *dimba*, growing vegetables, maize, and legumes on a small plot of land located near a water source, usually lowlands, riverbeds or marsh areas (Photo 5.2). The *dimba*

Photo 5.2 A *dimba* in June in the Northern Region of Malawi
Credit: Noelle LaDue

serves to sustain a household's livelihood, providing a source of food and income that complements more extensive production from rain-fed, upland cropping. Farmers used seed harvested from their upland field to plant in the *dimba*, and then harvested seed from the *dimba* to plant in the upland fields at the start of the next rainy season. The *dimba* was thus valued by farmers as a site of seed production, primarily for maize and legume crops, and as a method of seed saving. In particular, several farmers found that sourcing seed from the *dimba* ensured that they did not consume seeds before the next growing season, and avoided the risks of storage loss (e.g. pest damage) between growing seasons.

Not all farmers used the *dimba* to produce seed. The *dimba* was frequently used to sell vegetables to others who either did not have a *dimba* or whose *dimba* had not yet started producing. The earnings from these sales allowed some farmers to purchase seeds, thereby indirectly functioning as a seed sourcing pathway. Where food or income was a more pressing need for a household, the farmer might forgo seed production. Faless had oriented her farming choices towards preparing for the rainy season's maize crop, but prioritized the purchase of fertilizer over seed saving: 'Yes, we keep, but not really, because we always aim at selling so that we can find fertilizer.' She used earnings from her *dimba* to access fertilizer and found other pathways to source seeds.

*Dimba*s were not an option for all the farmers, however; land with access to water during the dry season was scarce in some villages, and poorer families were more likely to be excluded from this resource. Farmers without their own *dimba* plot were left with the option of sharing one with a relative or neighbour or renting land. Despite the constraints of income and food security experienced by poorer farmers, many aimed to sell enough of their rainy season crop to be able to rent a *dimba* during the next dry season.

Ganyu

In the Central Region, those farmers who had not saved seeds had most commonly obtained seed stock through *ganyu*, a form of casual or informal labour, usually working as temporary agricultural labourers on other farms. *Ganyu* provided access to seed through a number of pathways. In some cases, farmers purchased their seeds from income gained through *ganyu*, and in other cases farmers were paid in seed for their labour. Several farmers recounted times when they had received their *ganyu* payment in the form of maize seed, which they then sold to purchase different seed that they preferred. In the Northern Region, friends sometimes offered *ganyu* as a form of social support in exchange for seed, thereby allowing farmers to procure seed without paying the market price.

Farmers acknowledged that *ganyu* was a precarious way to obtain seed; this pathway was most often used by farmers who, at the beginning of the planting season, found themselves without sufficient seed stock or income to purchase seed. *Ganyu* is not always available and payment practices vary, with some farmers paid in low-quality seed or maize flour, limiting their seed options. *Ganyu* is most available at key stages of the agricultural calendar: namely, during field preparation, planting, and weeding. Farmers depending on seed sourced through *ganyu* must complete this work before they can sow their own crops; indeed, farmers in the Central Region often received *ganyu* payment in the form of seeds left over from the hiring farm's planting. The *ganyu* labourers only acquired seed to plant on their own farms after the ideal planting date, and therefore missed rain events critical to good crop germination and growth.

Many interviewees attributed poor yields to late planting due to *ganyu* seed sourcing, and faced food insecurity, with little or no seed stock for the next growing season. As Grace,[7] a 46-year-old farmer from the Central Region who grew maize, finger millet, groundnuts, and beans, explained, 'I can do *ganyu*, but how am I supposed to do *ganyu* when I have my own work to do?' Like Dorothy, many respondents viewed having a *dimba* and doing *ganyu* as competing livelihood strategies, and most preferred the former. The reason for their preference was elucidated by Dorothy, who had recently used earnings from her crop sales to purchase a *dimba*. In contrast to *ganyu*, where 'the money will finish' before long when earnings have to be spent on food and other goods, the *dimba* provides a reliable, continuous source of income, food, and seeds for the upcoming planting season.

Discussion and conclusions

The seed sourcing pathways available to farmers are shaped by the social, economic, and political conditions of their farming. Many of the farmers interviewed had limited autonomy in how they coped with barriers to seed sourcing such as low income or a lack of saved seeds. In our study, farmers told us that own-farm seed saving was a preferred method for sourcing seeds, but that it was difficult to save consistently in a context of widespread food insecurity. This study corroborates earlier research findings that seed saving by smallholders was the seed sourcing method that most maintained crop genetic diversity on both a farm and a community scale (Kloppenburg, 2010). Many farmers explained that they had been unable to recover crop species after losing a saved seed stock due to lack of money or low seed diversity available through markets and seed exchanges. Not all farmers agreed, however, that seed saving was the best sourcing pathway; this research provides a nuanced understanding of seed saving practice, capturing the trade-offs that make farmers reluctant to save seeds for certain crops or varieties. Our analysis highlights the knowledge and skill behind seed saving practices that success-fully preserve seed viability, and points to potential limits to seed sovereignty based on farmer-saved seed.

Dry-season gardening played an important role in maintaining or replen-ishing seed stock, providing further evidence of the *dimba* as a strategic asset to contend with food insecurity. The *dimba* became more widely used during the 1990s food crises but has continued to be omitted from official literature about irrigated land and food production in Malawi (Chinsinga, 2007). This chapter has further described the unique contributions of *dimba*s to seed saving; in particular, how that seed sourcing strategy avoids the problem of degradation and loss that can occur during seed storage. Greater understanding of the *dimba*'s critical role in mitigating food insecurity draws attention to how social inequalities are produced and compounded through unequal access to *dimba* land.

Context-specific interactions influence the development of local seed systems, while broader international policy and companies dictate the formal seed system. Dominant agricultural policy in Malawi, including recent state legislation that limits seed purchasing to certified seed, or predominantly hybrid varieties, encourages purchasing seed over alternative seed sourcing pathways (Malawi Government, 2018). Scarce income opportunities and unfavourable crop prices result in few farmers opting for a purchasing pathway, except to acquire a higher quality or diversity of seed than that available through other sourcing pathways. This study shows how an assessment of the benefits of seed sourced through a purchasing pathway must also examine how a farmer obtained the money used to buy the seed, as well as any inputs that may be needed for hybrid varieties. Farmers' accounts of the measures they took to earn this money, such as selling an asset or crop, or doing *ganyu*, illuminate how seed purchasing often required them to compromise their autonomy over other farm and household decisions. Although a farmer might value the traits of a hybrid variety, the trade-off of purchasing the seed – selling maize meant for subsistence, perhaps, or neglecting their own crops to do *ganyu* – was too costly for some.

Ganyu is a critical seed sourcing pathway for many farmers, yet respondents were aware of the danger that reliance on this pathway posed to their own agricultural production. This finding is consistent with previous findings that *ganyu*, as a livelihood strategy, might increase short-term food security but sacrifice stability in other ways due to reduced labour availability for the farmer's own food production (Bezner Kerr, 2005). Bezner Kerr (2005) found that households performing *ganyu* were frequently food insecure, and that reduced labour for their own production was a challenge, but not necessarily a cause of continued poverty, due to a net benefit from *ganyu*. Our study supports this finding: *ganyu* affected farmers' labour availability and, notably, the *timing* of planting, thereby influencing longer-term food security and subsequent seed sovereignty. Previous research has examined the dynamics of *ganyu* and how they change based on kin relationships and needs of the household. This study has gone on to elaborate the social dynamics that are associated with *ganyu* for seed provisioning, rather than overall livelihood outcomes (Peters, 1999; Whiteside, 2000; Bezner Kerr, 2005).

In communities with strong kinship networks, seed sharing offered the farmers in our study similar benefits to seed saving by ensuring that, when in need, they had access to seed, regardless of their financial situation. These results support Van Niekerk and Wynberg's findings (2017) in South Africa that exchanging seeds in local systems strengthened social cohesion and benefited farmers by providing 'insurance' in lean years. Differences in seed sharing reflected regional disparities in other forms of social support; in addition to sharing seed more than Central Region farmers, Northern Region farmers more commonly reported that they shared food with food insecure friends and neighbours. In both regions, seed exchange was a less frequently practised form of social support than food sharing. Further exploration of the social norms related to sharing seed as opposed to sharing food would shed light on the different roles these social practices play in maintaining longer-term seed availability versus short-term food security.

Few of the farmers sourced seeds through only one pathway. Our pathway model of analysis demonstrated the multiple ways that farmers interact with different seed sources to suit their needs. Pautasso et al. (2013) similarly found that farmers typically interact with seed systems on a continuum from formal to informal, supporting the idea that interactions for seed provisioning are necessarily a balancing act of decision-making and compromise to meet farm needs. Although they engaged with several sourcing pathways, most farmers interviewed had preferences, often linked to the social relations underlying seed sourcing and their implications. For most, the social relations underlying seed saving, *dimba*, and seed sharing, were preferred to those of purchasing and *ganyu*, and yet the last two could be complementary to the former three, providing greater quality or diversity, and possibly helping to recover seed stores when they had been reduced by crop loss. Seed saving, *dimba*, and seed sharing avoided the exploitative prices and payment schemes farmers encountered in seed and *ganyu* transactions. *Ganyu* and purchasing pathways were sometimes unavailable, with farmers not always able to find *ganyu* when they needed it, or local vendors not offering a sufficient quantity or variety of seed for a desired crop. Seed saving, *dimba*, and sharing ensured that seed would be available at planting time, except in the case of seed loss in the field or storage.

The farmers' seed sourcing pathways were linked to, and often constitutive of, their social, environmental, and material conditions, including social status and kinship networks, access to land, off-farm work opportunities, drought occurrence, access to *dimba* land, and market prices or dynamics. However, a farmer's decision to access different seed through a different source might signal a possible change in that individual's trajectory: that is, an effort to shape a different future. A number of farmers chose to invest income and energy into *dimba* cultivation, in the hope of extricating themselves from their reliance on *ganyu* for seed sourcing. Some farmers responded to a perceived growing threat of drought by purchasing hybrid maize seed, while others chose to save local varieties that were more resistant to storage problems, and still others preferred to diversify the risk of crop loss through a combination of seed sourcing pathways, varieties, and species. Crop diversification, through purchasing, sharing, or, as in this case, seed distribution by a non-profit organization, has been shown to lead to greater access to a diversity of foods for home consumption, higher-value crop sales and market flexibility, and resilience in dealing with crop loss (Jones et al., 2014; Madsen et al., 2021). Other studies have found that crop diversification encourages seed exchange, enhancing diversity and its benefits at a community level and reinforcing social ties (Deaconu et al., 2019). These benefits may then support farmers in autonomously sourcing seed through their preferred pathways and transitioning towards food sovereignty.

The critical role of seed as a technical requirement for farming meant that, to secure seed material for the next planting season, the respondent farmers often compromised control over key aspects of their farming and food production. Findings from this chapter confirm the critical importance of seed sovereignty in any future scenario setting out a just food system in smallholder production systems. La Via Campesina asserts that agroecological farming systems, in which crop diversity and the preservation of varieties and traits that are locally adapted are prioritized, will support farmers' autonomy over their seed sourcing. The agroecological intervention associated with this study helped to strengthen farmer seed and food sovereignty in this context (Madsen et al., 2021): seed distribution reintroduced diversity that had not been consistently accessible due to market prices, giving farmers the opportunity to cultivate a variety of crops that had previously been inaccessible. Farmers used the soil fertility benefits of intercropping legumes, in addition to compost application, crop residue incorporation, and crop rotation, to reduce reliance on synthetic fertilizers. Substituting agroecological management for purchased inputs reduced household expenses – a saving that, for some, meant an opportunity to restore seed stocks that had been sold in previous years to purchase fertilizer. Yet farmers did not always choose a seed saving pathway, finding that *dimba*, *ganyu*, seed sharing, and seed purchasing pathways offered distinct advantages, depending on their particular circumstances. This chapter has sought to explore the complex narratives of local seed systems and their contribution to food sovereignty, describing how farmers navigate the ever-present threat of scarcity to create their own best practices.

The farmers' stories reveal how seed sourcing is integrally linked to their livelihoods and well-being. Respondents were aware of the trade-offs implied in seed sourcing options; however, not all of them were able to source seed through their favoured pathway. The seed sourcing pathways they used often reflected varying levels of agency, which, in the process of accessing seed, were either reproduced or altered. Farmers' accounts of their struggles to access seed revealed how far removed many households were from a state of food sovereignty, even as the implications of these pathways exposed how strategies of seed sourcing actively undermined the possibility of a food-sovereign future. These findings support the food sovereignty movement's assertion that farmers' autonomy over their food and production systems is invariably shaped by the conditions under which they access seeds (Wittman et al., 2010).

Notes

All names are pseudonyms to protect respondent confidentiality. The research project was approved by Cornell University Institutional Review Board, Protocol # 1607006471.

1. Interview 56, July 2018, woman, Central Region
2. Interview 20, July 2018, woman, Central Region
3. Interview 37, June 2018, woman, Northern Region
4. Interview 8, June 2018, woman, Northern Region
5. Interview 12, June 2018, woman, Northern Region
6. Interview 14, June 2018, woman, Northern Region
7. Interview 22, July 2018, woman, Central Region

References

Bezner Kerr, R. (2005) 'Informal labor and social relations in northern Malawi: the theoretical challenges and implications of ganyu labor for food security', *Rural Sociology* 70(2): 167–187 <https://doi.org/10.1526/0036011054776370>.

Bezner Kerr, R. (2013) 'Seed struggles and food sovereignty in northern Malawi', *Journal of Peasant Studies* 40(5): 867–897 <https://doi.org/10.1080/03066150.2013.848428>.

Bezner Kerr, R. (2014) 'Lost and found crops: agrobiodiversity, indigenous knowledge, and a feminist political ecology of sorghum and finger millet in northern Malawi', *Annals of the Association of American Geographers* 104(3): 577–593 <https://doi.org/10.1080/00045608.2014.892346>.

Chibwana, C., Fisher, M. and Shively, G. (2012) 'Cropland allocation effects of agricultural input subsidies in Malawi', *World Development* 40(1): 124–133 <https://doi.org/10.1016/j.worlddev.2011.04.022>.

Chinsinga, B. (2007) 'Hedging food security through winter cultivation: the agronomy of dimba cultivation in Malawi', paper delivered at the *Education Development Conference 2007, National University of Ireland, Galway, 24–25 November*.

Chinsinga, B. (2010) *Seeds and subsidies: the political economy of input programmes in Malawi*, Working Paper 013, Future Agricultures Consortium <https:// citeseerx.ist.psu.edu/viewdoc/download?doi=10.1.1.174.4432&rep=rep1& type=pdf>.

Deaconu, A., Mercille, G., and Batal, M. (2019) 'The agroecological farmer's pathways from agriculture to nutrition: a practice-based case from Ecuador's highlands', *Ecology of Food and Nutrition* 58(2): 142–165 <https://doi.org/10. 1080/03670244.2019.1570179>.

Guei, R.G., Barra, A. and Silue, D. (2011) 'Promoting smallholder seed enterprises: quality seed production of rice, maize, sorghum and millet in northern Cameroon', *International Journal of Agricultural Sustainability* 9(1) 91–99 <https://doi.org/10.3763/ijas.2010.0573>.

Jones, A., Shrinivas, A., and Bezner Kerr, R. (2014) 'Farm production diversity is associated with greater household dietary diversity in Malawi: Findings from nationally representative data', *Food Policy* 46: 1–12 <https://doi. org/10.1016/j.foodpol.2014.02.001>.

Kloppenburg, J. (2010) 'Impeding dispossession, enabling repossession: biological open source and the recovery of seed sovereignty', *Journal of Agrarian Change* 10(3): 367–388 <https://doi.org/10.1111/j.1471-0366.2010.00275.x>.

Kydd, J. and Christiansen, R. (1982) 'Structural change in Malawi since independence: consequences of a development strategy based on large-scale agriculture', *World Development* 10(5): 355–375 <https://doi.org/ 10.1016/0305-750X(82)90083-3>.

Lentz, E.C. (2018) 'Complicating narratives of women's food and nutrition insecurity: domestic violence in rural Bangladesh', *World Development* 104: 271–280 <https://doi.org/10.1016/j.worlddev.2017.11.019>.

Madsen, S., Bezner Kerr, R., LaDue, N., Luginaah, I., Dzanja, C., Dakishoni, L., Lupafya, E., Shumba, L. and Hickey, C. (2021) 'Explaining the impact of agroecology on farm-level transitions to food security in Malawi', *Food Security* 13: 1–22 <http://dx.doi.org/10.1007/s12571-021-01165-9>.

Malawi Government (2018) *National Seed Policy*, Ministry of Agriculture, Irrigation and Water Development, Lilongwe <https://faolex.fao.org/docs/ pdf/mlw180417.pdf>.

Mayet, M. (2015) 'Seed sovereignty in Africa: challenges and opportunities', *Development* 58: 299–305 <https://doi.org/10.1057/s41301-016-0037-x>.

Méndez, V.E., Bacon, C.M., and Cohen, R. (2013) 'Agroecology as a transdisciplinary, participatory, and action-oriented approach', *Agroecology and*

Sustainable Food Systems 37(1): 3–18 <https://doi.org/10.1080/10440046.2
012.736926>.

Mngoli, M.B., Mkwambisi, D.D. and Fraser, E.D.G. (2015) 'An evaluation of
traditional seed conservation methods in rural Malawi', *Journal of International
Development* 27(1): 85–98 <https://doi.org/10.1002/jid.3052>.

Pautasso, M., Aistara, G., Barnaud, A., Caillon, S., Clouvel, P., Coomes, O.T.,
Delêtre, M., Demeulenaere, E., De Santis, P., Döring, T., Eloy, L., Emperaire, L.,
Garine, E., Goldringer, I., Jarvis, D, Joly, H.I., Leclerc, C., Louafi, S., Martin, P.,
Massol, F., McGuire, S., McKey, D., Padoch, C., Soler, C., Thomas, M. and
Tramontini, S. (2013) 'Seed exchange networks for agrobiodiversity conser-
vation: a review', *Agronomy for Sustainable Development* 33(1): 151–175
<https://doi.org/10.1007/s13593-012-0089-6>.

Peters, P.E. (1999) *Agricultural Commercialization, Rural Economy and Household
Livelihoods, 1990–1997: Final Report.* Harvard Institute for International
Development, Harvard, MA.

Silverman, D. (2005) *Doing Qualitative Research: A Practical Handbook*, 2nd edn,
SAGE Publications, London.

Van Niekerk, J. and Wynberg, R. (2017) 'Traditional seed and exchange systems
cement social relations and provide a safety net: a case study from KwaZulu-
Natal, South Africa', *Agroecology and Sustainable Food Systems* 41(9–10):
1099–1123 <https://doi.org/10.1080/21683565.2017.1359738>.

Wattnem, T. (2016) 'Seed laws, certification and standardization: outlawing
informal seed systems in the Global South', *The Journal of Peasant Studies*
43(4): 850–867 <https://doi.org/10.1080/03066150.2015.1130702>.

Whiteside, M. (2000) *Ganyu Labour in Malawi and its Implications for Livelihood
Security Interventions: An analysis of recent literature and implications for poverty
alleviation*, Overseas Development Institute, London.

Wittman, H., Desmarais, A. and Wiebe, N. (2010) 'The origins and potential of
food sovereignty', in H. Wittman, A. Desmarais, and N. Wiebe (eds), *Food
Sovereignty: reconnecting food, nature and community*, pp. 1–14, Fernwood,
Halifax, NS.

Box B 'When the rains come, I take my seeds and plant them': managing agricultural biodiversity at household level

Jaci van Niekerk

Photo B1 A household seed bank from Ingwavuma
Credit: Jaci van Niekerk

Introduction

The remote north-west corner of South Africa's KwaZulu-Natal province is home to rolling hills, free-roaming cattle, and scattered plots of land planted with a diversity of crops. The custodians of these crops are small-scale farmers, mostly middle-aged women, who learned the craft of cultivation and the art of seed saving from their parents and grandparents. In recent years the age-old practice of seed saving has become eroded, largely due to the introduction of 'modern' seed varieties such as hybrids. A local NGO working towards increasing agricultural biodiversity on smallholder farms, Biowatch South Africa, recognized the need to revive household seed saving as a complementary practice in programmes aimed at bolstering household level nutrition and food security. Integral to this was the requirement that seeds be saved for replanting and exchange in every household. This approach contrasts with interventions driven by development agencies and government programmes, which tend to favour more 'formal' arrangements such as the establishment of community seed banks or linkages with gene banks.

A research project carried out under the umbrella of the Seed and Knowledge Initiative saw the University of Cape Town partnering with Biowatch to investigate the importance

(Continued)

Box B Continued

and practicability of household seed saving. In November 2014, the team interviewed 40 small-scale farmers in Ingwavuma and Pongola, rural towns situated in north-west KwaZulu-Natal, close to the border with Eswatini, with responses revealing that a large variety of seeds were saved for replanting; traditional knowledge still informed many seed-saving practices; and that the act of seed saving fostered independence and social cohesion.

Diverse seeds are stored in multiple places

Despite a prolonged drought combined with unseasonal hot weather, which affected the harvest, farmers still managed to save seeds and vegetative propagation material from a range of crops: cucurbits, legumes, grains, vegetables, fruits, and others such as cassava, taro potatoes, coffee beans, and sunflowers. High levels of diversity were found within some species; for instance, farmers would typically save four or five types of maize or half a dozen types of beans. Seeds were generally kept in the kitchen, with maize, millet, and sorghum strung up above the fireplace to benefit from the storage-enhancing and pest-repelling qualities of the wood smoke. One farmer kept her seeds outdoors, in large metal drums in her vegetable garden, while another stored her highly prized seeds under her bed as she knew they would be safe from rodents there.

Seed preservation methods

To keep the seed viable until the next planting season, farmers employed a number of methods based on traditional knowledge, either passed on by elders or shared by other farmers involved in Biowatch projects. Some seeds were covered in ash from the fire (*umlota*) or, better yet, ash from burned aloe leaves (*inhlaba*), and others were mixed with whole or powdered dried citrus peels or *amahlamvu* (fragrant herbs that grow in the wild). In storage, seeds were placed in glass or plastic bottles as the pottery containers used in days gone by were no longer produced in the area.

Seed saving challenges

Saving seed is not without its challenges. Some crops may fall out of favour as the younger generation don't enjoy eating them, or climate change-induced disruptions in planting seasons may lead to failed harvests and thus a scarcity of seeds to save. A lack of fencing means that crops in the field are vulnerable to attack from chickens, goats, and cattle, and in the case of small grains such as sorghum and millet, wild birds might devour them before they are ready to harvest. A number of farmers remarked that unseasonal rains interfere with the post-harvest drying process and, once in storage, seeds may be destroyed by several creatures, including weevils, moths, and rodents. Excess moisture, often the result of storage in unsuitable containers, renders seeds sterile. Modernization also plays a role in seed preservation; for instance, some farmers no longer cook over a fire, and thus can't preserve their grains by smoking them.

Seed saving fosters preparedness and independence

A central finding of this study is the importance ascribed to being ready to plant one's own seed as soon as the rains come: in other words, timely crop production (Coomes, 2010). One farmer, for instance, remarked that she did not want to 'disturb' or rely on her neighbours when the planting season arrived. This level of independence translates into self-sufficiency, an attribute vital when farming in resource-limited conditions. Having one's own seed to plant saves precious cash reserves as there is no need to buy seed at markets or from agricultural dealers. Moreover, by saving one's own seed one is assured of its provenance; for instance, one farmer remarked that 'when saving seed, one is guaranteed that the seeds are not [chemically] treated'.

(Continued)

Box B Continued

The strengths of farmers' seed systems are identified as their 'availability, affordability, and timeliness' (Mkindi, 2015: 2). These attributes also apply to the household seed banks in Ingwavuma and Pongola, for not only do farmers save money by saving seed, but having one's own seed to plant in rain-fed agriculture systems circumvents issues related to access. Obtaining seed from 'formal' *ex situ* collections such as gene banks or community seed banks is likely to be governed by rules and regulations, whereas accessing seed from 'informal' sources such as fellow farmers, neighbours, and family members could well be contingent on 'community rules' such as expectations of reciprocity.

The social side of seed: enhanced food security increases social cohesion within households

Seed saving, and the enhanced food security it delivers, was credited for fostering social cohesion at household level. Sthapit et al. (2012: 19) found that farmers' seed systems in Nepal 'increased social cohesiveness as [they have] been managed through individual relationships'. In the words of one farmer who was teaching her grandchildren about this treasure of the home: 'It keeps the home happy when there is plenty of food [food grown from saved seed], no need to buy, no need to quarrel about money spent or money for food.' Another farmer underlined the importance of seed saving for household food security, stating: 'Even if there is drought and food runs out, we will survive because I have my seed.' Saving seed from a diversity of food crops has been shown to spread the risk of crop failure, thereby contributing to the security of household food supply channels (Richards et al., 2009). And, as noted by McGuire and Sperling (2016), a household does not require large quantities of seed in order to be food secure, since most seeds, especially grains, have a high multiplication rate.

Seed security: a backstop in times of privation

Being seed secure brings economic benefits for farmers who otherwise have limited opportunities to earn a living. Farmers practising agroecology do not incur costs for chemical fertilizers and pesticides, and if they produce a surplus, they can sell it, thereby generating an income. Linked to economic benefits and financial freedom, farmers in the study areas value their seed highly for reliable performance in low-input farming systems. The custodian farmers interviewed in Ingwavuma and Pongola see their household seed banks as a 'bank': having seed in the bank means they do not need money to farm. Poverty is frequently cited as a leading cause of agricultural biodiversity loss, but through practising seed saving, even the very poor can plant crops to feed their families (Sthapit, 2013; McGuire and Sperling, 2016).

A gendered perspective on seed custodianship

In traditional rural communities around the African continent, both men and women farm, but while men are generally responsible for keeping large livestock and cultivating cash crops, women are usually the ones in charge of feeding their families nutritious meals through maintaining a diversity of seed. Rural KwaZulu-Natal is a patrilineal society, and while women play important roles in reproductive, productive, and community-related activities, men have more authority, and are the ones who engage in community politics through traditional leadership structures (Mtshali, 2002). With their roles as rural women strongly circumscribed by tradition and culture, the lack of opportunity these farmers have had to further their formal education has translated into a deep connection to the land and comprehensive knowledge of crop production, seed management, and the provision of household food security. This accords with research carried out elsewhere in Africa, where women's knowledge and skills related to seed

(Continued)

Box B Continued

grant them status and respect within their families and communities, as they are not only the providers of food, but also supply seed for cultural activities and contribute to local economies too. One farmer in the study remarked that her kids respected her authority and she had dignity and status as she was seen as the head of the household by her extended family because she had the seeds.

The pernicious impact of modern seed varieties

External interventions can have significant impacts on local seed systems. Farmers associated with Biowatch reported that the agricultural extension that they received was firstly, minimal, and secondly, inappropriate, as extension officers recommended industrial agricultural practices which they did not wish to implement. In the past, extension officers had donated bags of hybrid seeds to the farmers; research elsewhere has shown the serious implications of this for local seed systems and economies. For instance, Pionetti (2005) found that commercial seeds appealed to men, who started growing them as cash crops to sell to companies or on formal markets. They began purchasing the inputs needed for commercial seed production, such as fertilizers and pesticides, and then, instead of cultivating food crops as before, they bought them. These purchases diminished the independence of male farmers, delivering them to a perpetual cycle of dependency and indebtedness. Women farmers were affected too, as cash crops displaced the diverse local food crops they once grew, and with them traditional knowledge, stripping women of their autonomy. This succession of events culminated in the gradual obliteration of the local non-monetary economy.

Threats to household seed saving: the role of traditional knowledge

Traditional knowledge held by rural women in KwaZulu-Natal plays an invaluable role in food production and post-harvest processes (Mtshali, 2002), and as the findings indicate, traditional methods of seed preservation such as smoke, ash, fragrant herbs, and citrus peels are successful deterrents to storage pests. Farmers find it more difficult, however, to combat threats to seed pre-harvest, such as birds that feed on small grains and free-roaming livestock that destroy crops due to a lack of fencing. Farmers commented that since the law dictated that children had to attend school, they were no longer available to drive birds and livestock away from the crops. Combined with a lack of money to fence off fields, the loss of children's assistance with farming considerably adds to the workload of small-scale women farmers.

Another form of traditional knowledge erosion detected in this study is related to changed contexts, often due to modernization, rendering existing traditional knowledge irrelevant. For example, farmers know that storing their seed over wood smoke will preserve the grains by making them less attractive to insects, but if they no longer cook over fires, this knowledge, although not lost, is no longer applicable. Altieri et al. (2015) maintain that one reason why the preservation of traditional knowledge related to farming practices is important is that it forms part of the social and human capital which will enable small-scale farmers to cope with and adapt to climate change-induced environmental change, a threat which is all too real in northern KwaZulu-Natal.

Conclusion

These findings show that even the most remote corners of South Africa are subject to change, be it environmental or induced by modernization. Small-scale farmers who wish to save their traditional seed have to negotiate failed harvests due to climate change, the replacement of wood fires with gas, and a paucity of suitable containers for seed saving.

Those who overcome such obstacles, whether by relying on traditional knowledge or by accepting support from NGOs, are richly rewarded. Being ready to plant one's own seed

(Continued)

Box B Continued

as soon as the rains come cultivates autonomy and self-reliance, instils knowledge about the provenance of the seed, and affords dignity, respect, and status. Underlining the value of seeds as social currency, being seed secure translates into enhanced household food security, which in turn fosters social cohesion.

In Africa, seed custodianship is generally within the purview of women, but this changes when modern seed varieties are introduced. The result is that men tend to step in and cultivate the cash crops these seeds were designed for, supplanting the diversity of traditional crops women plant to feed their households, and undermining women's roles as seed custodians.

In this study we aimed to fill a knowledge gap around household seed banks and probed the value of these to small-scale farmers in KwaZulu-Natal. The findings indicate that a significant amount of traditional knowledge accompanies seed saving at home, a knowledge set which is likely to be lost, should traditional seeds be replaced by commercial varieties. In fact, should household seed saving no longer be practised, small-scale farmers, their families, their communities, and agricultural biodiversity stand to lose immensely.

References

Altieri, M.A., Nicholls, C.I., Henao, A. and Lana, M.A. (2015) 'Agroecology and the design of climate change-resilient farming systems', *Agronomy for Sustainable Development* 35: 869–890 <https://doi.org/10.1007/s13593-015-0285-2>.

Coomes, O.T. (2010) 'Of stakes, stems, and cuttings: the importance of local seed systems in traditional Amazonian societies', *The Professional Geographer* 62(3): 323–334 <https://doi.org/10.1080/00330124.2010.483628>.

McGuire, S. and Sperling, L. (2016) 'Seed systems smallholder farmers use', *Food Security* 8: 179–195 <https://doi.org/10.1007/s12571-015-0528-8>.

Mkindi, A.R. (2015) *Farmers' seed sovereignty is under threat*, Policy Paper 3. Berlin: The Rosa Luxemburg Foundation. <https://www.rosalux.de/fileadmin/rls_uploads/pdfs/Standpunkte/policy_paper/PolicyPaper_03-2015.pdf>.

Mtshali, S.M. (2002) *Household Livelihood Security in Rural Kwa-Zulu Natal, South Africa*, PhD Thesis, Wageningen University, Wageningen, The Netherlands <https://library.wur.nl/WebQuery/wurpubs/fulltext/139851>.

Pionetti, C. (2005) *Sowing Autonomy: Gender and Seed Politics in Semi-Arid India*, London: IIED.

Richards, P., de Bruin-Hoekzema, M., Hughes, S.G., Kudadjie-Freeman, C., Kwame Offei, S., Struik, P.C. and Zannou, A. (2009) 'Seed systems for African food security: linking molecular genetic analysis and cultivator knowledge in West Africa', *International Journal of Technology Management* 45(1/2): 196–214 <http://dx.doi.org/10.1504/IJTM.2009.021528>.

Sthapit, B. (2013) 'Emerging theory and practice: community seed banks, seed system resilience and food security', in P. Shrestha, R. Vernooy and P. Chaudhary (eds), *Community Seed Banks in Nepal: Past, Present, Future*, Proceedings of a National Workshop, 14–15 June 2012, Pokhara, Nepal: LI-BIRD/USC Canada Asia/Oxfam/The Development Fund/IFAD/Bioversity International, 14–15 June 2012, Pokhara, Nepal. <https://www.doc-developpement-durable.org/file/Culture/Conservation-Graines-Semences-Vegetaux-Legumes-Refrigeration/banque-de-graines/Community_seed_banks_in_Nepal_past_present_&_future.pdf>.

Sthapit, B., Shrestha, P. and Upadhyay, M. (eds) (2012) *On-farm Management of Agricultural Biodiversity in Nepal: Good Practices*, Nepal: NARC/LI-BIRD/Bioversity International. <https://cgspace.cgiar.org/bitstream/handle/10568/104917/On-farm_management_of_agricultural_biodivesity_in_Nepal_Good_Practices_revised_edition_2012_1222_.pdf?sequence=3&isAllowed=y>.

Box C Are seed secure households also food secure?

Bulisani L. Ncube

Introduction

Smallholder farmers in Southern Africa rely on agriculture as their main source of food and livelihoods. Although agricultural production has increased both globally and in the region, the number of hungry people in Southern Africa that cannot access adequate, good quality food continues to increase (FAO, 2019). Smallholder farmers often suffer from a lack of appropriate seed as well as high levels of food insecurity. Interventions such as community seed production, seed aid, and input subsidies are used to address these concerns (Sperling and McGuire, 2010). However, the relationship between seed security and food security has not been sufficiently studied. The case study from eastern Zimbabwe aims to elucidate the factors that affect the relationship between seed security and food security. Specifically, it explores how the dimensions of seed security, which include availability, access, and quality, are related to those of household-level food security, which include dietary diversity and food consumption. The study thus asks whether households with adequate seed also have adequate food.

Study site and methods

The case study was conducted across Chikukwa and Chaseyama in the Chimanimani district. These areas were selected because of their different biophysical characteristics and agricultural potential. Chaseyama is located on the western side of Chimanimani, with the lowest and most erratic rainfall (between 300 mm and 450 mm per annum) in the district. The area is unsuitable for dry-land crop production (Campbell et al., 1995). Chikukwa is characterized by high rainfall of 1,000 mm per annum, with stable, deeply weathered, red-clay soils that are well suited to diversified cropping and high-value crops such as coffee, tea, and potatoes (Mugandani et al., 2012).

Methods used to collect data included a household survey of 227 farming households; interviews with 12 agro-dealers, local extension officers, and farmer groups; and individual life histories. The research was conducted in 2017.

Key findings

Informal seed systems play a dominant role

Informal sources such as own-saved seed, seed bought at local markets, and seed acquired through social networks played a dominant role when compared with formal sources such as seed sold by agro-dealers and distributed as aid. With the exception of maize, informal seed sources, at about 72 per cent, dominated seed supply. Eighty per cent of the maize planted was procured through formal seed sources such as agro-input sellers and seed aid. This confirms other studies that show informal channels as the main source of seed for small grains and legumes, providing more than 80 per cent of the seed grown in Zimbabwe (Mazvimavi et al., 2017). This underscores the importance of informal seed sources for smallholder farmers.

> My family struggles to obtain enough seed from agro-dealers, NGOs and the government seed aid. Due to the high cost of seed from agro-dealers, it is difficult to purchase enough quantities for my needs, while the government seed assistance comes late and is not guaranteed. The seed crops that I could not get include groundnuts, cowpeas, bambara groundnuts and finger millet. (Leo, case history interview, Chaseyama, September 2017)

(Continued)

BOX C ARE SEED SECURE HOUSEHOLDS ALSO FOOD SECURE?   111

Box C Continued

Seed availability, in terms of proximity and timeliness, does not always influence food security

The effects of timeliness and proximity were analysed to elucidate the relationship between seed availability and food security. Some researchers argue that there is a weak correlation because only a small proportion of the harvest is needed to meet sowing needs for the subsequent season (McGuire and Sperling, 2011). This study showed that farmers who obtained their seed from different sources, regardless of when, did not have observable differences in their food security status. A possible explanation is that the time difference between getting seed and planting period was not wide enough to affect crop productivity. Findings confirm that although timely seed availability affects seed security, it does not always have an impact on food security.

Farmers who obtained their seed from nearby sources had more experiences of food insecurity than farmers who obtained their seed from sources further away. Although obtaining seed from nearby sources was advantageous for farmers because of lower transaction costs and timeliness, it might have led to missed opportunities to travel to other areas and thus access food that was not available locally. Farmers from Chaseyama coped with drought and a lack of food availability by travelling to more distant locations to purchase food for their families. The findings suggest that factors such as proximity and timeliness, while important for ensuring seed availability, do not directly affect food security.

Links between seed access and food security are not evident

Studies have shown that access to food in rural settings is affected not only by decreased production, but also by loss of assets, lack of off-farm income, diminished purchasing power, and social marginalization (Mutea et al., 2019; Zhou et al., 2019). Households that perceived purchased seed to be expensive were therefore compared to those who perceived their seed to be affordable. The comparison revealed no differences in their food security outcomes.

These households were further assessed by comparing their food security with their access to assets and income. There were positive correlations between food security indicators and the number of assets owned, income sources, and quantities of seeds planted. On the other hand, the relationship between asset ownership, income sources, and seed security was inconclusive. This contrasts with other research that observed a strong relationship between access to seeds and ownership of household assets (Asfaw et al., 2011; Mesfin and Zemedu, 2018). However, those studies focused on accessing improved seed from formal sources, based on the ability to purchase seed. In contrast, for Chimanimani farmers, most seed sources included self-provision and social networks, which were predicated on social relations. A clear link between seed access and food security thus could not be determined.

> Though I source my seed from seed aid (sometimes) and purchases (from agro-dealers), this seed is rarely enough for my planting needs, especially when I would need to replant due to a drought. To ensure that my household has enough food we rear and sell the local chickens and some goats. (Leo, case history interview, Chaseyama, September 2017)

Seed quality is important for food security

The effect of seed quality on food security was explored, focusing on physical purity and seed germination. Results showed higher food security scores for seed-secure households than for those that had reduced seed security, with significant differences between the two categories. Thus, there was a direct relationship between the quality of seed and food security. A conclusion is that the quality of farmers' seed is essential to ensure that

(Continued)

Box C Continued

adequate food is produced. Quality seed leads to a healthy crop that ensures better crop yields, provided there is adherence to other, complementary measures of crop production. These findings have been corroborated by other researchers, such as Roy (2014) and Okello et al. (2017), demonstrating that the use of clean, good-quality planting materials translates into increased productivity and, subsequently, increased income.

Context matters

Seed security elements of availability, access, and quality affect the relationship between seed and food, as do the household and wider contexts in which farmers find themselves. These include farming practices, such as the level of crop production and diversity, and the number of years a household has been farming; economic and social assets; and agroecological characteristics.

Farmers and their families do not exist in a vacuum. They constantly interact with, and are moulded by, a broader socio-economic and political context, as well as erratic weather and a changing climate. It is critical to understand how these factors interact with, influence, and contribute to household seed and food security.

Conclusions

The relationship between seed security and food security is complex. Seed security does not necessarily equate to food security, nor does seed insecurity necessarily imply food insecurity. These findings are contrary to the assumptions of development actors who promote interventions based on seed provision as a linear path to ensuring better food access. Thus, the use of existing food-security assessments to extrapolate seed insecurity has critical flaws.

Those planning and implementing development programmes based on seed-related interventions should recognize that seed provision will not necessarily result in food security and might be counterproductive if it disrupts farmers' local seed systems. To ensure positive impacts for smallholders, attention should be given to the role of household and contextual factors in determining farmers' access to seed and food.

References

Asfaw, S., Shiferaw, B., Simtowe, F. and Hagos, M. (2011) 'Agricultural technology adoption, seed access constraints and commercialization in Ethiopia', *Journal of Development and Agricultural Economics* 3(9): 436–447 <https://academicjournals.org/journal/JDAE/article-full-text-pdf/F300BB68399>.

Campbell, B., Clarke, J., Luckert, M., Matose, F., Musvoto, C. and Scoones, I. (1995) 'Local-level economic valuation of savanna woodland resources: Village cases from Zimbabwe', *Hidden Harvest Project, Research Series* 2(3), Hot Springs Working Group, Sustainable Agriculture Programme of the International Institute for Environment and Development, London <https://pubs.iied.org/sites/default/files/pdfs/migrate/6002IIED.pdf>.

Food and Agriculture Organization of the United Nations (FAO) (2019) *Crop Prospects and Food Situation*, Quarterly Global Report, July, FAO, Rome. <https://www.fao.org/3/ca5327en/CA5327EN.pdf>.

McGuire, S.J. and Sperling, L. (2011) 'The links between food security and seed security: facts and fiction that guide response', *Development in Practice* 21(4–5): 493–508 <https://doi.org/10.1080/09614524.2011.562485>.

Mazvimavi, K., Murendo, C., Gwazvo, C., Mujaju, C. and Chivenge, P. (2017) *The Impacts of the El Niño-Induced Drought on Seed Security in Southern Africa: Implications for Humanitarian Response and Food Security*, ICRISAT, USAID and FAO <http://oar.icrisat.org/id/eprint/10279>.

(Continued)

BOX C ARE SEED SECURE HOUSEHOLDS ALSO FOOD SECURE?   **113**

Box C Continued

Mesfin, A.H. and Zemedu, L. (2018) 'Choices of varieties and demand for improved rice seed in Fogera District of Ethiopia', *Rice Science* 25(6): 350–356 <https://doi.org/10.1016/j.rsci.2018.10.005>.

Mugandani, R., Wuta, M., Makarau, A. and Chipindu, B. (2012) 'Re-classification of agro-ecological regions of Zimbabwe in conformity with climate variability and change', *African Crop Science Journal* 20(suppl 2): 361–369 <https://www.ajol.info/index.php/acsj/article/download/81761/71908>.

Mutea, E.N., Bottazzi, P., Jacobi, J., Kiteme, B., Ifejika Speranza, C. and Rist, S. (2019) 'Livelihoods and food security among rural households in the north-western Mount Kenya region', *Frontiers in Sustainable Food Systems* 3 <https://doi.org/10.3389/fsufs.2019.00098>.

Okello, J.J., Zhou, Y., Kwikiriza, N., Ogutu, S., Barker, I., Schulte-Geldermann, E., Atieno, E. and Ahmed, J.T. (2017) 'Productivity and food security effects of using certified seed potato: the case of Kenya's potato farmers', *Agriculture and Food Security* 6(25):1–9 <https://doi.org/10.1186/s40066-017-0101-0>.

Roy, B. (2014) 'Farmers' participatory quality seed production of field crops: a case study', *Journal of Crop and Weed* 10(2): 89–93.

Sperling, L. and McGuire, S.J. (2010) 'Persistent myths about emergency seed aid', *Food Policy* 35(3): 195–201 <https://doi.org/10.1016/j.foodpol.2009.12.004>.

Zhou, D., Shah, T., Ali, S., Ahmad, W., Din, I.U. and Ilyas, A. (2019) 'Factors affecting household food security in rural northern hinterland of Pakistan', *Journal of the Saudi Society of Agricultural Sciences* 18(2): 201–210 <https://doi.org/10.1016/j.jssas.2017.05.003>.

CHAPTER 6

Seed sovereignty, knowledge politics, and climate change in northern Ghana

Hanson Nyantakyi-Frimpong and Audrey Carlson

Introduction

The impacts of climate change are currently being felt in every region of the world. These impacts will be exacerbated in the future, with significant implications for sustainable agriculture, food security, and nutrition (FAO et al., 2018). In the African context, seed systems constitute one of the areas where agriculture-related climate impacts will be most pronounced (Niang et al., 2014; Sossou et al., 2019). Longer-term changes in temperature, precipitation, and the occurrence of extreme weather events will affect local seed systems, including seed selection, storage, production, distribution, and exchange (McGuire and Sperling, 2013; Niang et al., 2014; FAO, 2016). The need to bolster smallholder farmers' adaptation to these projected changes is now more pertinent than ever (Niang et al., 2014; FAO et al., 2018). However, the debate about what strategies and technologies are appropriate is often narrowly framed. Discussions and ensuing interventions frequently assume that scientists and experts from non-governmental organizations (NGOs) know what is best, and that seed users (e.g. Indigenous farmers) are merely involved in processes of adaptation and fine tuning (Juma, 2015).

The main objective of this chapter is to interrogate seed access and knowledge politics using Ghana as a case study and through the lens of climate change. The chapter examines climate change adaptation projects specifically linked to farmer seed systems, with a critical focus on who pursues these projects or gains from them, why and how they do so, and what their motivations and strategies are. The chapter also focuses on social justice, particularly for those who are potentially marginalized by seed-related interventions promoted by NGOs and government agencies.

In the Ghanaian context, a number of local and international NGOs are involved in seed interventions, most of them linked to the Alliance for a Green Revolution in Africa (AGRA). The work of AGRA entails a suite of technological interventions aimed at raising productivity among smallholder farmers on the assumption that increased yields are key to food security (Toenniessen et al., 2008; Moseley, 2016). In northern Ghana, AGRA works closely with partner NGOs (Asuru, 2015), as well as government institutions like the Savanna

Agricultural Research Institute (Martey et al., 2014). Aside from AGRA, other seed intervention projects in Ghana are also linked to the World Bank and Food and Agriculture Organization's (FAO) climate-smart agriculture (CSA) initiatives. As defined by the FAO, CSA involves technologies that increase the productivity of a given crop while simultaneously building resilience to climate change and reducing greenhouse gas emissions (Palombi and Sessa, 2013). Evidence for this chapter comes from one particular NGO project linked to both AGRA and CSA. For ethical reasons, we adopt a pseudonym for the NGO: the Seed Innovators Group.

Overall, the chapter contributes to understanding: how local farmers both engage with and are constructed by policy processes; whose realities and knowledge are used in the construction of acceptable seed systems adaptation strategies; and how institutions, both private and public, respond to the views of the local farmers they are meant to serve. The chapter argues that in order to enhance seed systems adaptation to climate-induced stresses in Africa, diverse views, not just those of mainstream scientists and NGO experts, must be accepted as legitimate and authentic. This argument is built based on in-depth empirical research and the literature on seed sovereignty and knowledge politics.

Seed sovereignty and knowledge politics

Seed sovereignty as a concept is defined as farmers' 'rights to save, breed and exchange seeds, to have access to diverse open source seeds which can be saved – and which are not patented, genetically modified, owned or controlled by emerging seed giants. It is based on reclaiming seeds and biodiversity as commons and public good' (Shiva, 2012; see also Bezner Kerr, 2013). Seed sovereignty serves as a key foundation to food sovereignty (Bezner Kerr, 2013; Adhikari, 2014; Kloppenburg, 2014), which asserts that people should be able to control the mechanisms and policies that shape their own food production, distribution, and consumption (Wittman et al., 2010). As emphasized by Kloppenburg (2014: 1225), 'it is difficult to imagine any form of "food sovereignty" that does not include a necessary and concomitant dimension of what might be called "seed sovereignty"'. Seed sovereignty also addresses some of the weaknesses inherent in the concept of seed security, defined as when farmers 'have sufficient access to quantities of available good quality seed and planting materials of preferred crop varieties at all times in both good and bad cropping seasons' (FAO, 2016: 1). Like seed security, seed sovereignty focuses on elements of seed availability and ready access to seeds. However, the critical distinction between seed security and seed sovereignty is the issue of 'local control over seeds', which includes farmers' rights to produce and maintain their own seeds (Bezner Kerr, 2013; Adhikari, 2014).

Knowledge politics relates to how ordinary citizens engage in critical scientific debates, interventions, and decisions that affect their lives

(Bezner Kerr et al., 2018). It is now recognized that local farmers have salient knowledge and critical perspectives that should be taken seriously as substantive inputs into the planning, design, and implementation of scientific interventions and development initiatives previously assumed to be the sovereign domain of expert scientific bodies. For example, anthropological, ecological, and historical research in low-income countries has frequently exposed major disjunctures between the knowledge and perspectives of land users and those underlying and reproduced through national and internationalized science and policy (e.g. Fairhead and Leach, 1996). This chapter puts a critical emphasis on how farmer participation and knowledge are both conceptualized in policy discourses.

Knowledge politics provides a framework in which to apply principles of social justice in environments of competing interests regarding science (Bezner Kerr et al., 2018). Both knowledge and its making can be seen as a good to be distributed, including all voices for whom the science will matter. In this framework, knowledge production is shared among a broader constituency of knowers. The chapter presents an argument for knowledge justice, meaning that an inclusive approach to making science is necessary for socially just outcomes. It also draws on the literature on political agronomy, which focuses on the knowledge politics involved in the problem and solution framing around some of the world's most contentious agrarian issues (Sumberg and Thompson, 2012).

Materials and methods

The case study setting

Fieldwork for this study was conducted in Ghana's Upper West Region (Figure 6.1). The region occupies an area of 18,476 km², has 11 districts, and is home to 702,110 people (Ghana Statistical Service, 2013). The yearly mean temperature is 18–22°C, with an average annual precipitation of 1,036 mm (Nyantakyi-Frimpong and Bezner Kerr, 2015a). Agriculture is the main source of income for more than 80 per cent of households (Ghana Statistical Service, 2013). The average farm size is 2.7 hectares (Chamberlin, 2007). Until recently, most farmers produced their own seeds year after year, or relied on other farmers for seeds. While NGOs have introduced several hybrid seed varieties to the region, farmers still value their landraces and continue to plant them, thereby contributing to the conservation of biodiversity. Most farming practices occur in a single rainy season of about five months, from late May to late October, with the main peak from June to July. Maize *(Zea mays)*, common beans *(Phaseolus vulgaris)*, finger millet *(Eleusine coracana)*, sorghum *(Sorghum bicolor)*, cowpea *(Vigna unguiculata)*, bambara beans *(Vigna subterranea)*, and groundnuts *(Arachis hypogaea)* are among the principal crops, with maize cultivated on more than 90 per cent of landholdings (Nyantakyi-Frimpong and Bezner Kerr, 2015b).

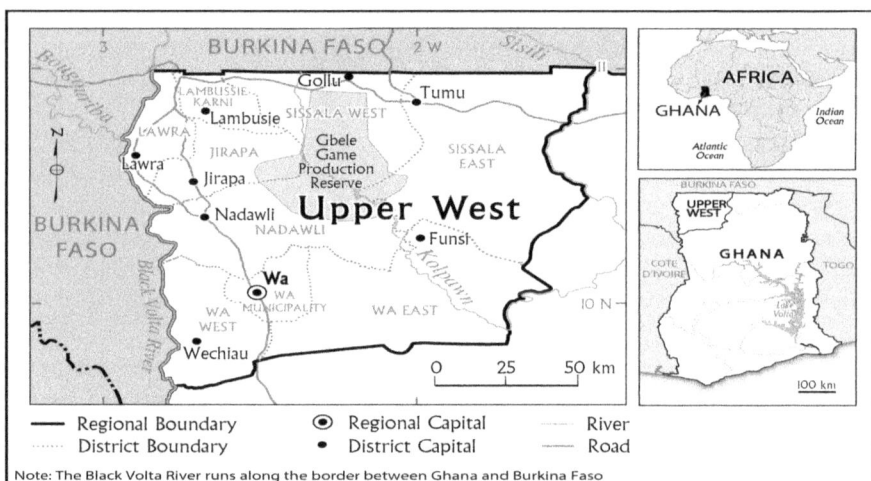

Figure 6.1 Map of the study area
Source: Map prepared by Karen van Kerkoerle

Over the last 10 years, many farmers in the region have lost most of their farmlands through land grabbing. The 1992 Ghanaian Constitution gave the government the power to take customary land for public use, subject to just and adequate compensation. Using this power for compulsory land acquisition, called 'eminent domain', the Ghanaian state has enclosed 316,400 hectares (3,164 km²) of farmlands for the purposes of gold mining in the study region. This land has been given as a concession to Azumah Resources Limited, a foreign-based mining company. Thousands of households have been affected by this land enclosure, leading to increasing levels of food and seed insecurity (Nyantakyi-Frimpong and Bezner Kerr, 2017). In comparison with the rest of Ghana, the Upper West Region has a greater proportion of poor and food-insecure households (Atuoye et al., 2019; Ghana Statistical Service, 2015). Closer interlinkages between food security and seed security (Madin, 2020) made the Upper West an ideal location for this case study.

Agricultural development in the Upper West Region has been severely transformed since neoliberalism began to take hold in Ghana in the early 1980s (Wiemers, 2015; Aryeetey and Kanbur, 2017; Awanyo and Attua, 2018). A snapshot of this transformation is critical to understanding the case study findings in this chapter. Like most low-income countries, Ghana was hit hard by the oil price shock in 1973, which led to a sharp increase in import expenditures and a decline in export proceeds (Konadu-Agyemang, 2000). This economic situation and a devastating drought in 1981 combined to plunge the Ghanaian economy into crisis (Awanyo and Attua, 2018). During this period, agricultural production stalled, industrial production faltered, and development goals were not met (Aryeetey and Kanbur, 2017). The government subsequently approached the World Bank and the

International Monetary Fund for an economic recovery loan (Awanyo and Attua, 2018). A loan of US$1.4 bn was granted, and with it came a set of conditionalities called structural adjustment programmes (SAPs) (Pearce, 1992; Hutchful, 2002). As elsewhere in Africa, SAPs in Ghana enforced spending cuts, currency devaluation, retrenchment, trade liberalization, and privatization (Konadu-Agyemang, 2000; Aryeetey and Kanbur, 2017).

Agriculture was one of various sectors in Ghana that were ferociously restructured (Wiemers, 2015). For example, state-run agricultural marketing boards, which prior to SAPs handled the sale of agricultural inputs like seeds, were all closed down (Pearce, 1992). Agricultural spending was deprioritized, including cuts to the provision of extension services. Currency devaluation also served to make agricultural inputs costly, especially as subsidies shrank (Konadu-Agyemang, 2000; Wiemers, 2015). Additionally, agriculture was restructured from domestic to export-oriented production (Pearce, 1992; Konadu-Agyemang, 2000). Overall, the net effect of the SAPs was that they subverted the state's capacity to invest in smallholder agriculture (Mohan, 2002).

Today, the scar of SAPs is still visible in Ghana, with their legacies more debilitating in the country's rural north (Konadu-Agyemang, 2000; Mohan, 2002; Nyantakyi-Frimpong and Bezner Kerr, 2015b; Aryeetey and Kanbur, 2017). For example, due to the privatization of markets, agriculture inputs such as seeds have become too expensive for smallholder farmers (Aryeetey and Kanbur, 2017). In the void created by the absence of the state in smallholder agricultural development, local and international NGOs have stepped in (Mohan, 2002; Dugle et al., 2015). Dugle et al. (2015) offer a comprehensive assessment of how these NGOs are formed and operated in the Upper West Region. International NGOs are mainly from Australia, Europe, and North America. Local NGOs, many of which are unregistered, include church-based bodies and those initiated by private individuals (Dugle et al., 2015). In terms of agriculture development, local and international NGO activities include integrated soil fertility management and the provision of seeds, fertilizers, and other farming inputs (Martey et al., 2014; Nyantakyi-Frimpong, 2019). As evidence from this chapter demonstrates, however, some of these NGOs are using a top-down and unsustainable approach to deliver seeds to farmers. It is important to clarify that not all NGO-initiated projects are problematic in Ghana; indeed, there have been several successful interventions (e.g. see Canadian Feed the Children, 2016). The analysis here focuses specifically on the NGO we are calling Seed Innovators Group. The overall agenda of this NGO is to assist farmers who have had their grain stocks severely depleted by successive poor harvests. Formed in 2008, with funding from several international sources (e.g. Canada and the United States), Seed Innovators Group's interventions have included the formation of seed banks and the promotion of high-yielding seed varieties. The organization describes these interventions as climate-smart, and believes that all farmers should be seed secure and have access to modern technology irrespective of characteristics like age, gender, and class. At the time of fieldwork, Seed Innovators Group was working with farmers in 15 rural communities in the Upper West Region. Before we present the research findings, the next section outlines the study methods.

Methods

This chapter is based on data collected in long-term fieldwork from 2012 to 2019 using a variety of methods: household surveys, in-depth interviews, focus group discussions, and a tracer analysis of seed flows among farm households. These different research methods complemented each other and allowed key issues to be addressed from several angles (Creswell and Plano Clark, 2017). They helped the researchers gain a deep understanding of trends and variability in climate, community perceptions on the occurrence of climate shocks, adjustment in farming practices in response to climate variability, and possible actions to respond to food production stresses. Data were collected in two villages, selected on the basis of contrasts in population size, remoteness, average landholding, mobile phone network coverage, climate-stress context, and livelihood activities (Table 6.1), all of which affect access to seeds.

The survey took place in 2012 and involved 404 households selected using systematic random sampling across both villages (see Nyantakyi-Frimpong, 2014).

Table 6.1 Key characteristics of the two study villages

Main points of comparison	Village 1	Village 2
District	Lawra District	Nadowli-Kaleo District
Estimated population (2018)	4,990	1,201
Total households (2018)	1,024	413
Average household size	7.6	8
Geographical location	Less remote	Highly remote and isolated
Average landholding	2.4 ha	0.6 ha
Road connectivity	Available, but not paved	Available, but not paved
Telephone communication	Yes	Partial coverage
Climate-stress context	Prolonged drought and dry spell	Flash flood, prolonged drought, and dry spell
Length of farming season	~ 5 months (April to August)	~ 5 months (April to August)
Mean annual rainfall	941 mm (1980–2012)	Data not available
Seed Innovators Group NGO presence	Yes – since 2008	Yes – since 2008
Principal livelihoods	Farming, livestock raising	Farming, livestock raising, fishing, and artisanal gold mining

Source: Compiled from field notes, 2012–2018.

Researchers walking along major village streets and footpaths sampled every fifth household. The survey questions were analysed descriptively. In-depth interviews ($n = 60$) were conducted with a subset of the survey respondents later in 2012. Out of these 60 in-depth interviews, 54 contained relevant information on seed access and sharing, and were retained for analysis in this chapter. In 2019, 48 of the 54 interview respondents were revisited for follow-up interviews to enable a better understanding of changes in seed access and sharing. A seed flow tracer analysis was conducted as part of the in-depth interviews in both 2012 and 2019. Emphasis was placed on farmers' explanations of seed transactions: why they had engaged in a transaction, with whom, and what the significance of the transaction was, among other factors. Three key crops (maize, beans, and groundnuts) were considered in the analysis because they dominate the farming system in northern Ghana. Eight focus group discussions were also carried out in 2012 with elderly men, elderly women, young men, and young women in each study community. In total, 75 people participated in these discussions.

For all qualitative methods of data collection, informants were selected through maximum variation sampling (Patton, 2014), to represent differences in age, gender, economic status, level of education, and farming experience. The dataset was analysed thematically. Thematic categories were identified through line-by-line coding (Miles et al., 2014; Patton, 2014). Overall, this dataset from long-term fieldwork permits a critical comparison of changes in seed selection and exchange practices over a seven-year period. Although the two study villages had contrasting characteristics (Table 6.1), the study findings were quite similar across both sites. The results are therefore presented as a single set, but with a clear distinction between the 2012 and 2019 datasets, where the strongest differences emerged.

Results

Climate impacts on seed access

Respondents from the 2012 household survey indicated that they had observed significant changes in climate variability. The climate factors most commonly mentioned by farmers were a decrease in total rainfall events, an increase in temperature, and an increased frequency of 'false starts', droughts, floods, and stronger winds (Nyantakyi-Frimpong and Bezner Kerr, 2015a). Farmers also mentioned significant irregularities in the onset and cessation of the planting rains. Analysis of regional and village-level rainfall data confirmed these perceptions. For example, village-level rainfall records from 1981 to 2012 showed that the planting rains had shifted markedly from an early start in mid-February to a late start in mid-April or mid-May over the preceding two decades (Nyantakyi-Frimpong and Bezner Kerr, 2015a). Farmers defined 'planting rains' as the accumulation of approximately 20–30 mm of rainfall, followed by a period of no more than 10 consecutive dry days over four weeks.

Table 6.2 Farmers' perceptions of climate variability and its impacts on seeds (2019)

Climate impacts on seeds	% of farmers reporting concern (n = 48)	
	Men (n = 27)	Women (n = 21)
Too much rainfall, leading to flooding and destroying seeds that are germinating	81	100
Too little rainfall, leading to seed germination failures	100	100
Delayed rainfall, leading to seed germination failures	96	90
Shorter farming seasons, leading to harvest shortfalls	100	95
New pests and diseases attacking seeds both in storage and in fields	67	67

Source: Fieldwork in 2019

During interviews conducted in 2019, farmers were again asked about their perceptions of climate variability, but with a greater focus on seed-related impacts. The most commonly mentioned impact was too little rainfall leading to seed germination failures (Table 6.2). In addition, farmers reported prolonged droughts, which shortened the farming season and led to harvest shortfalls. For example, one elderly female respondent said, 'The farming season has now reduced from seven to just four months, May, June, July, and August. This change has severely affected how much crops we can harvest from the field and how much we can set aside for seed saving.' In addition, low or poor yields were said to affect the quantity and quality of seeds farmers could save for subsequent planting seasons. Other factors were the emergence of new crop pests and diseases that attack seeds both in storage and in the fields. These farmer views are supported by recent case studies, confirming the growing incidence of new pests and diseases in Ghana's Upper West Region, including the fall armyworm *Spodoptera frugiperda* (Koffi et al., 2020; Nboyine et al., 2020). There were no marked differences in the responses provided by women and men (Table 6.2). Overall, farmers' responses revealed clear instances of food and seed insecurity, given recurring variations in climatic conditions.

Seed selection and exchange practices in 2012

Historically, the foundation of seed supply in the study area has been farmers selecting seed from previous harvests and saving it for the next planting season. One of the major goals of our analysis was to understand whether and how this had changed over time, given climatic and other stressors. Data for this assessment came from the seed tracer analysis, which was part of in-depth interviews conducted in both 2012 and 2019. During both interview periods, farmers were asked about their most common sources of seed for three major crops (maize, beans, and groundnut), and the proportion of seed they obtained from each source. In 2012, while farmers obtained seed from multiple

Table 6.3 Farmers' sources of seed during periods of stress (2012 and 2019)

	2012 (% of farmers who provided data, n = 54)			2019 (% of farmers who provided data, n = 48)		
	Maize	Beans	Groundnuts	Maize	Beans	Groundnuts
Farm saved	86	84	70	12	69	75
Local market	30	53	–	18	51	–
Agro-dealer	24	–	–	60	62	–
NGO project	25	15	10	76	70	79
Exchange with other farmers	70	71	72	49	22	69
Government research institutions	19	10	–	29	34	19

Note: This analysis was conducted with the same group of farmers.
Source: In-depth interviews and seed tracer analysis (2012 and 2019)

sources, the most dominant was the recycling of older varieties saved during harvest. This mode of seed acquisition was mentioned for all three crops: maize (86% of respondents), beans (84% of respondents), and groundnuts (70% of respondents) (Table 6.3).

Farmers noted that they were interested in saving their own seed varieties, especially those that adapted well to the changing environmental conditions (Nyantakyi-Frimpong and Bezner Kerr, 2015b). Selecting and saving seed was said to provide a sense of security and a chance to save money. Once seed is selected and safely set aside, a farmer can be assured that the seed for the next planting season is secured. Furthermore, the seed will be available when it is needed so that the farmer will not incur planting delays, which have become increasingly common in the region (Nyantakyi-Frimpong and Bezner Kerr, 2015b; Dapilah et al., 2020). A farmer can therefore avoid spending money and time acquiring seed at the last moment before planting, which is when prices typically increase and many households are struggling to raise the means necessary for land preparation and planting (Madin, 2020). Another important means of seed acquisition was through farmer-to-farmer exchanges; this was dominant for all three crops. Local markets also played a key role in providing seed for maize and beans, mentioned by 30 per cent and 53 per cent of respondents respectively.

For some of the farmers interviewed, their own maize seed was associated with a symbolic and cultural value. This aspect surfaced many times during individual interviews, and was also mentioned in focus group discussions. Results showed that seed is often inherited or passed on from parents to children when the latter start farming independently. Often, the seed has been in the family for many years, and has thus acquired an inherently affective or symbolic value. It links farmers to previous generations, and they feel compelled to pass it on to their descendants. Finally, saving seed was strongly associated with being 'a good farmer'. Thus, for many farmers in the study area, selecting and saving seed was not just a question of saving money, but a decision that had cultural, economic, and agroecological components. Although saving seed from one's own harvest is the backbone of local seed supply in the study area, farmers do acquire seed from other sources from time to time. Farmers' reasons for acquiring seed were identified and described during the focus group discussions and in-depth interviews with key informants, and were later quantified during the tracer study. Some farmers gave or exchanged seeds with others to ensure that particular varieties would persist beyond their farms and local environments and remain available to them later if needed. Seed transfer was typically accompanied by the transmission of information about crop varieties, agronomic requirements, yields, consumption qualities, and vulnerabilities to pests and disease (see also Madin, 2020).

Seed selection and exchange practices in 2019

Within the space of seven years, farmers' seed selection and exchange practices had changed considerably. Seed sources from farmers' own saving, as well as farmer-to-farmer exchanges, had both reduced compared to the findings from 2012 (Table 6.3). This finding was evident for all three seeds (maize, beans, and groundnuts) considered in the analysis. For example, the percentage of farmers who saved maize seed for subsequent planting fell from 86 per cent in 2012 to only 12 per cent in 2019. Similarly, farmer-to-farmer exchanges of beans fell from 71 per cent in 2012 to 22 per cent in 2019.

The most dominant seed sources were agro-dealers and NGO-based projects. The former involved trained retailers selling agriculture inputs such as seeds, synthetic fertilizers, and pesticides as part of AGRA (Toenniessen et al., 2008). The latter included NGO-sponsored village seed banks and other CSA initiatives, with funding mostly from international organizations (e.g. see Zundel, 2017; Nyantakyi-Frimpong, 2019). A small proportion of seeds came from government research institutions, which was a significant increase since the 2012 interviews. For example, in the 2012 interviews, no groundnut seed was reported to have come from government research institutions, compared to 19 per cent of reported cases in 2019.

There were substantial differences between seeds supplied through projects sponsored by Seed Innovators Group and the government and the varieties historically used by farmers. Seed sources from Seed Innovators Group projects were found to weaken traditional seed exchange relationships. In some cases, local seeds of particular maize varieties were found to be disappearing altogether. As a large literature demonstrates (e.g. Ricciardi, 2015; Violon et al., 2016), farmers' seed networks commonly supply material which farmers esteem highly, including varieties with traits not produced by formal breeding, such as tolerance to local climate stress. Thus, a weaker seed exchange relationship was reported to be detrimental to climate change adaptation.

As is evident in the World Bank publication *Ending Poverty and Hunger by 2030* (Townsend, 2015), the climate-smart agenda for food systems transformation is focused, in part, on seeds and seed systems. Some of the key elements are about promoting the adoption of drought-tolerant crop varieties such as maize. In the context of northern Ghana, where some of these programmes are being implemented, the net effect has been negative, with low yields and heightened susceptibility to crop diseases (Zundel, 2017; Madin, 2020).

Although a few of these CSA initiatives focus on reviving indigenous, drought-resilient seed varieties (e.g. see Canadian Feed the Children, 2016), the majority focus on agricultural modernization: for example, the substitution of local seed varieties with hybrids (Nyantakyi-Frimpong and Bezner Kerr, 2015b; Zundel, 2017). The field research revealed that many agriculture NGOs had projects spanning more than a decade, aimed at improving farmer access to hybrid seeds of various kinds. According to interviews and focus groups, a major challenge with these seeds was that they make farmers dependent on agrochemicals and other synthetic inputs such as fertilizers (Nyantakyi-Frimpong and Bezner Kerr, 2015b). Moreover, in the design of these projects, indigenous knowledge about seeds and farming practices was ignored, for the most part. As one study respondent revealed in an interview, 'As a way to increase corporate organizations' control over seeds, many of these organizations are generating a perception that traditional seeds are not important.' In explaining the effects of CSA interventions, another respondent added, 'Farmers' traditional seed-saving practices have been increasingly delegitimized through the various seed programmes in the [Upper West] region. With a lot of pressure and incentives to adopt these foreign seeds, farmers have less control on what they plant.' Overall, these findings suggested that farmers had less sovereignty over their own seed production.

The dramatic changes in farmer seed acquisition sources from 2012 to 2019 also raised equity questions. Different social groups were found to be excluded or marginalized from accessing seeds from sources such as agro-dealers, NGO-based projects, and government research institutions. For example, poorer farmers without adequate funds, as well as farmers without strong connections to village leaders and elites, were often excluded from these seed acquisition sources (Table 6.4). According to current national census data, more than 70 per cent of the study region's population lives in chronic poverty (Ghana Statistical Service, 2014), suggesting that a considerable group of farmers might be excluded from accessing seed from these sources. Other exclusionary practices included households not selected for project interventions by NGOs or government research institutions during experimental trials. Previous research suggests a strong link between farmers' relationships with village leaders and whether they were selected for community interventions (Nyantakyi-Frimpong, 2019). It could be argued that these seed sources and their exclusionary processes reinforce existing inequalities based on class, kinship relations, and other social networks. They therefore offer little

Table 6.4 Social groups excluded from access to seed through different sources

Seed sources	Social groups typically excluded or marginalized
Farmer saved	Food-insecure households; households experiencing land grabbing
Local market	Poorer farmers without adequate funds
Agro-dealer	Poorer farmers without adequate funds
NGO project	Households not selected by NGOs for project interventions
	Farmers without strong connections to village leaders and elites
Exchange with other farmers	Women
	Food-insecure households
Gov't research institutions	Households not selected for particular project interventions (e.g. experimental trials)
	Farmers without strong connections to village leaders and elites

transformative potential for smallholder farmers' food and seed sovereignty (Wittman et al., 2010; Shiva, 2012; Adhikari, 2014; Kloppenburg, 2014).

Discussion and conclusion

This chapter has demonstrated the changes in farmer seed acquisition practices in the context of northern Ghana. The analysis is drawn from seed-based interventions by one NGO, referred to in this chapter as Seed Innovators Group. Major limitations of Seed Innovators Group's interventions are that they are externally designed, top-down, and short-term in nature. Local farmers are not deeply involved in the design of these interventions, which, as explained earlier, include the formation of seed banks, as well as promoting the adoption of high-yielding seed varieties grown with synthetic fertilizer and pesticides. Despite a large body of literature showing that local farmers have salient knowledge that should be considered in agricultural interventions (Richards, 1985; Nyong et al., 2007; Sumberg and Thompson, 2012), indigenous knowledge has been ignored by Seed Innovators Group, as demonstrated in the interview accounts above. For example, the interviews revealed a perception among NGOs that traditional seeds are not important. These findings demonstrate the knowledge politics involved in some of the existing modern-day agrarian transformation initiatives in Ghana and other parts of Africa (Moseley et al., 2017; Bezner Kerr et al., 2018).

The proliferation of hybrid seeds as part of technologies for promoting CSA is deeply problematic. This high-external-input, market-based approach is ill-suited to addressing hunger in the context of northern Ghana: it reproduces and aggravates the same problems that have historically stalled agricultural development in the region, including class and gender inequalities and the lack of respect for indigenous knowledge (Nsiah-Gyabaah, 1994; Luginaah et al., 2009). Among the many limitations of this approach to transforming agriculture, one of the key concerns is the issue of cost and income inequalities

(Moseley et al., 2017). Given the nature of hybrid seed varieties, new seeds must be purchased annually, making this a recurring expenditure. Yet most poor farmers in northern Ghana do not have the funds for such recurring expenses (Dapilah et al., 2020; Nyantakyi-Frimpong, 2014). They are also not seen as creditworthy enough to receive agricultural loans. Thus, the use of hybrid seed varieties and related interventions has a class-based problem, which is a fundamental food and social justice issue. These interventions are inaccessible to the poorest of the poor, for whom food insecurity remains pervasive.

Ghana's agrarian political economy also needs to be considered in a context where farmers are increasingly being pushed to adopt seed technologies with huge capital investments. These seed technologies and their complementary inputs are unaffordable in Ghanaian markets already affected by structural adjustment reforms (Konadu-Agyemang, 2000). At the same time, smallholder access to credit has been firmly curtailed by the lingering impacts of structural adjustment implemented in the early 1980s (Pearce, 1992; Hutchful, 2002). Thus, expensive technologies are being promoted in a country where the local political economy offers little support for the small farmer.

The ecological implications of the findings here also merit some discussion. As noted in the findings, there are several ecological stressors affecting seeds, including too little rainfall, which leads to seed germination failures, and new pests and diseases that attack seeds both in storage and in the field. Hybrid seeds, often the varieties supplied through Seed Innovators Group and other NGOs, are often not hardy compared to local landraces (Schnurr, 2019). Although they are claimed to be climate-smart, they are highly susceptible to climate variability and change, including some of the farmer concerns noted in Table 6.2. It is therefore contradictory to promote these varieties as part of initiatives to address climate change and food and seed insecurity. Essentially, farmers are being supplied with seeds that are unable to address the ecological problems they face.

Finally, the findings here indicate that farmers have no seed sovereignty, given the ongoing rapid transformation of their seed systems. Farmers' seed-saving and exchange practices are being curtailed. Overall, there seems to be little local control over hybrid seeds, which undermines the ideals of seed sovereignty and the sustainable transformation of seed systems.

In conclusion, there are strong reasons why locally developed farmer seed varieties should be supported. There are many small-scale agroecological farming practices that promote the diversity of seeds, and these should be nurtured and supported (Bezner Kerr, 2013; HLPE, 2019). These local seed varieties are available to farmers without their needing to buy them or depend on external knowledge systems. Moreover, they are ecologically resilient seeds that can adapt to a changing climate along with many other challenges (Richards, 1985; Nyong et al., 2007; Niang et al., 2014). As several recent reports from a high-level panel of international experts have emphasized (e.g. HLPE, 2019), this is one of the most sustainable transformations of food systems that regions like Africa need: one based on agrobiodiversity, local resources, and knowledge.

Acknowledgements

Many thanks to Professor Rachel Wynberg for critical feedback on this chapter. Fieldwork activities were supported with funding from the Department of Geography and the Environment, University of Denver; a Public Good Fund grant from the University of Denver's Center for Community Engagement to advance Scholarship and Learning (CCESL) [Grant # 86847]; and an Internationalization Grant from the University of Denver.

References

Adhikari, J. (2014) 'Seed sovereignty: analyzing the debate on hybrid seeds and GMOs and bringing about sustainability in agricultural development', *Journal of Forest and Livelihood* 12(1): 33–46.

Aryeetey, E. and Kanbur, R. (eds) (2017) *The Economy of Ghana: Sixty Years After Independence*. Oxford University Press, Oxford.

Asuru, S.I. (2015) 'The new philanthropy, poverty reduction and rural development: a case study of Alliance for a Green Revolution in Africa (AGRA) in Ghana', *Jurnal Studi Pemerintahan* 6(1): 18–30 <https://doi.org/10.18196/jgp.2015.0003>.

Atuoye, K.N., Antabe, R., Sano, Y., Luginaah, I. and Bayne, J. (2019) 'Household income diversification and food insecurity in the Upper West Region of Ghana', *Social Indicators Research* 144(2): 899–920 <https://doi.org/10.1007/s11205-019-02062-7>.

Awanyo, L. and Attua, E.M. (2018) 'A paradox of three decades of neoliberal economic reforms in Ghana: a tale of economic growth and uneven regional development', *African Geographical Review* 37(3): 173–191 <https://doi.org/10.1080/19376812.2016.1245152>.

Bezner Kerr, R. (2013) 'Seed struggles and food sovereignty in northern Malawi', *Journal of Peasant Studies* 40(5): 867–897 <https://doi.org/10.1080/03066150.2013.848428>.

Bezner Kerr, R., Nyantakyi-Frimpong, H., Dakishoni, L., Lupafya, E., Shumba, L., Luginaah, I. and Snapp, S.S. (2018) 'Knowledge politics in participatory climate change adaptation research on agroecology in Malawi', *Renewable Agriculture and Food Systems* 33(3): 238–251 <https://doi.org/10.1017/S1742170518000017>.

Canadian Feed the Children (2016) *You Made CHANGE Happen!*, Project Wrap-Up Summary Report, CHANGE Project Stakeholder Learning Forum, Tamale, Ghana,17 May <https://canadianfeedthechildren.ca/downloads-projects/CHANGE-Project-Summary-of-Results.pdf>.

Chamberlin, J. (2007) *Defining smallholder agriculture in Ghana: who are small-holders, what do they do and how are they linked with markets?*, Background Paper No. GSSP 0006, Ghana Strategy Support Program, International Food Policy Research Institute, Washington, DC <http://ebrary.ifpri.org/utils/getfile/collection/p15738coll2/id/37971/filename/37972.pdf>.

Creswell, J.W. and Plano Clark, V.L. (2017) *Designing and Conducting Mixed Methods Research*, Sage Publications, Thousand Oaks, CA.

Dapilah, F., Nielsen, J.Ø. and Friis, C. (2020) 'The role of social networks in building adaptive capacity and resilience to climate change: a case study from northern Ghana', *Climate and Development* 12: 42–56 <https://doi.org/10.1080/17565529.2019.1596063>.

Dugle, G., Akanbang, B.A. and Salakpi, A. (2015) 'Nature of non-governmental organisations involved in local development in the Upper West Region of Ghana', *Ghana Journal of Development Studies* 12(1–2): 142–163 <https://doi.org/10.4314/gjds.v12i1-2.9>.

Fairhead, J. and Leach, M. (1996) *Misreading the African Landscape: Society and Ecology in a Forest-Savanna Mosaic*. Cambridge University Press, Cambridge.

Food and Agriculture Organization of the United Nations (FAO) (2016) *Seed Security Assessment: A Practitioner's Guide*, FAO, Rome <http://www.fao.org/3/i5548e/i5548e.pdf>.

FAO, IFAD, UNICEF, WFP and WHO (2018) *The State of Food Security and Nutrition in the World 2018: Building Climate Resilience for Food Security and Nutrition*, FAO, Rome <http://www.fao.org/3/i9553en/i9553en.pdf>.

Ghana Statistical Service (2013) *2010 Population and Housing Census: Regional Analytical Report, Upper West Region*, Ghana Statistical Service, Accra <https://www2.statsghana.gov.gh/docfiles/2010phc/2010_PHC_Regional_Analytical_Reports_Upper_West_Region.pdf>.

Ghana Statistical Service (2014) *Ghana Living Standards Survey Round 6 (GLSS6): Poverty Profile in Ghana (2005–2013)*, Ghana Statistical Service, Accra <https://www2.statsghana.gov.gh/docfiles/glss6/GLSS6_Poverty%20Profile%20in%20Ghana.pdf>.

Ghana Statistical Service (2015) *Ghana Poverty Mapping Report*, Ghana Statistical Service, Accra <https://www2.statsghana.gov.gh/docfiles/publications/POVERTY%20MAP%20FOR%20GHANA-05102015.pdf>.

High Level Panel of Experts on Food Security and Nutrition (HLPE) (2019) *Agroecological and other innovative approaches for sustainable agriculture and food systems that enhance food security and nutrition*, HLPE Report 14, Committee on World Food Security, Rome <http://www.fao.org/3/ca5602en/ca5602en.pdf>.

Hutchful, E. (2002) *Ghana's Adjustment Experience: The Paradox of Reform*, James Currey, Oxford.

Juma, C. (2015) *The New Harvest: Agricultural Innovation in Africa*, Oxford University Press, New York.

Kloppenburg, J. (2014) 'Re-purposing the master's tools: the open source seed initiative and the struggle for seed sovereignty', *Journal of Peasant Studies* 41(6): 1225–1246 <https://doi.org/10.1080/03066150.2013.875897>.

Koffi, D., Kyerematen, R., Eziah, V.Y., Agboka, K., Adom, M., Goergen, G. and Meagher, R.L. (2020) 'Natural enemies of the fall armyworm, *Spodoptera frugiperda* (J.E. Smith) (Lepidoptera: Noctuidae) in Ghana', *Florida Entomologist* 103(1): 85–90 <https://doi.org/10.1653/024.103.0414>.

Konadu-Agyemang, K. (2000) 'The best of times and the worst of times: structural adjustment programs and uneven development in Africa: the case of Ghana', *The Professional Geographer* 52(3): 469–483 <https://doi.org/10.1111/0033-0124.00239>.

Luginaah, I., Weis, T., Galaa, S., Nkrumah, M.K., Bezner Kerr, R. and Bagah, D. (2009) 'Environment, migration and food security in the Upper West Region of Ghana', in I.N. Luginaah and E.K. Yanful (eds), *Environment and Health in Sub-Saharan Africa: Managing an Emerging Crisis*, pp. 25–38, Springer, Dordrecht <https://doi.org/10.1007/978-1-4020-9382-1_2>.

Madin, M.B. (2020) *Climate Change and Seed Security Among Smallholder Farmers in Northern Ghana*, MA thesis, University of Denver <https://digital-ommons.du.edu/etd/1798/>.

Martey, E., Wiredu, A.N., Etwire, P.M., Fosu, M., Buah, S.S.J., Bidzakin, J., Ahiabo, B.D.K. and Kusi, F. (2014) 'Fertilizer adoption and use intensity among smallholder farmers in Northern Ghana: a case study of the AGRA soil health project', *Sustainable Agriculture Research* 3: 24–36 <https://doi.org/10.5539/sar.v3n1p24>.

McGuire, S. and Sperling, L. (2013) 'Making seed systems more resilient to stress', *Global Environmental Change* 23(3): 644–653 <https://doi.org/10.1016/j.gloenvcha.2013.02.001>.

Miles, M.B., Huberman, A.M. and Saldana, J. (2014) *Qualitative Data Analysis: A Methods Sourcebook*, Sage Publications, Thousand Oaks, CA.

Mohan, G. (2002) 'The disappointments of civil society: the politics of NGO intervention in northern Ghana', *Political Geography* 21(1): 125–154 <https://doi.org/10.1016/S0962-6298(01)00072-5>.

Moseley, W.G. (2016) 'The new green revolution for Africa: a political ecology critique', *The Brown Journal of World Affairs* 23: 177–190.

Moseley, W.G., Schnurr, M.A. and Bezner Kerr, R. (eds) (2017) *Africa's Green Revolution: Critical Perspectives on New Agricultural Technologies and Systems*, Routledge, Oxford.

Nboyine, J.A., Kusi, F., Abudulai, M., Badii, B.K., Zakaria, M., Adu, G.B., Haruna, A., Seidua, A., Osei, V., Alhassan, S. and Yahaya, A. (2020) 'A new pest, *Spodoptera frugiperda* (J.E. Smith), in tropical Africa: its seasonal dynamics and damage in maize fields in northern Ghana', *Crop Protection* 127: 104960 <https://doi.org/10.1016/j.cropro.2019.104960>

Niang, I., Ruppel, O.C., Abdrabo, M., Essel, A., Lennard, C., Padgham, J. and Urquhart, P. (2014) 'Africa', in V.R. Barros, C.B. Field, D.J. Dokken, M.D. Mastrandrea, K.J. Mach, T.E. Bilir, M. Chatterjee, K.L. Ebi, Y.O. Estrada, R.C. Genova, B. Girma, E.S. Kissel, A.N. Levy, S. MacCracken, P.R. Mastrandrea, and L.L. White (eds), *Climate Change 2014: Impacts, Adaptation, and Vulnerability. Part B: Regional Aspects. Contribution of Working Group II to the Fifth Assessment Report of the Intergovernmental Panel on Climate Change*, pp. 1199–1265, Cambridge University Press, Cambridge <https://www.ipcc.ch/site/assets/uploads/2018/02/WGIIAR5-PartB_FINAL.pdf>.

Nsiah-Gyabaah, K. (1994) *Environmental Degradation and Desertification in Ghana: A Study of the Upper West Region*, Avebury Publishers, Aldershot, UK.

Nyantakyi-Frimpong, H. (2014) *Hungry Farmers: A Political Ecology of Agriculture and Food Security in Northern Ghana*, PhD dissertation, University of Western Ontario, Canada <https://ir.lib.uwo.ca/etd/2276/>.

Nyantakyi-Frimpong, H. (2019) 'Visualizing politics: a feminist political ecology and participatory GIS approach to understanding smallholder farming, climate change vulnerability, and seed bank failures in Northern Ghana', *Geoforum* 105: 109–121 <https://doi.org/10.1016/j.geoforum.2019.05.014>.

Nyantakyi-Frimpong, H. and Bezner Kerr, R. (2015a) 'The relative importance of climate change in the context of multiple stressors in semi-arid Ghana', *Global Environmental Change* 32: 40–56 <https://doi.org/10.1016/j.gloenvcha.2015.03.003>.

Nyantakyi-Frimpong, H. and Bezner Kerr, R. (2015b) 'A political ecology of high-input agriculture in northern Ghana', *African Geographical Review* 34(1): 13–35 <https://doi.org/10.1080/19376812.2014.929971>.

Nyantakyi-Frimpong, H. and Bezner Kerr, R. (2017) 'Land grabbing, social differentiation, intensified migration and food security in northern Ghana', *The Journal of Peasant Studies* 44(2): 421–444 <https://doi.org/10.1080/03066150.2016.1228629>.

Nyong, A., Adesina, F. and Elasha, B.O. (2007) 'The value of indigenous knowledge in climate change mitigation and adaptation strategies in the African Sahel', *Mitigation and Adaptation Strategies for Global Change* 12(5): 787–797 <http://dx.doi.org/10.1007/s11027-007-9099-0>.

Palombi, L. and Sessa, R. (2013) *Climate-Smart Agriculture Sourcebook*, Food and Agriculture Organization of the United Nations, Rome <http://www.fao.org/3/i3325e/i3325e.pdf>.

Patton, M.Q. (2014) *Qualitative Research and Evaluation Methods: Integrating Theory and Practice*, 4th edn, SAGE Publications, Thousand Oaks, CA.

Pearce, R. (1992) 'Ghana', in A. Duncan and J. Howell (eds), *Structural Adjustment and African Farmers*, pp. 14–47, Overseas Development Institute, London.

Ricciardi, V. (2015) 'Social seed networks: identifying central farmers for equitable seed access', *Agricultural Systems* 139: 110–121 <https://doi.org/10.1016/j.agsy.2015.07.002>.

Richards, P. (1985) *Indigenous Agricultural Revolution: Ecology and Food Production in West Africa*, Westview Press, Boulder, CO.

Schnurr, M.A. (2019) *Africa's Gene Revolution: Genetically Modified Crops and the Future of African Agriculture*, McGill-Queen's University Press, Montreal.

Shiva, V. (2012) 'The seed emergency: the threat to food and democracy', *Al Jazeera English*, 6 February <https://www.aljazeera.com/opinions/2012/2/6/the-seed-emergency-the-threat-to-food-and-democracy>.

Sossou, S., Igue, C.B. and Diallo, M. (2019) 'Impact of climate change on cereal yield and production in the Sahel: Case of Burkina Faso', *Asian Journal of Agricultural Extension, Economics & Sociology* 37(4): 1–11 <https://doi.org/10.9734/AJAEES/2019/v37i430288>.

Sumberg, J. and Thompson, J. (eds) (2012) *Contested Agronomy: Agricultural Research in a Changing World*, Routledge, Oxford.

Toenniessen, G., Adesina, A. and DeVries, J. (2008) 'Building an alliance for a green revolution in Africa', *Annals of the New York Academy of Sciences* 1136(1): 233–242 <https://doi.org/10.1196/annals.1425.028>.

Townsend, R.F. (2015) *Ending Poverty and Hunger by 2030: An Agenda for the Global Food System*, World Bank Group, Washington, DC <https://openknowledge.worldbank.org/handle/10986/21771>.

Violon, C., Thomas, M. and Garine, E. (2016) 'Good year, bad year: Changing strategies, changing networks? A two-year study on seed acquisition in northern Cameroon', *Ecology and Society* 21(2) <https://doi.org/10.5751/ES-08376-210234>.

Wiemers, A. (2015) 'A "time of agric": Rethinking the "failure" of agricultural programs in 1970s Ghana', *World Development* 66: 104–117 <https://doi.org/10.1016/j.worlddev.2014.08.006>.

Wittman, H., Desmarais, A.A. and Wiebe, N. (eds) (2010) *Food Sovereignty: Reconnecting Food, Nature and Community*, Fernwood, Halifax, NS.

Zundel, T. (2017) *Climate-Smart Agriculture as a Development Buzzword: Framework for Flexible Development, or Greenwashing the Status Quo? Insights from Northern Ghana*, MA thesis, University of Guelph, Ontario <https://atrium.lib.uoguelph.ca/xmlui/handle/10214/10481>.

CHAPTER 7

Seedscapes of contamination: exploring the impacts of transgene flow for South African smallholder farmers

Rachel Wynberg and Angelika Hilbeck

Introduction

Genetically modified (GM) crops are deeply embedded in the rhetoric that, by analogy with the 'Green Revolution' of the 1960s and 1970s, a new 'Gene Revolution' is needed to save African agriculture. As with the Green Revolution, it comes with promises of higher yields, greater economic gains, and improved food security, premised on the argument that Africa was largely bypassed by the innovations and investments of this period, despite its much-contested benefits (Rosset et al., 2000; Singh, 2000; Patel, 2013). As a corollary, the hybrid seeds pushed under the Gene Revolution paradigm, genetically engineered to incorporate different traits for pest resistance, herbicide tolerance, and drought, are a microcosm of many of the harmful impacts of the Green Revolution. These include a reliance on expensive purchased seed and other inputs, the contamination of soils and water by agrochemicals, the increased planting of monocultures and reduced availability of diverse, nutritious food crops, and the displacement and reduced autonomy and agency of small farmers (Singh, 2000; Patel, 2013; DeFries et al., 2015; Davis et al., 2019).

Despite an increasing body of evidence indicating that these so-called 'first-generation'[1] GM crops fail to meet the needs of African smallholders (e.g. Witt et al., 2006; Dowd-Uribe, 2014; Fischer and Hadju, 2015; Schnurr, 2019; Marshak et al., 2021; Fischer, 2022), pressures for African governments to adopt GM crops have been relentless. For example, a series of USAID programmes aim to harmonize policy frameworks and strengthen capacity to manage biosafety. With support from the Rockefeller Foundation, the Gates Foundation, other government aid agencies, and the private sector, the Nairobi-based African Agricultural Technology Foundation actively promotes the uptake of GM crops among African smallholder farmers (Rock and Schurman, 2020). With the goal of 'getting to yes' (Schnurr and Gore, 2015), a swathe of supposedly 'neutral brokers' aims to advocate the uptake of GM crops in Africa – and to undermine any precautionary concerns. Despite these efforts, the uptake of GM crops across African landscapes has been low (Kedisso et al., 2022), likely due to what Rock and Schurman (2020) call 'the complex choreography' of sociopolitical, regulatory, and business conditions required for agricultural biotechnology projects to succeed. To date, only a handful of African countries have approved the commercial cultivation of GM crops – South Africa, Sudan, Ethiopia, Eswatini, Kenya, Malawi, and Nigeria – with only South Africa growing them at an industrial scale (Kedisso et al., 2022). Most countries have approved only one GM crop, insect-resistant cotton (Akinbo et al., 2021; Kedisso et al., 2022), largely as a non-food 'door opener'. Although South Africa remains the only African country growing a GM staple food crop (maize) commercially, recent developments in Kenya (GM maize) and Nigeria (GM cowpea) suggest that transgenic food crops are likely to become increasingly prevalent on the African continent, especially with the introduction of second-generation GM crops.

The possible accelerated adoption of GM crops in Africa has multiple implications for the millions of smallholders on the continent that produce food for their families. As Schnurr and Dowd-Uribe (2021) describe, these technologies, at least in their current first-generation form, are poorly suited to African smallholder systems and their 'lofty projected benefits' are unlikely to be realized by most smallholder farmers. Moreover, through transgene flow – the contamination by GM traits of non-GM varieties of the same crop (*Nature Biotechnology*, 2002) – they also pose direct threats to the diverse, nutritious, and resilient seeds that have been nurtured over generations and form the food basis for most inhabitants of the African continent. A growing body of research is revealing that transgene flow is widespread, even in remote situations previously thought to be impervious to cross-contamination (Quist and Chapela, 2001; Fitting, 2011; Bøhn et al., 2016; Bourgou et al., 2020; Fernandes et al., 2022). In this regard, important lessons are emerging from South Africa. As a minimally processed, milled raw material, industrial GM maize is used directly in the human food chain in South Africa, especially in poorer communities, where the porridge *uphuthu*, or *pap*, is typically consumed daily as the bulk of the meal.[2]

South Africa has produced GM maize, cotton, and soybeans for more than 20 years, and the outcrossing and spread of transgenic traits into open-pollinating varieties of smallholder farmers are now well described (Iversen et al., 2014) – mainly for maize, which also represents the focus of this chapter. Several studies have revealed the contamination of smallholder fields and seed systems from varied on-farm and off-farm pathways (Iversen et al., 2014; Price and Cotter, 2014; Kganyago, 2020). Combined, these events facilitate and accelerate the flow of transgenic material along a multitude of pathways to non-GM varieties and present serious barriers for farmers who wish to maintain the use of local varieties and grow a diverse mixture of non-GM maize and other crop varieties. The use of GM crops also has profound implications for farmers who wish to pursue agroecological farming, significantly curtailing possibilities, limiting farmers' choices, and raising questions about impacts on agrobiodiversity, food security, and farming practices such as seed saving and seed selection (Jacobson and Myhr, 2013).

The deep connections between traditional seed systems and culture are well known (Van Niekerk and Wynberg, 2017), and while several papers reveal how farmers' seed systems have been contaminated by transgenic gene flow, few explore the social, ecological, cultural, and psychological implications for smallholder farmers. The inspiration for this chapter emerged from a series of workshops with and visits to communities in KwaZulu-Natal, South Africa, who farm agroecologically. In 2017, the preliminary testing of traditional maize seed of these smallholder farmers, presumed to be local farmer varieties farmed on an agroecological basis, revealed that 5 of the 42 samples unquestionably contained transgenes. This result was a shock for farmers; many had carefully kept their traditional seed apart from hybrid seed, and proudly identified themselves as GM-free farmers. Several, in fact, had been key instigators of protests against Monsanto and the use of GM seed in their communities (Photo 7.1). With the support of the NGO Biowatch South Africa and the two authors, a process was conceptualized to help farmers think through reasons for their crop having become contaminated and

Photo 7.1 Ingwavuma agroecology farmers join the international call to 'March Against Monsanto' May 2015
Credit: Biowatch South Africa

ways in which they could manage the impacts and reduce negative effects. A four-day 'caravan' was convened in January 2018, bringing together farmers from across KwaZulu-Natal and scientists, researchers, gene bank managers, and NGOs from South Africa, eSwatini, Lesotho, Zimbabwe, and Zambia. Collectively, this group engaged on these issues in farmers' fields, in small, group-facilitated discussions, and in plenary. This chapter presents the results of that process, drawing also on unpublished research from elsewhere in the country. It begins by describing the history of GM crop adoption in South Africa, before presenting an overview of the contemporary use of such crops in the country. The chapter then proceeds to explain why contamination matters for smallholder farmers in the agro-environments of which they are a part, before concluding with a set of possible approaches to be adopted in response to contamination.

The history of GM crop adoption in South Africa

Genetically modified crops first became prominent on the South African agricultural landscape in 1992, when the apartheid government approved Monsanto's field trials for Bt insect-resistant transgenic cotton. At the time, there were no regulatory frameworks in place for these novel crops, which remained untested, and oversight was through a voluntary group of scientists, the South African Genetic Experimentation Committee, which had close ties to industries promoting the development and marketing of GM crops and seeds. As the decade progressed, multinational companies such as DuPont, Bayer, Cargill, BASF, and Monsanto were experiencing rapid growth through trade liberalization, advances in biotechnology, and the granting of patents on genetically engineered organisms. A suite of mergers and acquisitions cemented the position of these industrial gene giants, and by 2008 just six companies controlled virtually the entire market for GM seeds (Howard, 2009), a trend that, as Greenberg (Chapter 8, this volume) explains, has continued to intensify. There was immense pressure to commercialize new products and to open new markets in Africa, given European reservations about the technology (Wynberg, 2003; Bowman, 2015). South Africa, with its relatively sophisticated infrastructure and research capacity, provided an ideal launch pad to do so. A model evolved whereby multinational biotechnology companies – all with their origins in agrochemicals – typically financed research and partnered with local research facilities to develop and promote GM crops. This system laid a crucial foundation for the rapid adoption of GM crops in South Africa.

Faced with a plethora of policy imperatives, the newly elected post-apartheid government played a largely passive role in determining policy on genetic engineering. More pressing issues dominated the policy arena, allowing civil servants and those with vested interests to introduce laws and policies in more peripheral areas without following due process. One such law was the Genetically Modified Organisms (GMO) Act 15 of 1997, promulgated only *after* the first commercial planting of a GM crop in South Africa. In contrast to other laws and policies at the time, the GMO Act was promulgated without a policy in place and without a comprehensive public participation process. Structures set up to implement the GMO Act similarly excluded public-interest groups, but access to the state by major seed companies continued, often through industry-funded public science that actively promoted genetic modification.

The increased lobbying from seed companies, a flood of permit applications for GM crop plantings, and the need for South Africa to engage in international negotiations for a biosafety protocol under the UN Convention on Biological Diversity were not without consequence. The stance of the government in regulating GM crops soon shifted from one of 'convenient neglect' towards one representing all the characteristics of a country strongly promotional of their uptake (e.g. Paarlberg, 2000). This reflected the strong pro-business stance of the African National Congress government, which had come to power not only with immense popular support but also with substantial backing from large capital (Sitas, 2010; Marais, 2011; Du Toit, 2022). Although this capital was not necessarily linked to companies promoting GM crops, it indicated the post-apartheid government's accommodating position regarding business engagement in policy formulation and decision-making (Du Toit, 2022).

GM crop adoption in South Africa

South Africa is today a country well known for its highly promotional approach to GM crops. Not only was it the first African country to commercialize GM crops, but it was also the first in the world to produce a GM subsistence crop and staple food: insect-resistant and herbicide-tolerant white maize (Gouse et al., 2005; Wynberg and Fig, 2013). With 2.7 million hectares of the country's land currently under cultivation to GM maize (an estimated 2.17 million ha, equalling 85 per cent of the country's production), soybean (494,000 ha, 95 per cent of production), and cotton (9,000 ha, 100 per cent of production), it constitutes the largest area devoted to GM crops in Africa and the ninth largest in the world (ISAAA, 2018). White maize comprises about 58 per cent of this amount, produced as a food staple, while yellow maize constitutes about 42 per cent, produced largely for animal feed (Masehela et al., 2021).

Since the commercialization of the first GM crop in South Africa in 1995, a total of 27 general release permits have been granted for commercial planting in the country, with 10 GM crops granted trial approval (Masehela et al., 2021). GM crops developed and approved for field trials have included sugar cane, potato, canola, cassava, wheat, grapevine, and groundnut. As in the rest of the world, almost all field trials and commercial releases comprise mainly two traits: tolerance to herbicides, and pest resistance through incorporation of genes from *Bacillus thuringiensis* (Bt). Stacked traits, which combine two or more genes of interest into a single plant, are increasingly incorporated into GM seed, particularly maize (Mawasha, 2020).

The past 30 years of GM adoption in South Africa have revealed sharp differences in their uptake and impacts, reflecting the country's dual agrarian structure, which comprises both large-scale commercial farms, typically under white ownership, and small-scale subsistence farms, concentrated in the former communal 'homelands', where the majority of black South Africans were forced to live under apartheid laws. While the apartheid legacy of this dualized system is unique to South Africa, the colonial shaping of its agricultural landscape is similar to many other countries in the Global South, especially with regard to historical patterns of maize industrialization and modernization and the marginalization of smallholder farmers (Bernstein, 1996). In many ways, South Africa thus embodies the development paradigm forecast for the rest of the continent and beyond.

A sharp division exists between technologically advanced and capital-intensive, large-scale agriculture in former white areas (on 85 per cent of agricultural land), and marginalized, smallholder subsistence farming in the remaining 15 per cent of (generally poor) agricultural land by approximately 2.4 million smallholder farmers (Greenberg, 2013). In practice, both types of farming systems co-occur, sometimes adjacent to each other in the same region.

Large-scale commercial farming typically produces commodity crops for domestic markets (feed and food) or export, while smallholder farmers produce largely for subsistence and local and domestic markets (Aliber and Hart, 2009; Greenberg, 2013). In each system, maize is a key crop but is farmed with entirely different objectives, markets, values, cultivation methods, and meanings (Marshak et al., 2021). On commercial farms, maize is produced industrially as a commodity crop for food, feed, and export, with high levels of fertilizer, herbicide, and pesticide inputs.

In contrast, for many smallholder subsistence farmers, maize is viewed as the 'pillar of the household' and is valued inter alia as a staple food, for the variety of ways in which it can be prepared, for its taste and resistance to drought, pests, and diseases, and for the strong traditions and customs that accompany the crop (Van Niekerk and Wynberg, 2017). Although not indigenous to the region, maize has become embedded in farming systems and cultures across many regions in Africa over several centuries, replacing, or displacing, traditional staples such as sorghum or millet, but also assimilating within existing seed systems and being passed down from generation to generation (McCann, 2009; Van Niekerk and Wynberg, 2017).

Smallholders typically grow maize as a primary staple crop, with many farmers reliant on open-pollinated varieties (OPVs) and/or local varieties, although, in South Africa, hybrid and GM seeds are also planted. In addition to affordability, crop choices are often influenced by the successes and failures of fellow farmers and neighbours, which in turn are strongly dependent on the type of agricultural advice received. Over the past 30 years, the much-depleted state extension support services increasingly have been replaced by multinational seed companies and agro-dealers, which, except for a few NGOs, are the *de facto* sources of agricultural advice for many farmers, especially large-scale producers (Fischer and Hadju, 2015; Mahlase, 2017). As Vanessa Black describes in Box G, those state-based extension services that exist are, moreover, heavily influenced by the training received at tertiary education institutions which typically promote modern crop varieties and denigrate farmer varieties.

Estimates of the number of smallholders that have adopted GM crops in South Africa are difficult to calculate due to the limited availability of smallholder data, the diverse categorizations of smallholder farmers, and the aggregation of seed sale data (Gouse et al., 2016). What is clear, however, is that the uptake has been extremely low. Gouse et al. (2008, in Gouse et al., 2016) have estimated that in 2008 approximately 10,500 smallholder farmers in South Africa planted GM maize, comprising less than 0.5 per cent of the country's 2 million small-scale maize farmers.

The limited number of smallholder farmers who have taken up GM crops in South Africa typically were incentivized to do so through government and seed industry-sponsored campaigns, designed and implemented to promote the uptake of GM cotton and maize (Witt et al., 2006; Assefa and Van den Berg, 2010; Gouse, 2012; Fischer and Hadju, 2015; Mahlase, 2017). Monsanto, for example, disseminated Bt insect-resistant maize to smallholder farmers via their 'Seeds of Success' programme, which at first distributed free GM seed in a package alongside fertilizers and herbicides (Gouse et al., 2016). Through government-supported maize development projects such as the Massive Food Production Programme, smallholders were similarly supplied with free or subsidized Bt maize and fertilizer (Gouse et al., 2005; Jacobson, 2013; Jacobson and Myhr, 2013; Fischer and Hajdu, 2015; Mahlase, 2017). Over time, most, if not all, of these projects collapsed, especially as subsidies were withdrawn and farmer debt increased, but GM seed retains an unintentional presence in many smallholder fields in South Africa.

GM contamination in smallholder agriculture in South Africa: why it matters

The low level of GM maize adoption among smallholder farmers in South Africa belies the extent to which their crops and farming systems are affected by GM seed. Maize is a wind-pollinated outcrossing plant with no known biological barriers to gene flow within the *Zea mays* species complex (Ellstrand and Hoffman, 1990; Ellstrand et al., 1999; Quist and Chapela, 2001; Johnston et al., 2004; Aheto et al., 2013). Therefore, gene exchange between all types of maize is common, a fact that underpins the widely recorded outcrossing and spread of transgenic traits into the seed of non-GM OPVs and landraces in South Africa (Jacobson and Myhr, 2013; Iversen et al., 2014).

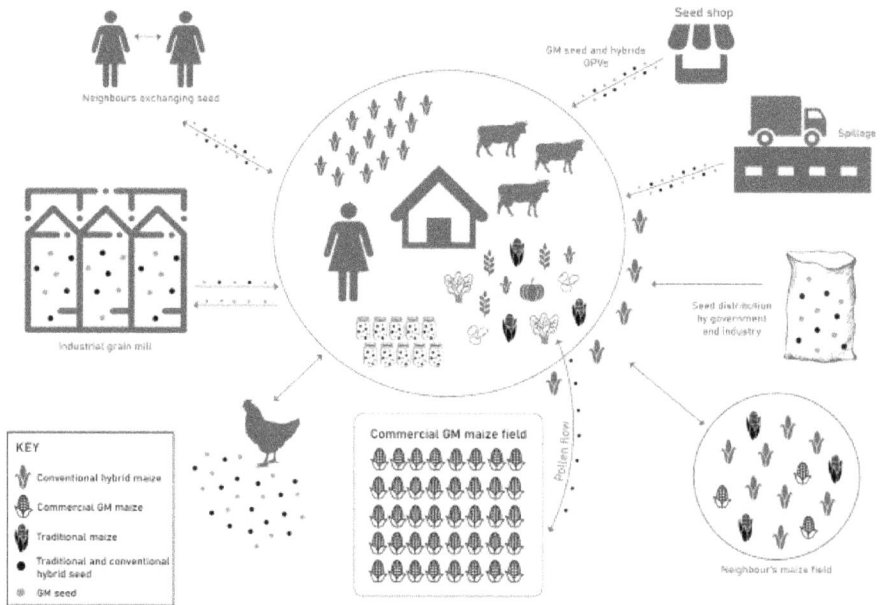

Figure 7.1 Pathways for the contamination of local seed systems by GM seed

This contamination occurs through several pathways (Bøhn et al., 2016; Schnurr, 2019), all usually beyond the farmers' knowledge and capacity to detect: pollen flowing between adjacent GM and non-GM maize fields, seed recycling and sharing, the sale of incorrectly labelled seed, and the distribution of 'free' or subsidized seed by government agencies and seed companies (see Figure 7.1).

Pollen flow between maize fields is a major source of contamination in both smallholder fields and industrial plantations, despite requirements for farmers to plant a so-called 'refuge' area of non-Bt maize alongside their Bt crop in order to delay the widely documented resistance to stem-borer (e.g. Tabashnik et al., 2013; Tabashnik and Carrière, 2019). A study involving 105 commercial farmers farming 87,778 hectares of maize in the main highveld maize production region of South Africa revealed that 99.8 per cent of farmers allowed no spatial separation between the Bt field and the required buffer (Kruger et al., 2012). Unsurprisingly, low levels of compliance are also evident among South African smallholders, where small fields, combined with low levels of awareness among both farmers and extension workers of the distinctions between GM and other hybrid seeds, make refuge requirements near impossible to achieve (Assefa and Van den Berg, 2010; Jacobson and Myhr, 2013; Van den Berg et al., 2013; Witt, 2018; Marshak et al., 2021). High levels of pollen flow between fields are thus inevitable, especially in contexts where small-scale subsistence and large-scale commercial farmers exist alongside each other. For example, as Kganyago (2020) describes, the vast, industrialized maize monoscapes that characterize the North West province are a world apart from the diverse cropping systems farmed by smallholders in the small North West village of Sespond. While Sespond villagers may not grow GM maize, their farms are located adjacent to industrial GM plantations in a complex agricultural landscape and their traditional and open-pollinated varieties of maize will almost inevitably be exposed to transgene flow.

The prevalence of seed recycling, sharing, and replanting among smallholder farmers provides a further avenue for contamination. Research across several South African provinces, in KwaZulu-Natal (Mahlase, 2017; Marshak et al., 2021), the North West (Kganyago, 2020), and the Eastern Cape (Iversen et al., 2014), has revealed how small-scale farmers are not able to detect different maize varieties in their seed systems, with farmers often unsure of the cultivar or seed company, and of whether the seed is OPV, conventional hybrid, or GM. Remarked one farmer in Kganyago's (2020) study: 'In the olden days when we only knew the seeds our fathers grew in their fields, we understood where the seeds came from and how they adapted to the environment.' As a result, locally recycled maize varieties often intermingle with GM maize seed that is unknowingly accepted as a gift from families and friends, reused from chicken feed, acquired from agro-dealers with unclear or incorrect information, or donated through government subsidy programmes (Iversen et al., 2014; Witt, 2018). Combined, these diverse pathways of contamination accelerate the flow of transgenic material to non-GM varieties, while breaching the biosafety requirements indicated in the approval and licensing conditions of GM crops. They also confirm the problems of applying management measures designed for large-scale commercial farming to small-scale, subsistence agriculture.

Transgenic contamination has profound implications for both small-holders and the farming systems that they nurture. In addition to impacts of introgression on agrobiodiversity, food security, and farming practices such as seed saving and seed selection, contamination undermines the centrality of traditional seed storage and exchange systems. These systems not only maintain and enrich crop genetic diversity and social cohesion, but also increase the resilience and autonomy of small-scale farmers through reducing their dependence on commercial seed (Kloppenburg, 2010; Helicke, 2015; McGuire and Sperling, 2016). Such violations raise questions about the laws designed to protect farmers from transgene contamination, including liability for unintended contamination and its legal consequences. Despite a strong articulation of the right to food in the South African Constitution, small-scale farmers are situated in an uncomfortable legal space which has been strongly promotional of industrial agriculture and the interests of the commercial seed sector, while neglecting the rights of small-scale farmers to save and plant seeds of their choice (Wynberg et al., 2012). Such rights extend also to the consumption of traditional maize. Kganyago (2020) explains how a lack of local milling and storage facilities led farmers in the village of Sespond in the

North West province of South Africa to send their traditional maize harvest to an adjacent mill for *both* processing and storage. Although farmers consciously chose to plant traditional maize because of its preferred taste and nutritional qualities, the lack of separation facilities at the mill meant that the processed maize flour (*bupi*) they received back was not what they had planted and was perceived to be of an inferior quality. Similarly, while farmers received back a similar volume of stored seed from the mill to that which they had deposited, it was typically not the traditional maize they had originally sent for storage, which opened up yet another pathway for contamination of their seed.

In the 'caravan' that the authors accompanied in KwaZulu-Natal, agroeco-logical farmers who were reviving their traditional varieties and actively avoiding GM maize spoke of the deep trauma they experienced when learning that their maize had tested positive for GM proteins. Some pronounced that they would 'tear up their fields', while others felt affronted that their right to plant and eat traditional maize had been obliterated through contamination. Such sentiments are not, however, confined to farmers who intentionally avoid GM crops and pursue an agroecological approach to farming. Farmers in Sespond, for example, grew a mix of GM, open-pollinated, and traditional maize varieties and intentionally planted these crops in different areas of their fields, for different purposes. However, the contamination of their traditional seed, passed down for generations, left them feeling despondent, ignorant, and insecure (Kganyago, 2020).

Despite the paucity of research on this topic, the deep connections between traditional seed systems and culture are well known (Van Niekerk and Wynberg, 2017), and the contamination of traditional seed is likely to have significant effects on local farming systems and on farmer psychology, well-being, and morale. Moreover, as farmers lose touch with the qualities of the seed they are planting, seed becomes anonymous, and patterns of seed and knowledges disintegrate. As Marshak et al. (2021) describe, farmers may perceive their own knowledge to be inadequate, leading to the loss of socio-ecological agency and farmer disempowerment.

More insidious impacts are also possible. For example, as Box 7.1 describes, the occurrence of introgressed GM traits in OPVs or landraces grown in small-scale farming systems may lead to farmers unknowingly or accidentally selecting the Bt trait in their maize varieties. This may have unpredictable consequences for the genetic diversity of their seeds (see also Lohn et al., 2020). Thus, knowledge about the dynamics of transgene flow and the seed selection behaviour of farmers is of fundamental importance for small-scale farmers who prefer to cultivate and select GM-free OPVs and/or landraces.

There is also a risk of trans-border transgene flow, given that South Africa currently stands alone as a producer of GM crops in the Southern African region. Indeed, many GM farming activities straddle the borders of South Africa and eSwatini, Lesotho, Zimbabwe, and Mozambique. This is likely to have multiple socio-ecological, economic and trans-border cultural political implications for affected smallholder communities.

Box 7.1 Understanding transgene flow

Maize (*Zea mays*) is a wind-pollinated outcrossing plant with no known biological barriers to gene flow among the species complex. Therefore gene exchange between all types of maize is common. This is important in the context of GM crop production. If transgene flow occurs from GM hybrid maize to the OPVs and landraces that are maintained, owned, and cherished by smallholders, it will have multiple implications at many levels. However, evaluating the consequences of gene flow, regardless of whether transgenes are involved, is a challenge because it is difficult to predict the ecological and evolutionary effects of genes that are expressed in different, potentially new genetic backgrounds and agroecological contexts.

In new research relevant for South Africa (Erasmus et al., 2019; Lohn et al., 2020), researchers found that a GM trait expressed in GM hybrid maize widely grown in South African industrial agricultural systems could successfully outcross to South African OPVs. This GM trait induced the production of an insecticidal toxin from the bacteria *Bacillus thuringiensis* (Bt), which is intended to kill certain caterpillar pests. The OPV or landrace maize plants that had accidentally received the Bt transgene produced the Bt toxin as in the original GM maize hybrids. This transgene introgression led to consistent but highly variable concentrations of the Bt toxins in what now became GM OPV plants. While the measured Bt toxin concentrations fluctuated unpredictably, even under controlled experimental conditions, the research confirmed that the Bt toxins newly produced in the GM OPV plants affected targeted pests, primarily caterpillars, that attack all maize plants. Maize varieties from a commercial hybrid breeding programme and those produced through farmer selection of OPVs responded similarly to transgene introgression.

Confirmation of bioactivity in outcrossed GM OPVs suggests that farmers will likely be influenced to select this GM trait. This is because in areas where caterpillar pest damage is prevalent, farmers will unknowingly select healthy-looking plants or cobs, which in turn will be more likely to carry the insecticidal trait. The strong insecticidal trait conferred by GM OPVs can conceal and underplay other traits selected over many years. This analysis suggests that transgene flow into OPVs and landraces may have significant consequences for the evolution of maize populations and the seed diversity cultivated by small-scale farmers. It may also influence the genetic diversity of varieties cultivated and maintained by small-scale farmers. Moreover, because the Bt toxin initially exerts a pest control effect which could be perceived as beneficial, pest resistance could develop. Pest resistance is widely documented to evolve quickly against Bt toxins. This could have negative livelihood impacts given that farmers unknowingly have been growing Bt OPVs and landraces and inadvertently selecting for increased prevalence of the Bt transgene and its toxins in their seed stock. Farmers could thus suffer increased crop losses through an increase in resistant pests. They may also have less diversity of non-Bt seed stock to remedy the situation.

Very little is currently known regarding the spread of Bt transgenes and the selection behaviour of maize seeds in smallholder farming systems. However, given the complexities described in this chapter, it is likely that a programme to systematically sample and monitor smallholder farmers' fields is needed. Such an initiative could help determine to what degree the crossing of non-Bt and Bt maize plants may affect their interactions with pest insects and the subsequent selection behaviour of farmers.

Responding to contamination

The realities and permanence of transgene contamination are clearly profound. While the bigger, systemic issues of transforming food and agricultural systems remain at the forefront, on the ground a different set of practical actions is needed to support smallholder farming communities in managing contamination and mitigating its negative consequences. This section of the chapter reports on the outcome of a four-day dialogue among farmers and other actors, in the hope that it provides some tools that can be developed over time.

Community dialogue and support

Discussion of solutions with farmers emphasized the importance of community dialogue and support. Contamination can be traumatic for farmers who have consciously chosen to avoid planting GM crops. One farmer said that after she discovered that her seeds were contaminated, she felt 'so much pain over the night' and thought of going to speak to the *induna*[3] to ask for another field. Remarked one farmer, 'When people farm in the village and community, we farm close together and there is potential for a lot of cross-pollination and therefore there is a need to come up with solutions.'

Farmers spoke strongly about the need to put responsive, healing strategies in place, with one female farmer drawing a powerful analogy: 'When you have a sick child you don't just run away and dump them but you try and make them healthy again.' Curative suggestions to do this included dialogues with neighbours and engaging with traditional leaders and other community members to discuss the implications of growing GM maize. This had already yielded helpful experiences. For example, a farmer from KwaNgwanase said that his seeds were contaminated but it did not make him worried or sad as he saw it as a lesson. He had selected seed from the boundaries of his field, approximately one kilometre away from his younger brother's maize field. His brother had been using herbicides and (probably) GM seed. When the farmer's traditional seed was shown to include GMOs, he started a dialogue with his brother. There were no arguments; instead, he told his brother about the health impacts of synthetic fertilizers and agrochemicals. In the end, his brother decided to start adopting agroecological practices. Contamination, it was suggested, 'can help you find ways of getting better seeds and ways of preventing further contamination'. Less positive experiences were reported elsewhere. One farmer explained how she had gone to speak to her neighbour about planting GM maize but been told to leave as her 'space was her own and she makes her own decisions'.

Low levels of awareness

Low levels of awareness among farmers were a key concern – both about the nature of GM contamination and about the ways in which their seed can become contaminated. There were also low levels of awareness about GM crops in general. This points to the need for a concerted awareness-raising and capacity-development effort focused not only on the basic facts about GM crops, but also on ways in which farming practices can be adjusted or managed to avoid contamination. Also clear is the importance of extending awareness-raising beyond farmers to include actors such as NGOs that support farmers and are involved in advocacy work, farmers in neighbouring countries that might be affected by transgene flow, government officials, and policymakers. As described by Vanessa Black in Box G, this points also to the pivotal role played by agricultural extension officers. Without such interventions, small-scale farmers will continue to be unintended casualties of transgene flow.

Farming practices

The farmer consultations revealed a number of strategies that could be adopted or strengthened in order to avoid GM contamination, and gave farmers the opportunity to express their concerns. Farmers discussed avoiding cross-pollination in their fields by spacing out the planting times of different varieties. This strategy is also employed by farmers who know that their neighbours are planting GM maize. Sourcing strategies were discussed, including the notion of 'trusted suppliers'. There was lively discussion about different approaches to seed selection: to ensure that GM seed was not unintentionally selected, and to set in place ways in which selection could reduce contamination. One farmer from Ingwavuma explained how she had established a special seed plot, where 'the plants are tall and the cobs are long'. She was also experimenting with planting different densities of maize plants, and had seven different varieties, all planted at different times to avoid cross-pollination. 'It is very important to plant your own seeds', remarked another farmer, 'and to stagger planting among farmers.' He explained, 'If you plant in October or November then the other farmers will plant after you. In this way there won't be any contamination. By the time the rain comes your seeds will be the first ones to grow. They will be the first ones to tassel and there won't be any contamination'.

Mitigation strategies were also discussed, to prevent farmers from simply 'pulling up their fields' when detecting contamination. Importantly, farmers voiced their concern about multiple forms of 'contamination' – not just transgene contamination, but also the effects of chemical agricultural inputs on health, soils, and water. A concern about 'knowledge contamination' was also raised, where traditional ways of knowing are now having to respond and adapt to the impacts of transgene flow.

Community monitoring, research, advocacy, and policy

Adapting farming practices, initiating dialogues, and raising awareness are important strategies to manage change, but for farmers to do this effectively they need to know what is happening in their fields. Farmers expressed concern about the implications of GM contamination for the blessing of their traditional seed, and how testing would help put their minds at rest for the blessing ceremony, and help them manage their selection, planting,

Photo 7.2 Ntombithini Ndwandwe from the Zimele Project demonstrating the GM 'strip test' to Enoch Dlamini from the Africa Cooperative Action Trust (ACAT), eSwatini. Ingwavuma, January 2018
Credit: Rachel Wynberg

and storage practices. To date, farmers have lacked the resources, technical know-how, and capacity to detect the presence or absence of GMOs in their crops and seeds. Currently, antibody-based strip tests are the quickest and simplest way of detecting the novel proteins produced by transgenes and were the method of choice used during our 'caravan' testing project. Farmers quickly learnt this testing method and carried out many tests on their own seed samples (Photo 7.2). However, this method fails to detect transgenes that may be present but which are not sufficiently functioning or producing altered forms or fragments of the novel proteins. Other methods, like the polymerase chain reaction (PCR) tests, are necessary and more precise to detect transgenes at the DNA level. However, these more sophisticated tests require high levels of training, punctilious chemistry and biochemistry, and time-critical access to high-development urban centres, as well as expensive and voluminous equipment, toxic and hazardous materials, and sophisticated infrastructure for the interpretation and resolution of results. This puts them out of the reach of most farmers. However, exciting new approaches are under development to enable farmers to self-monitor their crops. For example, Ignacio Chapela, a professor at the University of California, Berkeley, and one of the scientists responsible for first describing the contamination of traditional maize landraces in their Mexican centre of origin (Quist and Chapela, 2001), and his colleagues have developed a new and affordable GM testing technology which, once launched, promises to resolve many of the challenges posed by PCR testing and could empower farmers to assess their own maize and make decisions for themselves (Bekta and Chapela, 2016). The method, based on an isothermal DNA amplification reaction (loop-mediated isothermal amplification reaction, or LAMP) is described as 'turtle methodology' and includes concepts, biochemical reactions, instruments, and procedures that enable field-based detection of genetic contamination in a manner that can be controlled locally and can reveal GMO presence within two hours. Such approaches, combined with possibilities to develop wider programmes that enable communities to monitor soil and water health, could be powerful tools for community action and response.

The value of research, advocacy, and policy emerged as cross-cutting themes across these discussions. In practice, many farmers are already active researchers, experimenting with different varieties and methods and bringing in embedded, visceral, tacit, and ecological ways of knowing to their farming practices. Bringing these knowledges and skills to the fore, and providing additional technical support where needed, emerged as a strong call. Particular research needs were identified to deepen understanding about the ways in which seeds are stored, separated, and selected, and the range of planting practices used both on their farms and on adjacent farms. Drawing on this information, farmers believed that strategies could be developed to minimize transgene flow.

Questions were asked about the role of the gene bank in repatriating uncontaminated seed to farmers, and a discussion with gene bank managers yielded exciting possibilities for seed restoration programmes to reintroduce farmers' traditional varieties and strengthen on-farm conservation. Concerns were also expressed about the extent to which farmers might be held legally liable for GM crops unintentionally planted in their fields, and the need for legal and policy work on transgene flow as it relates to farmers' rights, and potential infringements of these rights. The importance of linking this work to ongoing debates around seed policy and biosafety in the region was highlighted as critical, especially with the revision of plant breeder's rights laws that criminalize the unauthorized use of protected varieties.

Conclusion

Genetically modified crops have a myriad of consequences for smallholder farmers on the African continent and beyond, but little attention has been paid to the threats posed to farmer-managed seed systems by the contamination of GM seed. Although transgene flows are known to be widespread, the implications of this contamination for smallholder farmers are less well understood. Several African countries – all of which have significant smallholder-based farming systems centred on the production of maize and its consumption as a staple food – are now poised to allow the commercial planting of GM maize. The experiences of smallholder farmers in South Africa are thus likely to portend something similar for farmers elsewhere in the region. Findings suggest that transgene contamination of local seed systems and farmers' fields in South Africa is common, but under-reported. Contamination takes place through several pathways, all usually beyond the farmers' knowledge and capacity to detect: pollen flowing between adjacent GM and non-GM maize fields, seed recycling and sharing, the sale of incorrectly labelled seed, and the distribution of 'free' or subsidized seed by government agencies and seed companies. Contamination negatively impacts agrobiodiversity, food security, farming practices, and traditional seed saving and exchange systems, and has also resulted in the deskilling of farmers, who no longer know what they are planting and perceive their own knowledge to be inadequate. The contamination of traditional seed is also likely to have significant effects on farmer psychology, well-being, and morale. Farmer-led strategies to manage contamination are crucial, but setting them in place requires awareness-raising, as well as technical support and advice about the ways in which farming practices can be adjusted or managed to avoid contamination, and methods farmers can use to conduct their own research and testing. Ongoing dialogues within farming communities to enable landscape- and community-level interventions are a critical part of this strategy. Such actions might not only help mitigate the negative impacts of GM contamination, but also strengthen the agency of farmers and local farming communities to secure productive and healthy food systems.

Notes

1. First-generation GM crops are those that are insect-resistant or herbicide-tolerant versions of commodity crops – canola, cotton, maize, and soybean – designed to be used in large-scale, industrial agriculture. Second-generation GM crops typically focus on staple crops used by smallholder farmers such as cowpea, brinjal, sorghum, millet, and cooking banana and may result from public-private partnerships that potentially circumvent restrictive intellectual property rights.
2. In North and South American countries GM maize is also grown at a vast scale, but in contrast to South Africa, it is mostly exported as feed for animals or energy production such as biogas and ethanol, with negligible amounts of the overall volumes processed into industrial food stuffs such as high-fructose corn syrup. Like South Africa, Mexico consumes maize as a staple food, but it has not approved cultivation of GM maize. The Mexican government is currently considering a ban on the import of GM maize to avoid contamination of the food and feed chain through unintended gene flow, and prevent negative impacts on this global centre of origin of maize diversity (Deslandes, 2022).
3. A Zulu or Xhosa headman or chief.

References

Aheto, D.W., Bøhn, T., Breckling, B., Van den Berg, J., Ching, L.L. and Wikmark, O. (2013) 'Implications of GM crops in subsistence-based agricultural systems in Africa', in B. Breckling and R. Verhoeven (eds), *GM-Crop Cultivation: Ecological Effects on a Landscape Scale*, pp. 93–103, Peter Lang, Frankfurt.

Akinbo, O., Obukosia, S., Ouedraogo, J., Sinebo, W., Savadogo, M., Timpo, S., Mbabazi, R., Maredia, K., Makinde, D. and Ambali, A. (2021) 'Commercial release of genetically modified crops in Africa: interface between biosafety regulatory systems and varietal release systems', *Frontiers in Plant Science* 12 <https://doi.org/10.3389/fpls.2021.605937>.

Aliber, M., and Hart, T.G. (2009) 'Should subsistence agriculture be supported as a strategy to address rural food insecurity?', *Agrekon* 48(4): 434–458 <https://doi.org/10.1080/03031853.2009.9523835>.

Assefa, Y. and Van den Berg, J. (2010) 'Genetically modified maize: adoption practices of small-scale farmers in South Africa and implications for resource poor farmers on the continent', *Aspects of Applied Biology* 96: 215–223.

Bekta, A. and Chapela, I.H. (2016) 'Efficiency of a fluorescent, non-extraction LAMP DNA amplification method: toward a field-based specific detection of maize pollen grains', *Aerobiologia* 32: 481–488 <https://doi.org/10.1007/s10453-016-9420-z>.

Bernstein, H. (1996) 'The political economy of the maize filière', *The Journal of Peasant Studies* 23(2–3): 120–145 <https://doi.org/10.1080/03066159608438610>.

Bøhn, T., Aheto, D.W., Mwangala, F.S., Fischer, K., Bones, I.L., Simoloka, C., Mbeule, I., Schmidt, G. and Breckling, B. (2016) 'Pollen-mediated gene flow and seed exchange in small-scale Zambian maize farming, implications for biosafety assessment', *Scientific Reports* 6(1): 1–12 <https://doi.org/10.1038/srep34483>.

Bourgou, L., Kargougou, E., Sawadogo, M. and Fok, M. (2020) 'Bt cotton seed purity in Burkina Faso: status and lessons learnt', *Journal of Cotton Research* 3(30) <https://doi.org/10.1186/s42397-020-00070-4>.

Bowman, A. (2015) 'Sovereignty, risk and biotechnology: Zambia's 2002 GM controversy in retrospect', *Development and Change* 46(6): 1369–1391 <https://doi.org/10.1111/dech.12196>.

Davis, K.F., Chhatre, A., Rao, N.D., Singh, D., Ghosh-Jerath, S., Mridul, A., Poblete-Cazenave, M., Pradhan, N. and DeFries, R. (2019) 'Assessing the sustainability of post-Green Revolution cereals in India', *Proceedings of the National Academy of Sciences* 116(50): 25034–25041 <https://doi.org/10.1073/pnas.1910935116>.

DeFries, R., Fanzo, J., Remans, R., Palm, C., Wood, S. and Anderman, T.L. (2015) 'Metrics for land-scarce agriculture', *Science* 349(6245): 238–240 <https://doi.org/10.1126/science.aaa5766>.

Deslandes, A. (2022) 'Mexico prohibits planting of GM corn, but stops short of banning imports', *Diálogo Chino*, 3 March <https://dialogochino.net/en/agriculture/mexico-gm-ban-planting-corn-imports/>.

Dowd-Uribe, B. (2014) 'Engineering yields and inequality? How institutions and agro-ecology shape Bt cotton outcomes in Burkina Faso', *Geoforum* 53: 161–171 <https://doi.org/10.1016/j.geoforum.2013.02.010>.

Du Toit, P. (2022) *The ANC Billionaires: Big Capital's Gambit and the Rise of the Few*, Jonathan Ball, Johannesburg.

Ellstrand, N.C. and Hoffman, C.A. (1990) 'Hybridization as an avenue of escape for engineered genes: strategies for risk reduction', *BioScience* 40(6): 438–442 <https://doi.org/10.2307/1311390>.

Ellstrand, N.C., Prentice, H.C., and Hancock, J.F. (1999) 'Gene flow and introgression from domesticated plants into their wild relatives', *Annual Review of Ecology and Systematics* 30: 539–563 <https://doi.org/10.1146/annurev. ecolsys.30.1.539>.

Erasmus, R., Pieters, R., Du Plessis, H., Hilbeck, A., Trtikova, M., Erasmus, A. and Van den Berg, J. (2019) 'Introgression of a cry1Ab transgene into open pollinated maize and its effect on Cry protein concentration and target pest survival', *PLoS ONE* 14(12): e0226476 <https://doi.org/10.1371/journal. pone.0226476>.

Fernandes, G.B., Silva, A.C.D.L., Maronhas, M.E.S., Santos, A.D.S.D. and Lima, P.H.C. (2022) 'Transgene flow: challenges to the on-farm conservation of maize landraces in the Brazilian semi-arid region', *Plants* 11(5): 603 <https://doi.org/10.3390/plants11050603>.

Fischer, K. (2022) 'Why Africa's new green revolution is failing: maize as a commodity and anti-commodity in South Africa', *Geoforum* 130: 96–104 <https://doi.org/10.1016/j.geoforum.2021.08.001>.

Fischer, K., and Hajdu, F. (2015) 'Does raising maize yields lead to poverty reduction? A case study of the Massive Food Production Programme in South Africa', *Land Use Policy* 46: 304–313 <https://doi.org/10.1016/ j.landusepol.2015.03.015>.

Fitting, E. (2011) *The Struggle for Maize: Campesinos, Workers, and Transgenic Corn in the Mexican Countryside*. Duke University Press, Durham, NC.

Gouse, M. (2012) 'GM maize as subsistence crop: the South African smallholder experience', *AgBioForum* 15(2): 163–174.

Gouse, M., Pray, C.E., Kirsten, J. and Schimmelpfennig, D. (2005) 'A GM subsistence crop in Africa: the case of Bt white maize in South Africa', *International Journal of Biotechnology* 7(1–3): 84–94 <https://doi.org/10.1504/ IJBT.2005.006447>.

Gouse, M., Sengupta, D., Zambrano, P. and Zepeda, J.F. (2016) 'Genetically modified maize: less drudgery for her, more maize for him? Evidence from smallholder maize farmers in South Africa', *World Development* 83: 27–38 <https://doi.org/10.1016/j.worlddev.2016.03.008>.

Greenberg, S. (2013) *The Disjunctures of Land and Agricultural Reform in South Africa: Implications for the Agri-Food System*, Working Paper 26, Institute for Poverty, Land and Agrarian Studies, University of the Western Cape, Bellville <http://hdl.handle.net/10566/4489>.

Helicke, N.A. (2015) 'Seed exchange networks and food system resilience in the United States', *Journal of Environmental Studies and Sciences* 5(4): 636–649 <https://doi.org/10.1007/s13412-015-0346-5>.

Howard, P.H. (2009) 'Visualizing consolidation in the global seed industry: 1996–2008', *Sustainability* 1(4): 1266–1287 <https://doi.org/10.3390/ su1041266>.

International Service for the Acquisition of Agri-biotech Applications (ISAAA) (2018) *Global Status of Commercialized Biotech/GM Crops in 2018: Biotech*

Crops Continue to Help Meet the Challenges of Increased Population and Climate Change, ISAAA Brief No. 54, ISAAA, Ithaca, NY <https://www.isaaa.org/resources/publications/briefs/54/>.

Iversen, M., Grønsberg, I.M., Van den Berg, J., Fischer, K., Aheto, D.W. and Bøhn, T. (2014) 'Detection of transgenes in local maize varieties of small-scale farmers in Eastern Cape, South Africa', *PLoS ONE* 9(12): e116147 <https://doi.org/10.1371/journal.pone.0116147>.

Jacobson, K. (2013). *From Betterment to Bt Maize: Agricultural Development and the Introduction of Genetically Modified Maize to South African Smallholders*, Doctoral thesis, Swedish University of Agricultural Sciences, Uppsala <http://pub.epsilon.slu.se/10406/>.

Jacobson, K. and Myhr, A.I. (2013) 'GM crops and smallholders: biosafety and local practice', *The Journal of Environment & Development* 22(1): 104–124 <https://doi.org/10.1177/1070496512466856>.

Johnston, J., Blancas, L., and Borem, A. (2004) 'Gene flow and its consequences: a case study of Bt maize in Kenya', in A. Hilbeck and D.A. Andow (eds), *Environmental Risk Assessment of Genetically Modified Organisms Vol. 1: A Case Study of Bt Maize in Kenya*, pp. 187–209, CAB International, Wallingford, UK <https://doi.org/10.1079/9780851998619.0187>.

Kedisso, E.G., Karim, M., Joseph, G. and Koch, M. (2022) 'Commercialization of genetically modified crops in Africa: opportunities and challenges', *African Journal of Biotechnology* 21(5): 188–197 <https://doi.org/10.5897/AJB2021.17434>.

Kganyago, M.C. (2020) *Understanding Farmer Seed Systems in Sespond, North West Province*, Master's thesis, University of Cape Town <http://hdl.handle.net/11427/32486>.

Kloppenburg, J. (2010) 'Impeding dispossession, enabling repossession: biological open source and the recovery of seed sovereignty', *Journal of Agrarian Change* 10(3): 367–388 <https://doi.org/10.1111/j.1471-0366.2010.00275.x>.

Kruger, M., Van Rensburg, J.B.J. and Van den Berg, J. (2012) 'Transgenic Bt maize: farmers' perceptions, refuge compliance and reports of stem borer resistance in South Africa', *Journal of Applied Entomology* 136(1–2): 38–50 <https://doi.org/10.1111/j.1439-0418.2011.01616.x>.

Lohn, A.F., Trtikova, M., Chapela, I., Van den Berg, J., Du Plessis, H. and Hilbeck, A. (2020) 'Transgene behavior in *Zea mays* L. crosses across different genetic backgrounds: segregation patterns, cry1Ab transgene expression, insecticidal protein concentration and bioactivity against insect pests', *PloS ONE* 15(9): e0238523 <https://doi.org/10.1371/journal.pone.0238523>.

Mahlase, M.H. (2017) *Exploring the Uptake of Genetically Modified White Maize by Smallholder Farmers: The Case of Hlabisa, South Africa*, Master's thesis, University of Cape Town.

Marais, H. (2011) *South Africa Pushed to the Limit: The Political Economy of Change*, Bloomsbury Publishing, London.

Marshak, M., Wickson, F., Herrero, A. and Wynberg, R. (2021) 'Losing practices, relationships and agency: ecological deskilling as a consequence of the uptake of modern seed varieties among South African smallholders', *Agroecology and Sustainable Food Systems* 45(8): 1189–1212 <https://doi.org/10.1080/21683565.2021.1888841>.

Masehela, T.S., Rhodes, J.I., Groenewald, H., Poole, C.J., Van den Berg, J., Gouse, M., Skowno, A.L., Barros, E., Seymour, C.L., Mandivenyi, W.G. and Van der Colff, D. (2021) *An Initial Assessment of Impacts on Biodiversity from GMOs Released into the Environment in South Africa*, South African National Biodiversity Institute, Department of Forestry, Fisheries and the Environment, Pretoria.

Mawasha, J.L. (2020) *An Assessment of South Africa's Non-Genetically Modified Maize Export Potential*, MSc (Agricultural Economics) dissertation, University of Pretoria, Pretoria <http://hdl.handle.net/2263/72932>.

McCann, J. (2009) *Maize and Grace: History, Corn, and Africa's New Landscape, 1500–1999*, Boston University Press, Boston, MA.

McGuire, S. and Sperling, L. (2016) 'Seed systems smallholder farmers use', *Food Security* 8(1): 179–195 <https://doi.org/10.1007/s12571-015-0528-8>.

Nature Biotechnology (2002) 'Going with the flow', Editorial, *Nature Biotechnology* 20(6): 527 <https://doi.org/10.1038/nbt0602-527>.

Paarlberg, R.L. (2000) *Governing the GM Crop Revolution: Policy Choices for Developing Countries*, Food, Agriculture, and the Environment Discussion Paper 33, International Food Policy Research Institute, Washington, DC <https://www.ifpri.org/publication/governing-gm-crop-revolution>.

Patel, R. (2013) 'The long green revolution', *The Journal of Peasant Studies* 40(1): 1–63 <https://doi.org/10.1080/03066150.2012.719224>.

Price, B. and Cotter, J. (2014) 'The GM Contamination Register: a review of recorded contamination incidents associated with genetically modified organisms (GMOs), 1997–2013, *International Journal of Food Contamination* 1(5) <https://doi.org/10.1186/s40550-014-0005-8>.

Quist, D. and Chapela, I.H. (2001) 'Transgenic DNA introgressed into traditional maize landraces in Oaxaca, Mexico', *Nature* 414(6863): 541–543 <https://doi.org/10.1038/35107068>.

Rock, J. and Schurman, R. (2020) 'The complex choreography of agricultural biotechnology in Africa', *African Affairs* 119(477): 499–525 <https://doi.org/10.1093/afraf/adaa021>.

Rosset, P., Collins, J. and Lappé, F.M. (2000) 'Lessons from the Green Revolution: Do we need new technology to end hunger?', *Tikkun Magazine* 15(2): 52–56 <https://web.archive.org/web/20001202181100/http://www.twnside.org.sg/title/twr118c.htm>.

Schnurr, M.A. (2019) *Africa's Gene Revolution: Genetically Modified Crops and the Future of African Agriculture*, McGill-Queen's University Press, Montreal.

Schnurr, M.A. and Gore, C. (2015) 'Getting to "yes": governing genetically modified crops in Uganda', *Journal of International Development* 27(1): 55–72 <https://doi.org/10.1002/jid.3027>.

Schnurr, M.A. and Dowd-Uribe, B. (2021) 'Anticipating farmer outcomes of three genetically modified staple crops in sub-Saharan Africa: insights from farming systems research', *Journal of Rural Studies* 88: 377–387 <https://doi.org/10.1016/j.jrurstud.2021.08.001>.

Singh, R.B. (2000) 'Environmental consequences of agricultural development: a case study from the Green Revolution state of Haryana, India', *Agriculture, Ecosystems & Environment* 82(1–3): 97–103 <https://doi.org/10.1016/S0167-8809(00)00219-X>.

Sitas, A. (2010) *The Mandela Decade 1990–2000: Labour, Culture and Society in Post-Apartheid South Africa*, Unisa Press, Pretoria.

Tabashnik, B.E. and Carrière, Y. (2019) 'Global patterns of resistance to Bt crops highlighting pink bollworm in the United States, China, and India', *Journal of Economic Entomology* 112(6): 2513–2523 <https://doi.org/10.1093/jee/toz173>.

Tabashnik, B.E., Brévault, T. and Carrière, Y. (2013) 'Insect resistance to Bt crops: lessons from the first billion acres', *Nature Biotechnology* 31(6): 510–521 <https://doi.org/10.1038/nbt.2597>.

Van den Berg, J., Hilbeck, A. and Bøhn, T. (2013) 'Pest resistance to Cry1Ab Bt maize: field resistance, contributing factors and lessons from South Africa', *Crop Protection* 54: 154–60 <https://doi.org/10.1016/j.cropro.2013.08.010>.

Van Niekerk, J. and Wynberg, R. (2017) 'Traditional seed and exchange systems cement social relations and provide a safety net: a case study from KwaZulu-Natal, South Africa', *Agroecology and Sustainable Food Systems* 41(9–10): 1099–1123 <https://doi.org/10.1080/21683565.2017.1359738>.

Witt, H. (2018) *Policy Impacts: The Impact of Government Agricultural and Rural Development Policy on Small-holder Farmers in KwaZulu-Natal*, Biowatch Research Paper, Biowatch South Africa, Durban <https://biowatch.org.za/download/research-paper-policy-impacts-the-impact-of-government-agricultural-and-rural-development-policy-on-smallholder-farmers-in-kwazulu-natal/?wpdmdl=511&refresh=6526616cb507b1697014124>.

Witt, H., Patel, R. and Schnurr, M. (2006) 'Can the poor help GM crops? Technology, representation & cotton in the Makhathini flats, South Africa', *Review of African Political Economy* 33(109): 497–513 <https://doi.org/10.1080/03056240601000945>.

Wynberg, R. (2003) 'Biotechnology and the commercialisation of biodiversity in Africa', in B. Chaytor and K.G. Gray (eds), *Environment & Policy, vol. 36: International Environmental Law and Policy in Africa*, pp. 83–102, Springer Science+Business Media, New York <https://doi.org/10.1007/978-94-017-0135-8_5>.

Wynberg, R. and Fig, D. (2013) *A Landmark Victory for Justice: Biowatch's Battle with the South African State and Monsanto: The Inside Story*, Biowatch South Africa, Durban <https://biowatch.org.za/download/a-landmark-victory-for-justice/>.

Wynberg, R., Van Niekerk, J., Williams, R. and Mkhaliphi, L. (2012) *Policy Brief: Securing Farmers' Rights and Seed Sovereignty in South Africa*, Biowatch South Africa, Durban, and the Environmental Evaluation Unit, University of Cape Town <https://biowatch.org.za/download/policy-brief-securing-farmers-rights-and-seed-sovereignty-in-south-africa/>.

Box D Small-scale farmers – the guardians of plant plasticity: reflections from Dr Melaku Worede, former director of the Ethiopian gene bank

Melaku Worede

The first Seed and Knowledge Initiative (SKI) seminar was held at Mont Fleur in the winelands outside Cape Town, South Africa, in 2014. Those who attended were captivated by the wisdom of one of the keynote speakers, Dr Melaku Worede. A retired plant geneticist from Ethiopia, Dr Worede was known and respected around the world for his tremendous contribution to the conservation of agricultural biodiversity and is credited with helping to restore food security in Ethiopia after years of devastating drought-induced famine.

Dr Worede played a key role in establishing the first African gene bank, the Plant Genetic Resources Centre in Addis Ababa, serving as its director from 1979 to 1993. Under his leadership, gene bank staff collected and safely stored a considerable amount of Ethiopia's seeds and plant materials, in the process establishing one of the world's foremost genetic conservation systems. A distinguishing feature of Dr Worede's work was that he used genetic science to support small-scale farmers, and, rather than impose Western science on farmers, he valued their knowledge and age-old practices. Through connecting farmers with scientists, he was instrumental in breaking down barriers of supposed 'superiority' between the knowledge bases of those working respectively in the laboratory and in the field.

In 1989 Dr Worede was presented with the Right Livelihood Award (known as 'the Alternative Nobel Prize') for 'preserving Ethiopia's genetic wealth for the benefit of all humanity'. Throughout his career was very active at the international level: he served as the first Chair of the African Committee for Plant and Genetic Resources, and was instrumental in setting up the African Biodiversity Network.

Putting farmers first

The following section is an adaptation of Dr Worede's talk, 'Putting farmers first', which he delivered at the seminar. It represents his reflections on a long and distinguished career.

Farmers are known to spread the risk of farming over locations, seasons, and diversity – a concept known as 'plasticity'.[1] In the past, the marketplace was the central hub for farmers, the place where farmers were able to cross-fertilize ideas, and exchange views and materials – important foundational elements for on-farm plasticity. With the number of local markets declining in Ethiopia, community seed banks have in some instances come to fulfil this role of knowledge and material exchange hubs.

Community seed banks in Ethiopia are comprehensive and dynamic: they espouse the thought of doing better without compromising what is already there. With their seeds safely backed up in a community seed bank, farmers in Ethiopia are able to maintain their own choices. Saving seed and storing it at a community seed bank is a matter of sovereignty and also has a gendered aspect as it mostly involves women farmers. Responsible for the bulk of the farm work, including seed selection, women farmers play important roles at both household and community level.

The linkages between community seed banks and farming households foster a type of 'conservation system'. This system is not necessarily better than what each farmer is doing individually, but its potency is enhanced due to its multiplicity of connections

(Continued)

Box D Continued

which feed into a complex network. Knowledge is enriched through collaboration, and when scientists are included in farming networks, enhancement, or value adding, is made possible. In this context, enhancement may refer to raised productivity, better-quality seed or crops, or, in the case of an undesirable trait, replacement with a desired trait.

Agrobiodiversity is best conserved through use by local farmers and I strongly support the free exchange of germplasm among small-scale farmers, as this would enrich seed stocks and knowledge, especially if networks included farmers from further afield.

I prefer the term 'farmers' varieties' to 'landraces' as the former acknowledges the important role of farmers in their development. Farmers always know best how to grow their varieties, and also know how to determine the optimum population size and growing space for crops. In relation to growing space, I believe that farmers' varieties should not be grown over a large area, as this would compromise the gene complex inherent in these materials. If cultivation across a large area continued, the variety would ultimately be lost.

During the famine in the mid-1980s agrobiodiversity was severely impacted by food aid which arrived in the form of (much-needed) grain. Many farmers were forced to consume their seed stocks due to hunger, and when conditions were favourable enough to cultivate again, they ended up planting the grain which had been donated. My colleague Dr Regassa Feyisa and I came across this phenomenon while we were collecting seeds for the gene bank. We were alarmed, because farmers would normally never, ever consume their own seed – unless circumstances forced them to do so.

Initially Dr Feyisa and I collected seed to store in the gene bank, but, questioning the usefulness of this *ex situ* form of conservation, and cognizant of the enormous amount of knowledge held by farmers, we decided to work alongside farmers, sharing their expertise and also learning from those on the ground. At the same time, experts from abroad were suggesting that farmers adopt external inputs and improved seed varieties to raise their productivity. A representative from the FAO even suggested that teff (*Eragrostis tef*), a staple crop, be abandoned as it was deemed to be 'useless' due to its low yields and lack of nutritional value.

The Ethiopian gene bank had a different perspective, believing that farmers were the best conservers and users of both the wild gene pool and farmers' varieties. While gene banks are able to collect representative samples and store them under controlled systems, that doesn't include the other dynamics found on the ground outside the laboratory. In a formal gene bank one might store something that one doesn't want to lose, but it would not be representative enough of the diversity that is in the field. I believe that farmers and gene banks can work together in a mutually supportive way. This would work better in a farmer-led system than a scientist-led system, as the majority of the knowledge base is on the ground.

In closing, I want to underline the importance of small-scale farmers' knowledge once more, and I would label them the original 'plant breeders' as they employed selective breeding to raise yield, improve quality, and promote diversity long before formal plant breeding became an established discipline.

Note

1. In the plant science world, 'plasticity' is defined as the ability of an organism to change its phenotype in response to different environments, an important characteristic which enables sessile plants to adapt to rapid changes in their surroundings (Laitinen and Nikoloski, 2019).

Reference

Laitinen, R.A.E. and Nikoloski, Z. (2019) 'Genetic basis of plasticity in plants', *Journal of Experimental Botany* 70(3): 739–745 <https://doi.org/10.1093/jxb/ery404>.

Box E Forging links between gene banks and African smallholder farmers: opportunities and constraints

Kudzai Kusena and Jaci van Niekerk

What is a gene bank?

In the early 1960s, the growing realization that anthropogenic activity was driving agrobio-diversity loss at an alarming rate led to multiple strategies aimed at halting the erosion of plant genetic resources (Frankel and Hawkes, 1975). Several international treaties, including the Convention on Biological Diversity, the Food and Agriculture Organization's Undertaking on Plant Genetic Resources, and its successor, the International Treaty on Plant Genetic Resources for Food and Agriculture (ITPGRFA), specify conservation strategies for plant genetic resources which are vital for food and agriculture, including the establishment of gene banks (Hawkes et al., 2000).

There are two broad categories of plant resource conservation: *ex situ* and *in situ*. *Ex situ* refers to the conservation of genetic resources outside their natural habitat, for instance in gene banks or botanical gardens. *In situ* conservation, on the other hand, relates to the preservation of genetic resources within the cultivated ecosystems where they developed their distinctive features – for example, on a farm (Engelmann and Engels, 2002).

As a widely used *ex situ* conservation approach, gene banks function at global, regional, and national levels and the concept has been extended to the local level by including community seed banks, where community-managed seed collections are conserved and the exchange of local genetic resources is enhanced (Khoury et al., 2010; Sthapit 2012). *Ex situ* collections are envisaged as the solution to the restoration of the diversity of plant genetic resources in the event of losses incurred as a result of both rapid and slow onset disasters such as floods, fires, earthquakes, drought, and climate change.

The primary objectives of gene banks are:

- to collect representative samples from specific crop populations;
- to conserve these samples away from the evolutionary pressures of their natural or on-farm habitat.

Types of gene banks

There are several types of gene bank, but this commentary focuses on public gene banks. Table E.1 provides an overview of the different types of public gene banks found around the world.

Table E.1 The most common types of public gene banks

Global gene banks

The largest global gene bank – the Svalbard Global Seed Vault – is situated on a Norwegian island within the permafrost of the Arctic Circle. The vault stores duplicates of seed samples from the world's crop collections and currently holds more than 1.1 million seed samples from 91 gene banks (www.seedvault.no). The Svalbard vault provides a backup for gene banks where seeds can be retrieved and restored to farmers in the event of catastrophic or incremental loss.

Photo E.1 The Svalbard Global Seed Vault
Credit: Fredrik Naumann-Panos; https://time.com/doomsday-vault/

(Continued)

Box E Continued

CGIAR gene banks

Strategically located in centres of crop diversity, 11 international gene banks operate under CGIAR. These gene banks are mandated by the ITPGRFA to conserve crops and trees and make them available on behalf of the global community. Users are mainly researchers and plant breeders from developing countries, with 80% of material distributed to national agricultural research stations, universities, and national gene banks, 8% to farmers, farmer organizations, and NGOs, and 7% to the commercial seed sector (Halewood et al., 2020).

Photo E.2 A gene bank staff member holds packets of rice samples in the CGIAR AfricaRice gene bank in Mbe, Côte d'Ivoire
Credit: Neil Palmer/Crop Trust

Regional gene banks

Regional gene banks act as duplicate repositories for national collections. Member states deposit their collections for safety duplication so that, in the event of a loss at the national level, regional gene banks can restore materials by retrieving duplicates. An example is the Southern African Development Community (SADC) Plant Genetic Resources Centre located in Lusaka, Zambia.

Photo E.3 The SADC Plant Genetic Resources Centre
Credit: www.spgrc.org.zm

National gene banks

National gene banks conserve seeds in active as well as safety duplicate collections. Active collections house germplasm stored under short-term storage conditions, with resources available for distribution to farmers, researchers, and breeders. Safety duplicates are kept under long-term storage conditions, and materials are not readily available for distribution. Materials in safety duplication are only retrieved if the active collections are depleted and need to be multiplied or regenerated.

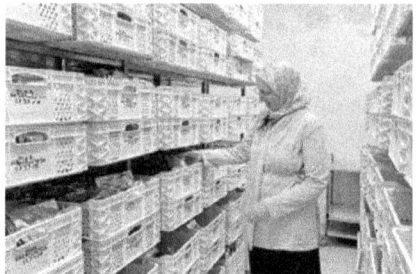

Photo E.4 A staff member in the seed storage room at Morocco's national gene bank
Credit: Nelissa Jamora

(Continued)

Box E Continued

Community seed banks

The community seed bank concept mimics the function of national gene banks. Primarily driven by civil society and farmers, this is a local approach to the conservation and sustainable use of plant genetic resources. The development of community seed banks has been instrumental in organizing farmers and linking them to gene banks at national, regional, and global levels.

Photo E.5 A community seed bank in Malawi
Credit: Rachel Wynberg

How do gene banks function?

Gene banks are designed to collect, document, conserve, characterize,[1] and distribute plant genetic resources. Before collecting specimens, gene banks carry out eco-geographic surveys to identify and target certain plant populations, be they farmers' varieties or wild crop relatives.

Three types of collection missions commonly used by gene banks:

- *General.* Gene banks go to farmers to collect samples they deem fit to be included in their collections.
- *Gap filling.* Gene banks target unique genetic resources not already captured in their collections.
- *Rescue.* Gene banks set out to collect threatened plant or crop populations.

Besides seeds, gene banks also collect tissue culture, or actively growing plants, though seeds are most commonly used in gene banks in sub-Saharan Africa. Seed samples are stored in freezers at sub-zero temperatures ($-20°C$) and less than 10 per cent humidity, conditions which reduce the biochemical activities of seeds and allow them to survive long-term storage. Seeds stored in this way can subsist for more than 15 years without losing viability.

During a collection mission, a questionnaire is used to collect what gene banks call passport data. Passport data record the geographical information of the seed, the types of soil in which it is grown, and economic and social uses. Thus, gene banks generate and hold essential information about their collections, critical to carrying out seed distribution functions.

Gene banks and smallholder farmers

As described, much of a gene bank's functionality hinges on collaboration with plant breeders and researchers. There are instances, however, of gene banks successfully linking with small-scale farmers. At a global level this has been limited, but there have been examples of withdrawals, such as seed deemed at risk from the ongoing war in Syria. Gathered through rescue missions, the seed was stored in the Svalbard Global Seed Vault until samples were withdrawn to be multiplied in Lebanon and Morocco. Subsequently,

(Continued)

Box E Continued

larger quantities of the seed were returned to the vault. Another example is the Andean community, which received potato accessions from the Svalbard vault, because the community had lost the variety.

Data from CGIAR-managed gene banks indicate that there is little (8 per cent) direct restoration of seed diversity to farmers and that those collections that do reach farmers are likely to be of limited genetic diversity and supplied at a cost, a factor which may reduce their accessibility.

As a rule, regional gene banks do not link directly with small-scale farmers, although this may be facilitated indirectly via national gene banks and their connections with community seed banks.

At a national level, there is much more scope for gene banks to link with small-scale farmers. This largely relates to the promotion of the conservation of farmers' varieties and crop wild relative populations. Around the world, these efforts have paid off in times of disaster such as the 2015 earthquakes in Nepal and the aftermath of Cyclones Idai and Kenneth in Zimbabwe, Mozambique, and Malawi in 2019. In these cases, national gene banks were directly involved in restoring and rebuilding lost crop diversity in smallholder farming systems.

Links between national gene banks and smallholder farmers are frequently brokered via community seed banks, as illustrated in an example from Zimbabwe, where the NGO Community Technology Development Organization arranged for farmers associated with their seed bank to access and deposit seeds in gene banks, including the Svalbard Global Seed Vault. The community seed bank concept is multipronged, as it also promotes on-farm seed conservation and local seed exchanges. In Zimbabwe, community seed banks are linked with seed fairs, farmer field schools, participatory plant breeding, and participatory variety selection: activities that promote the protection of locally adapted seed varieties and support the continued diffusion of seeds in local seed networks.

Challenges facing interaction between gene banks and smallholder farmers

While gene banks play an important role in conserving and distributing plant genetic resources, a number of critical challenges, described below, limit their utilization by smallholder farmers.

Bureaucracy and cost constrain linkages between smallholder farmers and gene banks

On paper it is possible for smallholder farmers to connect with large gene banks such as those operating at regional level, but in reality this is a complex and time-consuming procedure with associated costs. Therefore it may not suit smallholder farmers who need germplasm urgently for the next planting season, or do not have cash reserves to spend on seed. For instance, a farmer from Malawi may want to access seed from the SADC Plant Genetic Resources Centre in Lusaka. The request would have to be made via a local community seed bank, which would approach the national gene bank to request a copy from the regional gene bank. Acquiring the germplasm would have to follow the same bureaucratic procedure in reverse.

The genetic quality of conserved collections decreases over time

Gene banks withdraw samples at intervals during the seed storage cycle to test germination rates and to regenerate and multiply seeds. This process leads to the inbreeding of conserved collections and an inferior genetic make-up. Crop generations of a gene bank collection are thus likely to be of lower genetic quality than the original collection, with negative outcomes for smallholder farmers.

(Continued)

Box E Continued

Low quantities of seed are not smallholder-friendly

Gene banks keep limited seed quantities, and while this may suit researchers and plant breeders, it is not appropriate for production by smallholder farmers. To overcome this practical barrier, gene banks need to undertake seed multiplication. However, it can take several seasons before the seeds are plentiful enough for smallholder farmers to use.

During the seed restitution efforts following Cyclones Idai and Kenneth and the earthquakes in Nepal, it was noted that quantities of seeds stored in gene banks were too small to effectively respond to disasters of such magnitude. Furthermore, the process of multiplying seeds was laborious and costly.

Gene banks are designed to deliver improved varieties, but these are more useful to plant breeders than smallholder farmers

Crop breeders and researchers are mainly interested in elite germplasm for crop development, a need which gene bank collections are designed to fulfil. Many crop varieties developed and released in commercial markets or formal seed systems in sub-Saharan Africa come from materials selected from gene banks, in particular, the CGIAR centres. This underscores the fact that gene banks tend to benefit researchers, crop breeders, and the development of formal seed systems while smallholders receive less attention. In some cases, improved crop varieties are developed for smallholder farmers, but multiple factors such as access, availability, and cost are barriers which limit their benefits.

Gene bank storage 'freezes' seed evolution

The long-term conservation of gene bank collections in freezers arrests the evolutionary pressures that are exerted on seeds in continuous cultivation. Releasing such collections to smallholder farmers risks introducing undesired and weak genes into local seed systems, thereby undermining local seed resources that have been carefully selected and developed by farmers over years.

Conclusion

Gene banks' *in situ* and on-farm conservation activities provide an opportunity to promote and advance smallholder farmers' seed systems. The benefits brought by the increased participation of gene banks in local seed systems are mutual because seeds get to evolve in farmers' fields, and gene banks can continuously update their collections. Community seed banks can play an invaluable role in realizing such benefits.

Several strategies have emerged to improve the linkages between *ex situ* and on-farm genetic diversity, such as farmer field schools, participatory plant breeding and variety selection, emergency seed intervention, seed restoration, integrated seed systems, and community seed banking (Westengen et al., 2017). These strategies deserve further investigation, because the diversity maintained by smallholder farmers may well hold the key to the development of future climate-adapted crops.

Note

1. Characterization involves recording and compiling data on important characteristics that distinguish one species from another (Bioversity International, 2007).

References

Bioversity International (2007) *Guidelines for the Development of Crop Descriptor Lists*, Bioversity Technical Bulletin Series 12, Bioversity International, Rome, Italy.
Engelmann, F. and Engels, J.M.M. (2002) 'Technologies and strategies for ex situ conservation', in J.M.M Engels, V. Ramanatha Rao, A.H.D. Brown and M.T. Jackson (eds), *Managing Plant Genetic Diversity*, pp. 89–103, CABI Publishing, Wallingford.

(Continued)

Box E Continued

Frankel, O.H. and Hawkes, J.G. (1975) *Crop Genetic Resources for Today and Tomorrow*, International Biological Programme 2, Cambridge University Press, Cambridge.

Halewood, M., Jamora, N., Noriega, I.L., Anglin, N.L., Wenzl, P., Payne, T., Ndjiondjop, M.N., Guarino, L., Kumar, P., Yazbek, M., Muchugi, A., Azevedo, V., Tchamba, M., Jones, C.S., Venuprasad, R., Roux, N., Rojas, E. and Lusty, C. (2020) 'Germplasm acquisition and distribution by CGIAR genebanks', *Plants* 9(10): 1296 <https://doi.org/10.3390/plants9101296>.

Hawkes, J.G., Maxted, N. and Ford-Lloyd, B.V. (2000) *The Ex Situ Conservation of Plant Genetic Resources*, Springer Science & Business Media, Dordrecht <https://doi.org/10.1007/978-94-011-4136-9>.

Khoury, C.K., Laliberté, B. and Guarino, L. (2010) 'Trends in *ex situ* conservation of plant genetic resources: a review of global crop and regional conservation strategies', *Genetic Resources and Crop Evolution* 57: 625–639 <https://doi.org/10.1007/s10722-010-9534-z>.

Sthapit, B. (2012) 'Emerging theory and practice: community seed banks, seed system resilience and food security', in P. Shrestha, R. Vernooy and P. Chaudhary (eds), *Community Seed Banks in Nepal: Past, Present, Future, Proceedings of a National Workshop, 14–15 June 2012, Pokhara, Nepal*, pp. 16–40, Local Initiatives for Biodiversity, Research and Development (LI-BIRD) Pokhara, Kaski, Nepal.

Westengen, O.T., Hunduma, T. and Skarbø, K. (2017) *From Genebanks to Farmers: A Study of Approaches to Introduce Genebank Material to Farmers' Seed Systems*, Noragric Report No. 80, Department of International Environment and Development Studies, Noragric, Faculty of Landscape and Society, Norwegian University of Life Sciences, Aas, Norway <http://dx.doi.org/10.13140/RG.2.2.25699.76327>.

PART 2

Privatizing profit, socializing costs

CHAPTER 8

Corporate expansion in African seed systems: implications for agricultural biodiversity and food sovereignty

Stephen Greenberg

Global context of corporate concentration in seed

Commercial seed markets have, over the past four decades, morphed from a base of small-scale or family-owned businesses to large global multinational corporations, integrating on a biotechnology-seed-agrochemical[1] techno-logical platform with extremely high barriers to entry. A first wave of concentration occurred in the late 1970s to mid-1980s, primarily in the hands of the petrochemical sector, which sought to combine its fossil fuel interests – linked to synthetic fertilizers as well as industrial pesticides – into a package with seeds. Later, pharmaceutical companies took the lead. These manufactured pesticides and considered a similar strategy of combining seed and agrochemicals into a package (Mayer and Runyon, 2016). In 1980 the extension of intellectual property rights to living organisms in the US, and consequently elsewhere, led to massive investment in biotechnology, resulting in the integration of these three elements of the currently dominant platform. This was linked to the rise of the leveraging of proprietary knowledge as the main source of wealth creation in the capitalist economy (Stephan et al., 2006: 108).

The result was the commercialization of herbicide-tolerant and insect-resistant genetically modified (GM) crops in 1995. Most significant was Monsanto's package of Roundup Ready (herbicide-tolerant) GM seed and glyphosate herbicide that set the trend for the industry. This boosted the earlier declining profitability of agrochemicals, and stimulated a merging of these sectors on a common technological platform, leading to a second wave of consolidation in the 1990s. Seed companies were acquired as a source of genetic materials under private ownership. Between 1996 and 2013, the 10 largest seed corporations absorbed nearly 200 seed companies and purchased equity stakes in dozens more (Howard, 2016: 112).

By 2015, three of the largest corporations (Monsanto, DuPont Pioneer, and Syngenta) controlled 55 per cent of the global commercial seed market and another three (Syngenta, Bayer CropScience, and BASF) 51 per cent of the agrochemicals market (ETC Group, 2015) (Figure 8.1). The 'Big Six' mega

BASF
US$7.2 bn, 7%

Bayer-Monsanto
US$29 bn, 30%

ChemChina
US$18.3 bn, 19%

Dow-DuPont
US$18.6 bn, 19%

Other
US$24 bn, 25%

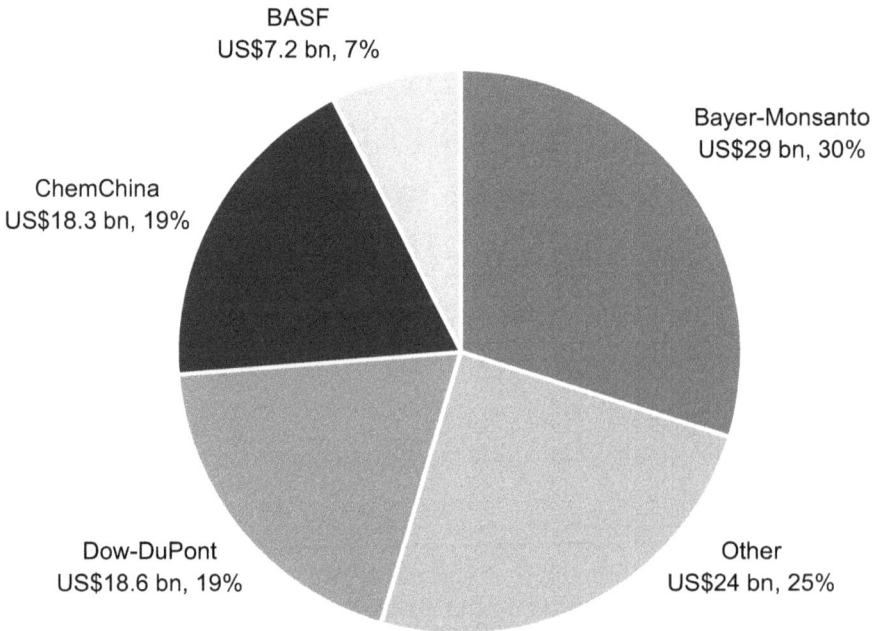

Figure 8.1 Seed and agrochemical markets at the time of the third wave mergers
Source: ETC Group, 2016: 6

corporations, namely, BASF, Bayer, Dow, DuPont, Monsanto, and Syngenta, together controlled 75 per cent of the global agrochemicals market, 63 per cent of the commercial seed market, and over 75 per cent of all private sector research and development (R&D) in the sector. In 2013 the combined agricultural R&D budget of the Big Six was 20 times bigger than the total expenditure of CGIAR on crop-oriented research and breeding, including gene bank conservation (ETC Group, 2015: 4). The 'four-firm' concentration ratio (CR4), a reputable measure of economic concentration, assumes an oligopoly[2] if the combined market share of the four largest firms in a given industry is over 40 per cent (see also Bonny, 2017; Clapp, 2018).

These constant acquisitions led to a third wave of consolidation in 2017, with three mega-mergers (Figure 8.2). Regulatory agencies globally approved two acquisitions – Bayer-Monsanto for around US$66 bn and ChemChina-Syngenta for US$43 bn – and a 'merger of equals' between Dow and DuPont. The main requirement of the acquisitions and merger was some limited divestiture of assets, of which BASF was the main beneficiary. In 2019 DowDuPont was broken into three separate companies: Dow (industrial chemicals for intermediate uses), DuPont de Nemours (finished products, growth sectors), and Corteva Agriscience (biotech-seed-agrochemicals). Corteva had a market value of US$21 bn in mid-2019 (Bromels, 2019). In 2021 a merger of ChemChina and Sinochem, a far larger Chinese chemical

TIMELINE OF MERGERS AND ACQUISITIONS

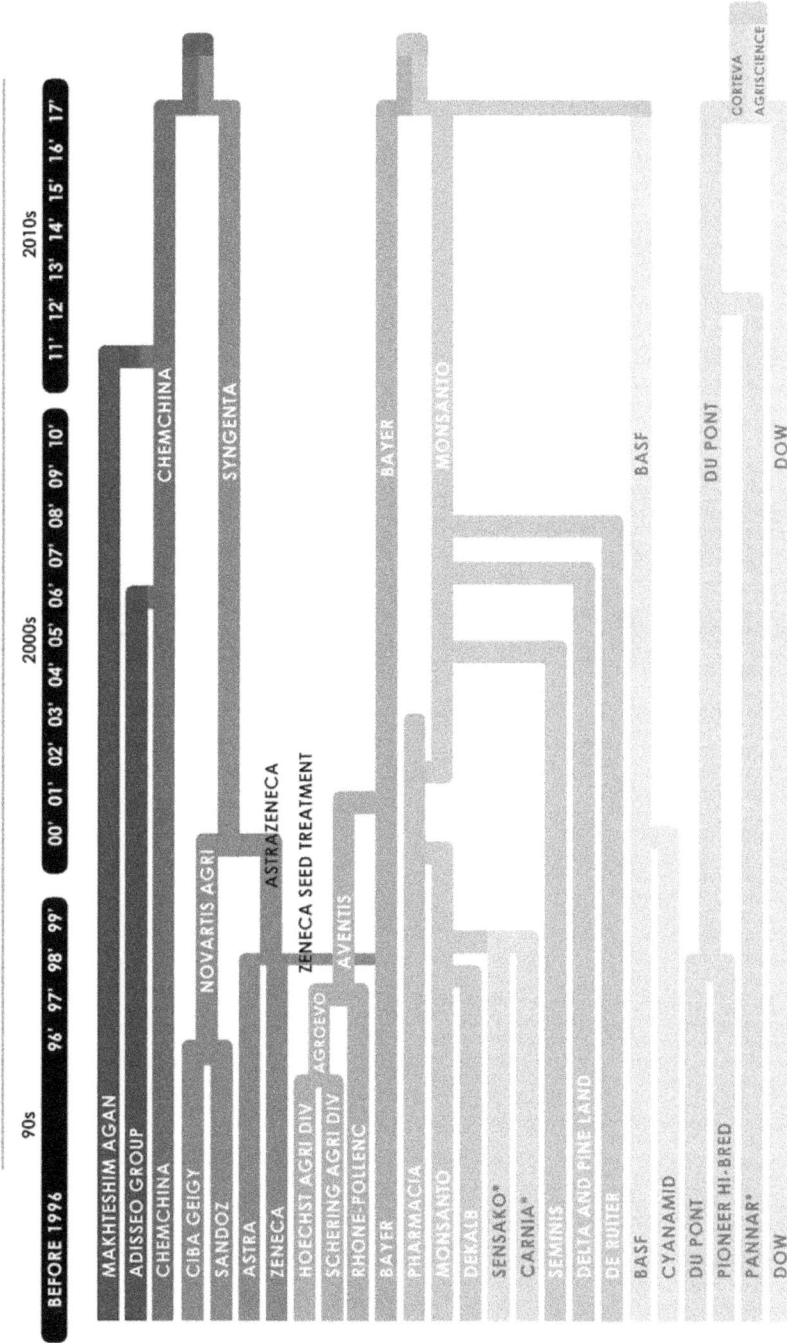

Figure 8.2 Timeline of major mergers and acquisitions, from 1990s to 2017
Source: ACB, 2017

conglomerate, was approved by the Chinese government. The two corporations became wholly owned subsidiaries of a new holding company operated by China's State-owned Assets Supervision and Administration Commission (Bloomberg, 2021).

Corporate concentration in agricultural inputs is linked to wider interconnected developments in the late capitalist global economy, in particular the rise of intellectual property as a profit centre, financialization, and a range of new technologies from genomics to agricultural production in the field.

Intellectual property

Intellectual property (IP) underpins the contemporary capitalist mode of accumulation as the latest area of capitalist enclosure and value extraction (Stephan et al., 2006). Africa remains 'trapped' in the first two 'dynasties' (land and labour), long ago characterized by commodification and diminished returns (Stephan et al., 2006: 115). The cheapening of land and natural resources, labour, food, and energy is an integral part of the capitalist world praxis to construct a 'Cheap Nature' which is central to the endless accumulation of capital (Moore, 2017). The cheapening is twofold: reducing the costs of working for capital, directly and indirectly; and cheapening in the sense of treating as unworthy of dignity and respect (Moore, 2017: 7). The value of natural resources and the ongoing activities of smallholder producers and Indigenous communities in maintaining and adapting biodiversity for human use is systematically denigrated as backward and obsolete. These resources are then quantified based on capitalist metrics, commodified, and extracted at low cost to the acquiring party.

Multinational investment in agricultural inputs is premised on wealth extraction on the basis of proprietary IP rather than the material resource itself, whether seed or agrochemicals. IP ownership does not necessarily mean exclusive use by the owner, only exclusive use rights. Indeed, the licensing out of IP for royalties is a corporate profit centre. Examples of this are GM traits or processes, or agrochemical formulations. In South Africa, Monsanto owns the vast majority of maize GM traits, which are then licensed to its competitors such as Pioneer (Corteva) for a fee. These traits and the associated processes are then used to produce other patented or IP-protected GM varieties. The same is true for agrochemicals. Patents and the related licensing agreements have helped integrated biotech-seed-agrochemical corporations establish their market dominance, not just over the patented product but also over complementary products through extensive licensing agreements.

The dominant biotech-seed-agrochemicals corporations have essentially 'formed ... a cartel with a web of cross-licensing agreements' that 'secure the dominance of [a] few corporate giants' (Wager, 2016). Prolific cross-licensing of genetic traits makes it harder for new and smaller companies to enter the market because they have to pay multiple and often expensive licensing fees. Smaller companies that cannot afford licensing fees often opt to be bought out, thereby producing further concentration in germplasm ownership. Bryant et al. (2016) refer to this web of cross-licensing as 'non-merger mergers', in that they effectively enable competitors to lock up markets without the regulatory approval required for formal mergers.

Finance and ownership

Vast financial interests lie behind corporate concentration in the biotech-seed-agrochemicals sector. Even where commercial farmers and some share-holders have expressed concern about the mergers, especially their potential impacts on input prices and the decline in innovation and choice, the mergers have been pushed through (Bjerga and McLaughlin, 2016; Farm Aid, 2018). A closer look at the shareholders behind the merging contestants reveals that the world's genetic resources rest in the hands of global financial firms. BlackRock and other asset management firms such as Vanguard and State Street are among the largest shareholders in every one of the biotech-seed multinationals (BlackRock, n.d.; Roumeliotis and Stone, 2015; MultiWatch, 2016: 23; Werner, 2016; MarketScreener, 2017).

Corporate expansion and concentration in seed systems in sub-Saharan Africa

Contextual background

Most seed in Africa is still maintained, adapted, and shared directly by farmers in their fields, networks, and systems. According to recent research (Sperling et al., 2021) drawing on 10,209 transactions in Africa, 36 per cent of seed for all crops was sourced from farmers' own stocks, and another 16 per cent from social networks. Another 30 per cent, sourced from local markets, was in turn sourced from many places including local farmers themselves. A total of 82 per cent of seed thus came from these three sources. In contrast, 2 per cent of seed was from agro-dealers and just over 6 per cent from government, including input subsidy programmes.

Despite the historical and ongoing role of African farmers in maintaining, adapting, and employing agricultural biodiversity, this role was not formally recognized when their countries gained independence. Most countries adopted modernization approaches, farmers' own practices were disregarded, and there was a strong emphasis on the role of the state. The public sector was tasked with agricultural R&D. African governments worked with the institutes that eventually formed CGIAR to establish state-owned breeding and seed production, with distribution also through the state (Buhler et al., 2002; Lynam, 2011).

This system was effective in providing some R&D and farmer support, but started becoming costly, especially in the context of the global debt crisis from the early 1970s. Africa was caught in a debt trap. International Monetary

Fund and World Bank structural adjustment programmes were imposed as a condition for receiving financial aid, with a focus on investing in profitable ventures to allow debt repayment. Conditions included trade and financial liberalization, deregulation, privatization, and export orientation, including for R&D, seed production, extension, market support, and financing (Cheru, 1989; Klein, 2007; Lynam, 2011).

Despite privatization, the private sector only ventured into lucrative crops. These included crops with regional economies of scale such as maize and, more recently, soya. The private sector also operated 'closed' value chains for other lucrative crops such as cotton, coffee, and tobacco, where companies organize the whole chain, including inputs, production methods, and outputs. As a result of the failure of the private sector to fill the space opened by privatization, national and CGIAR agricultural research institutes remain an important source of new seed and plant varieties in Africa, often working in consortia with private breeding companies (Access to Seeds Foundation, 2019: 38). While public sector breeders were able to produce potentially useful varieties for a range of agroecological contexts, they lost the capacity to multiply and get these varieties to farmers.

Structural adjustment programmes led to a reduction in state support to farmers, resulting in some critical services (e.g. extension and technical support) being inadequately provided. The 1990s are considered the lost decade for African agriculture, as there was limited investment, and by and large agriculture was neglected and forgotten. However, the early 2000s saw renewed interest in African agriculture arising from a global raw materials commodities boom, linked to rapid global expansion, especially in China. According to Moore (2015: 226), the costs of the 'Four Cheaps' – labour, food, energy, and raw materials – rose simultaneously from 2003, producing a 'signal crisis' indicating the exhaustion of the accumulation regime based on the low cost of these inputs. The main physical commodity frontiers (spaces for new expansion) were exhausted, while the mass of capital continued to rise, requiring more and more fuel. Efforts to arrest this falling ecological surplus, such as agricultural biotechnology, have been insufficient to improve intrinsic yields (Moore, 2015: 270).

The new African Green Revolution

The original Green Revolution was in the North, based on mechanization, synthetic agrochemicals and fertilizers, and plant breeding. Its extension to the South followed in the 1950s and 1960s, especially in Asia and to some extent Latin America, which massively increased yields of staple crops (Conway, 1997; Otsuka and Muraoka, 2017). The Green Revolution also generated significant social and ecological problems and ultimately did not respond effectively to the challenge of persistent high global food insecurity (George, 1976; Holt-Giméenez et al., 2006; Kumbamu, 2020). This second green revolution failed in Africa for many reasons, including poor adaptation of varieties to diverse socio-ecological and production contexts, an over-reliance on a narrow technological fix without considering political and market systems, social dynamics, and top-down development processes.

From the mid-2000s, a renewed push to extract value from labour and natural resources took the form of the so-called new African Green Revolution. Staples include privatization, private investment protection, the ability to repatriate profits, low or no taxes, conversion of land to private title with an effective institutional structure that can uphold private ownership rights, government subsidies, and the use of public resources to support opportunities for private profit.

This new thrust has multiple sponsors. Private philanthropic institutions the Bill & Melinda Gates Foundation and the Rockefeller Foundation jointly launched the Alliance for a Green Revolution in Africa (AGRA) in 2006. The Rockefeller Foundation has a long history of sponsoring Green Revolution activities through CGIAR. Many governments, multilateral institutions such as the Food and Agriculture Organization of the United Nations (FAO), and other philanthropies and diverse donors have joined them in numerous overlapping partnerships, programmes, and networks (Brooks et al., 2009; Morvaridi, 2012; Ignatova, 2017; Bergius and Buseth, 2019; Kumbamu, 2020). These include USAID Feed the Future, as well as British, European, Chinese, Indian, and other Green Revolution programmes and investments in Africa. The African Union's Comprehensive Africa Agricultural Development Programme provides a cloak of legitimacy, with a profound impact on the structure of governance and institutions in agriculture at national and regional levels.

Institutional arrangements are modelled on multilateral public–private partnerships with other donor support and a strong corporate thrust. In essence, this is state-facilitated expansion of global multinationals. Diverse activities include institution building; legal and policy work (e.g. IP and plant variety protection) including regional and continental harmonization; education, training, and technical capacity building; commercial seed breeding, production, and distribution; markets and value chains; and infrastructure development.

Corporate expansion in African seed systems

There was a sharp rise in the number of private seed companies in Africa from the 1990s following deregulation and privatization. Results of a nine-country survey showed an increase from 33 companies in 2002 to 332 in 2012 in the selected countries (ACB, 2015: 22). Five West African countries (Burkina Faso, Ghana, Mali, Niger, Nigeria) all had single monopoly state-owned seed companies before 2007, but by 2014 had a combined 114 seed enterprises in operation, with Nigeria alone accounting for almost two-thirds of these. In East Africa, Uganda, Tanzania, and Kenya had around 40 seed enterprises in operation, rising to 179 in 2014 (many of which were vegetable seed importers in Kenya) (AGRA, 2017: 63).

Table 8.1 Layers of companies in the private seed industry

Segment	Examples of companies
Farmer seed system (70–80% of overall supply)	Farm level and networked seed saving, adaptation, production, and sharing
Small and medium enterprises	Hundreds, for example NAFASO (Burkina Faso), Faso Kaba (Mali), Alheri (Niger), Kamano, Kamasika Seed Growers' Association, Afriseed (Zambia), Meru Agro (Tanzania), Dryland (Kenya), Pearl (Uganda), Funwe Farm, Peacock (Malawi), Phoenix, Nzara Yapera (Mozambique), Champion (Zimbabwe)
Large national companies	Zamseed, Tanseed, Kenya Seed Co., Value (Nigeria), NASECO, Victoria (Uganda), Demeter (Malawi)
Regional companies	Seed Co, Klein Karoo, Hygrotech, Capstone, Starke Ayres
Second-tier multinationals	Vilmorin, East-West, Rijk Zwaan
Top-tier multinationals	Corteva, Bayer-Monsanto, ChemChina-Syngenta

The private seed industry consists of numerous layers of companies (Table 8.1), including the largest global multinationals; a secondary group of European and Asian multinationals, many with a focus on horticulture; African regional multinationals; large national seed companies (a mix of state-owned, formerly state-owned, and private); and a group of newly emerging local seed enterprises, many of which have received support from AGRA. More recently a new category of farmer-owned seed enterprises, operating in commercial markets, has emerged, as described below.

Even though the commercial share of overall African seed supply is relatively small, at about 20 per cent to 30 per cent, the major biotech-seed-agrochemical corporations and other global, regional, and domestic companies are all present in selected markets in Africa, in alignment with wider Green Revolution interventions and programmes. Markets are usually only defined in terms of large-scale commercial seed markets in selected crops. Market figures do not incorporate farmer seed because this is not traded through formally constituted commercial markets, which do not register any demand for these seeds. High levels of concentration characterize national commercial seed markets in the main crops (see TASAI, 2018 for more details), with Ethiopia, Kenya, Nigeria, and Tanzania standing out.

Multinational corporations dominate in major field crops such as maize and soya, and in horticulture. Corteva Agriscience, Monsanto, and Syngenta all have a physical presence in many countries in Africa, having acquired several domestic seed companies in the past few decades.

Aside from South Africa, which permits the commercial cultivation of GM maize, soya, and cotton, 'an increasing number of countries are allowing for further research into GM food crop varieties as well as cotton, notably Nigeria (cowpea, rice, sorghum, cassava, maize), Ghana (cowpea, rice), Burkina Faso (cowpea), Kenya (banana, rice, maize, cassava, sweet potato), Uganda (banana, maize, rice, potato, cassava), Tanzania (maize), Mozambique (maize) and Malawi (banana, soybean, cowpea)' (Access to Seeds Foundation, 2019: 42).

In the second tier of multinationals, French seed company Vilmorin & Cie, the world's fourth largest by market share, has also expanded its operations on the continent, acquiring South African company Link Seed in 2012, and an equity stake in Zimbabwe's Seed Co in 2014, which stood at 29 per cent in 2018 (Seed Co, 2019: 67).

East-West Seed conducts R&D and produces vegetable and some flower seeds. East-West Seed started its expansion into Africa through Tanzania, forming a company and initiating the Afrisem breeding programme with Rijk Zwaan in 2008, later expanding to other countries such as Nigeria (De Ocampo-De Guzman, 2022). Other second-tier multinationals focusing on tropical vegetable seed are Rijk Zwaan and Technisem from France.

The next layer comprises regional and large domestic seed enterprises. There has been some expansion across national boundaries into the region in the past 10 years. Zimbabwe's Seed Co is the largest indigenous regional seed company in sub-Saharan Africa (Seed Co, 2019). It started out as a private company and is listed on the stock exchange. Seed Co International was successfully launched on the Botswana Stock Exchange in 2018, incorporating all non-Zimbabwean operations. It focuses mainly on hybrid maize seed, but also includes wheat, soya, and, to a lesser extent, barley, sorghum, and groundnut seeds. Seed Co has a gene bank which maintains over 5,000 proprietary germplasm accessions in field crops and vegetables. It has a footprint in 15 African countries, with active operations in 7, mostly in Southern and East Africa, and some expansion into West Africa, with support from AGRA. Government farm input subsidy programmes (FISPs) in Zambia, Zimbabwe, and Malawi are among its key markets. It has planned market acquisitions and capital expenditure programmes in Tanzania, Nigeria, Zambia, Botswana, and Kenya. Seed Co had a turnover of US$73 m in 2019. This is about 2 per cent of the estimated value of the commercial seed market in Africa, as indicated above. This suggests that first- and second-tier multinationals occupy the bulk of the commercial market, not only in breeding but also in the production of seed.

Joining Seed Co at a regional level are a group of South African companies including Klein Karoo, Hygrotech, Capstone Seeds, and Starke Ayres. Klein Karoo is a subsidiary of Zaad Holdings, which is owned by Zeder Investments, the agricultural investment arm of PSG Group (owned by the Mouton family) in South Africa. Zaad also owns Agricol, which could also expand into the region in future. Klein Karoo produces field crop, pasture, and vegetable seed for African and global markets, with offices and/or facilities around the world, including South Africa and Zimbabwe (Zaad Holdings, n. d.). Hygrotech has expanded into Zambia and East Africa, mainly in vegetable seed. Hygrotech's FertAgChem division was launched in 2011 to produce and supply fertilizer and agrochemicals.

Seed Co was one of just five African companies with independent breeding capabilities in 2015 (ACB, 2015). Of the other four, Pannar and MRI have been taken over by the dominant multinationals, while Zamseed and Kenya Seed Co. are in the category of large domestic enterprises with national scope and some potential to expand. Zamseed was formerly the monopoly state-owned seed producer in Zambia. Following liberalization in the early 1990s, it was privatized and competed against foreign entrants to the commercial seed market. Kenya Seed Co. is a state-owned corporation operating in Kenya and Uganda. Tanseed is another privatized former state-owned enterprise that continues to play an important national seed production role. In Ethiopia, the public sector still plays the main role in seed breeding, production, and distribution, through (sub)regional state-owned seed enterprises.

Other domestic companies with national scope started out as private enterprises, such as Demeter Seed in Malawi, Victoria Seeds and NASECO in Uganda, East African Seed in Kenya, and Value Seeds in Nigeria. These companies focus on field crops, again mostly maize and soya but with some diversification in different places into millet, sorghum, wheat, rice, cowpea, groundnut, sunflower, and other seeds, as well as vegetatively propagated crops such as sweet potato and cassava. In some countries, especially in Southern Africa, these companies rely on the FISPs as an important market. As state or private companies from this tier establish a regional presence, they are liable to become prime targets for acquisition by larger corporations.

Results of a survey released in 2019 of 99 production sites of mainly large and medium companies in Southern, East and West Africa showed that 39 per cent involved smallholder farmers in seed production (Access to Seeds Foundation, 2019: 39).

The next layer consists of new, smaller seed enterprises that emerged after liberalization. Many of these received significant support from AGRA through its Program for Africa's Seed Systems (PASS) from 2006. PASS had four pillars: tertiary education (plant breeding); R&D and the release of new crop varieties; the creation of private seed companies for the production and sale of seed; and agro-dealers to distribute Green Revolution inputs. In pillar two, 600 new varieties were registered through AGRA-sponsored activities as from 2006, with grants to national breeding programmes in 18 countries. Of these new varieties, 53 per cent were maize and 36 per cent beans, cassava, sorghum, wheat, sunflower, and groundnuts. Other crops with fewer releases were soya, cowpea, pigeon pea, rice, lablab, and teff (AGRA, 2017: 176). In the third pillar, AGRA supported 112 private seed enterprises, 7 public sector entities, 10 agricultural research systems, and 14 farmer associations, cooperatives, and NGOs in 18 countries (AGRA, 2017: 64). Most grant beneficiaries were start-up seed enterprises that had been in existence for less than two years. These enterprises were given a one-off grant spanning two years to help them increase seed production, enhance farmer awareness, and connect with agro-dealer networks for dissemination (AGRA, 2017: 67). Total production was around

600,000 tonnes of seed from 2007 to 2016. Maize, wheat, and rice predominated, but a range of other crops were also produced.

Farmer-owned enterprises are another type of small and medium enterprise, for instance Champion Farmer Seed Cooperative in Zimbabwe. This commercial enterprise was established in 2016 with the support of the Community Technology Development Trust. Its initial focus was on small grain crops (sorghum and pearl millet), but more recently it has diversified into cowpeas, sugar beans, maize (open-pollinated and hybrid varieties) and groundnuts. Smallholder farmers produce the seed and own the company. They started with 90 farmers and in 2018–19 they were working with about 700 farmers, with a target of 170 tonnes of seed (Personal interview with W. Zonge, Sales and marketing agronomist, Champion Seeds, 23 May 2019, Harare).

Social, economic, and ecological implications

This section briefly reflects on three interrelated areas of concern, namely agricultural biodiversity and dietary and nutrition diversity; production systems; and the privatization of public policy.

Agricultural biodiversity and dietary diversity

The industry argument is that the formal system brings in diversity with new varieties and combinations. While it may be true that the number of registered varieties of selected crops increases as a result of commercial activity, this does not consider the impacts on:

- varieties or populations that are not registered;
- the narrowing of genetic diversity within the crop because many modern varieties are slight adaptations of one another from a common genetic base;
- narrowing genetic diversity in related crops (e.g. so-called 'neglected and underutilized species').

These deficiencies are a result of a narrow investment and promotion focus on relatively few crops and varieties, driven by a profit motive. Although farmers have developed and cultivated thousands of different and genetically unique populations, today only 150 plant species are widely cultivated, and just 12 provide three-quarters of the world's plant-based food. These 'mega-crops' include rice, wheat, and maize, along with sorghum, millet, potatoes, and sweet potatoes (Fowler and Mooney, 1990).

Hybrids and GM seed exacerbate the tendency towards monocultures at scale, resulting in low plant biodiversity in zones of production. They also bring with them toxic agrochemicals that destroy soil biodiversity in the field and poison water supplies, killing life in the water. The result is genetic erosion and increasing dependence on relatively few plant varieties, with species loss and reduction of diversity, as well as a gradual breakdown of processes that maintain the evolution of diversity (Mahmood et al., 2016; Tsatsakis et al., 2017).

Crop and seed diversity relate directly to dietary diversity as a key component of food and nutrition security. Limited agricultural biodiversity (especially from monocultural production) automatically limits dietary diversity both for producing households and for consumers or users more widely.

Production systems

The modernization project orients production towards a large scale and a focus on monocrops using a technological package of hybrid seed, synthetic fertilizers, and agrochemicals. The impact this has on the agrarian structure is profound. At the top there is an expansion of a relatively thin layer of commercializing producers. They are encouraged by market forces and policy interventions to expand their area under production, leading to land encroachment and dispossession (Dawson et al., 2016). A small proportion of the newly landless peasantry are converted into mostly low paid labourers in commercial production (Smalley, 2013). Others are forced onto other land, creating more crowded conditions for those remaining, leading to out-migration to urban areas with limited opportunities for stable employment or income in the context of a huge oversupply of manual labour. Long-term consequences include the disintegration of farming households and communities and the destruction of the peasantry, with limited urban employment to absorb the dispossessed (as has happened in Europe and elsewhere) (Mburu, 1986; Ozden and Enwere, 2012; Dawson et al., 2016). Industrial and manufacturing markets are already globally saturated, and subject to highly uneven terms of trade in global trade regimes overseen by the World Trade Organization, for example tariff-free imports (Madeley, 2002; Tilzey, 2006; FAO, 2015).

Farming households and communities are further locked into the cash economy. Cash is required for transport, education, health care, and energy, pushing farmers to produce bulk 'commodity' crops for commercial markets. Upon adoption of this model, agricultural inputs are added as a cost requiring cash, with a price now on seed, soil fertility, and pest management. The noose gets tighter around the necks of the African peasantry and smallholder farmers.

Privatization of public policy

Corporate influence over national governments has been ascendant since the start of the neoliberal era and structural adjustment across Africa as a whole, even if uneven at country level. Deregulation and privatization have facilitated rising corporate power. Neoliberal states in the US, Europe, and elsewhere worked in partnership with private corporations and philanthropies to engage with African states on policies and laws. A strong emphasis of the new Green Revolution is on the institutional, technical, and policy dimensions, including the introduction of restrictive commercial plant variety protection and seed policies and laws (Correa et al., 2015), and the channelling of public sector and donor resources to subsidizing Green Revolution inputs (ACB, 2016; Jayne et al., 2018).

Alternatives and potential solutions

Alternatives arise from popular activity in Africa under the umbrella of food sovereignty (AFSA, 2016). Food sovereignty brings the political dimension into technical processes of agrifood production, allocation, and distribution. A fundamental pillar is democratic control and governance throughout the food system. Food sovereignty is rooted in democratic producer organization, and realized through action learning processes driven by producers and including multiple stakeholders such as intermediate and end users, and social and technical systems support.

Agroecology is the technical, material basis for food sovereignty, grounded in ecologically and socially sustainable production. There is a huge literature on agroecology, covering many diverse techniques, organizational forms, results, values, and strategies (e.g. Wibbelman et al., 2013; Cacho et al., 2018; Bergez et al., 2019).

Farmer seed systems can be considered an integral component of agroecology, alongside soil and water conservation and management, and pest management (Almekinders and Louwaars, 1999). 'Farmer seed is defined by a process of production, and is conserved and multiplied by farmers in the same field as it is cultivated … This seed constitutes populations, not varieties. Evolving selection allows us to choose changing characteristics every year. We must characterise the farmer seed. Where does the seed come from, what are its origins, and which are the parents' (ACB, 2019).[3]

Conclusion

This chapter has highlighted the increasing concentration of economic power and control in the hands of a few multinational corporations throughout the food system, including seed and other agricultural input supply. Development has been privatized, with states generally aligning themselves towards (subsidized) markets for delivery. Farmer seed systems have been neglected even though they continue to generate the majority of seed on the continent. The Green Revolution is the dominant development discourse in African agriculture today, and reinforces these dual trends of concentration and marginalization. The Green Revolution has wide-ranging impacts and implications for agrarian structure, social inequality, ecological damage (of water, soil, or biodiversity), dietary diversity, democracy, and accountability.

An alternative discourse and practice have arisen in response, based on food sovereignty and agroecology, with farmer seed systems as an integral part of this movement.

The private capture of state systems is a significant challenge to resource-poor farmers and farming communities, since external technical and other support may be required at times. If this does not come from the state, it usually comes from the private sector with strings attached, pushing farmers into adopting defined practices and inputs. The state is a contradictory entity. Part of its original purpose was to organize the smooth functioning of society, through the provision of services, infrastructure, and so on. But we now exist within the late capitalist, corporate, profit-making context. Liberal representative democracy appears to have exhausted its historical role in recent years, with financial power purchasing political power, from the US down to the smallest and weakest state, with few exceptions. Corruption, secrecy, and deceit are the norm of political behaviour. However, states are not monolithic. Individuals and even units may have a progressive agenda and some authority to define programming.

Food sovereignty and agroecology movements could benefit from seeking out and working with such people within the state. There are multiple roles for the state and public sector, including participatory R&D and extension, the facilitation of farmer-to-farmer networks, training, communications, multi-stakeholder planning, and budgets to support. There are vast public and donor resources available, and demands should still be made of them, even if this is not always successful. This is critical to institutionalizing programmes, rather than relying on temporary, donor-funded NGO projects to support the development of farmer seed systems, agroecology, and food sovereignty. At the same time, autonomous action beyond the state is required to secure a material basis for an alternative, even if the state does not contribute – or even obstructs. Examples of such actions are already under way, such as local seed-saving networks; revival, production, and use of indigenous and underutilized crops; producers linking with users for diverse, healthy food production; and seed activities connecting with wider movements for agroecology and social justice. As part of a systematic response to deepening inequality and dispossession, it is also necessary to resist encroachment upon, commodification, and privatization of seed systems by corporations.

Notes

1. Here 'agrochemicals' refers to pesticides (herbicides, insecticides, and fungicides) and plant-growth regulators. Although synthetic fertilizers can also be categorized as agrochemicals, that is a separate market and is not dealt with here.
2. In an oligopoly, a market or industry is dominated by a small number of large sellers. Oligopolies can lead to collusion, less choice, and higher prices for consumers.

3. Comment by Guy Kastler, Confédération Paysanne/La Via Campesina, at an ACB workshop on quality controls in farmer seed systems in Africa, Zanzibar, 21–23 August 2019.

References

Access to Seeds Foundation (2019) *Access to Seeds Index 2019: Synthesis Report* <www.accesstoseeds.org/app/uploads/2019/06/Access-to-Seeds-2019-Index-Synthesis-Report.pdf>.

African Centre for Biodiversity (ACB) (2015) *The Expansion of the Commercial Seed Sector in sub-Saharan Africa: Major Players, Key Issues and Trends*, ACB, Johannesburg.

ACB (2016) *Farm Input Subsidy Programmes (FISPs): A Benefit For, or a Betrayal Of, SADC's Small-Scale Farmers?* ACB, Johannesburg.

ACB (2017) *The Three Agricultural Input Mega-Mergers: Grim Reapers of South Africa's Food and Farming Systems*, ACB, Johannesburg.

ACB (2019) *Changing the Discourse, Policy and Practice on Farmer Seed Systems in Africa*, ACB, Johannesburg.

Alliance for Food Sovereignty in Africa (AFSA) (2016) *Agroecology: The Bold Future of Farming in Africa*, AFSA, Addis Ababa.

Alliance for a Green Revolution in Africa (AGRA) (2017) *Seeding an African Green Revolution: The PASS Journey*, AGRA, Nairobi <https://agra.org/wp-content/uploads/2018/02/PASS-Book-web.pdf>.

Almekinders, C. and Louwaars, N. (1999) *Farmers' Seed Production: New Approaches and Practices*, Intermediate Technology Publications, London.

Bergez, J.E., Audouin, E. and Therond, O. (eds) (2019) *Agroecological Transitions: From Theory to Practice in Local Participatory Design*, Springer, Cham, Switzerland.

Bergius, M. and Buseth, J. (2019) 'Towards a green modernisation development discourse: the new green revolution in Africa', *Journal of Political Ecology* 26: 57–83 <http://dx.doi.org/10.2458/v26i1.22862>.

Bjerga, A. and McLaughlin, D. (2016) 'Farmers head to D.C. to protest agribusiness consolidation', *Bloomberg*, 9 September <www.bloomberg.com/news/articles/2016-09-09/farmers-head-to-d-c-to-protest-agribusiness-consolidation-wave>.

BlackRock (no date) 'Introduction to BlackRock' [website] <www.blackrock.com/sg/en/introduction-to-blackrock>.

Bloomberg (2021) 'China approves restructuring of Sinochem and ChemChina', 31 March <www.bloomberg.com/news/articles/2021-03-31/china-approves-restructuring-of-sinochem-and-chemchina>.

Bonny, S. (2017) 'Corporate concentration and technological change in the global seed industry', *Sustainability* 9: 1632 <http://dx.doi.org/10.3390/su9091632>.

Bromels, J. (2019) 'After the DowDuPont split: an investor's guide to the 3 new companies', *The Motley Fool*, 17 August <www.fool.com/investing/2019/08/17/after-the-dowdupont-split-an-investors-guide-to-th.aspx>.

Brooks, S., Leach, M., Lucas, H. and Millstone, E. (2009) *Silver Bullets, Grand Challenges and the New Philanthropy*, STEPS Working Paper 24, STEPS Centre, Brighton.

Bryant, H., Maisashvili, A., Outlaw, J. and Richardson, J. (2016) *Effects of Proposed Mergers and Acquisitions among Biotechnology Firms on Seed Prices*, Working Paper 16–2, Research conducted for the Agricultural and Food Policy Center, Department of Agricultural Economics, University of Texas, September <www.afpc.tamu.edu/research/publications/675/WP_16-2.pdf>.

Buhler, W., Morse, S., Beadle, A. and Arthur, E. (2002) *Science, Agriculture and Research: A Compromised Participation?* Earthscan, London.

Cacho, M., Giraldo, O., Aldasoro, M., Morales, H., Ferguson, B., Rosset, P., Khadse, A. and Campos, C. (2018) 'Bringing agroecology to scale: key drivers and emblematic cases', *Agroecology and Sustainable Food Systems* 42(6): 637–665 <http://doi.org/10.1080/21683565.2018.1443313>.

Cheru, F. (1989) *The Silent Revolution in Africa: Debt, Development and Democracy*, Zed Books, London.

Clapp, J. (2018) 'Mega-mergers on the menu: corporate concentration and the politics of sustainability in the global food system', *Global Environmental Politics* 18(2): 12–33 <http://dx.doi.org/10.1162/glep_a_00454>.

Conway, G. (1997) *The Doubly Green Revolution: Food for All in the 21st Century*, Penguin Books, London.

Correa, C., Shashikant, S. and Meienberg, F. (2015) *Plant Variety Protection in Developing Countries – A Tool for Designing a Sui Generis Plant Variety Protection System: An Alternative to UPOV 1991*, APBREBES, Bonn <www.apbrebes.org/files/seeds/ToolEnglishcompleteDez15.pdf>.

Dawson, N., Martin, A. and Sikor, S. (2016) 'Green revolution in sub-Saharan Africa: implications of imposed innovation on the wellbeing of rural smallholders', *World Development* 78: 204–218 <http://dx.doi.org/10.1016/j.worlddev.2015.10.008>.

De Ocampo-De Guzman, L.M. (ed.) (2022) *2021 Our Year in Review: Growing Forward*, East-West Seed, Nonthaburi, Thailand <www.eastwestseed.com/reports>.

ETC Group (2015) *Breaking Bad: Big Ag Mega-Mergers in Play – Dow + DuPont in the Pocket? Next: Demonsanto?*, Communiqué 115, 15 December <www.etcgroup.org/content/breaking-bad-big-ag-mega-mergers-play>.

ETC Group (2016) 'The Monsanto-Bayer tie-up is just one of seven: mega-mergers and big data domination threaten seeds and food security' [News release], 15 September <www.etcgroup.org/content/monsanto-bayer-tie-just-one-seven-mega-mergers-and-big-data-domination-threaten-seeds-food>.

Farm Aid (2018) 'Farmers overwhelmingly oppose Bayer-Monsanto merger' [blog], 8 March <www.farmaid.org/issues/corporate-power/farmers-overwhelmingly-oppose-bayer-monsanto-merger/>.

Food and Agriculture Organization of the United Nations (FAO) (2015) *The State of Agricultural Commodity Markets 2015–16 – Trade and Food Security: Achieving a Better Balance between National Priorities and the Collective Good*, FAO, Rome <www.fao.org/3/i5090e/i5090e.pdf>.

Fowler, C. and Mooney, P. (1990) *Shattering: Food, Politics and the Loss of Genetic Diversity*. University of Arizona Press, Tucson, AZ.

George, S. (1976) *How the Other Half Dies: The Real Reasons for World Hunger*, Penguin Books, Harmondsworth.

Holt-Giméenez, E., Altieri, M. and Rosset, P. (2006) *Ten Reasons why the Rockefeller and the Bill and Melinda Gates Foundations' Alliance for Another Green*

Revolution Will Not Solve the Problems of Poverty and Hunger in sub-Saharan Africa, Food First Policy Brief 12, Food First, Oakland, CA <http://foodfirst. org/wp-content/uploads/2013/12/PB12-Ten-Reasons-Why-AGRA-Will-not-Solve-Poverty-and-Hunger-in-Africa.pdf>.

Howard, P. (2016) *Concentration and Power in the Food System*, Bloomsbury Publishing, London.

Ignatova, J.A. (2017) 'The "philanthropic" gene: bio-capital and the new green revolution in Africa', *Third World Quarterly* 38(10): 2258–2275 <http://doi. org/10.1080/01436597.2017.1322463>.

Jayne, T., Mason, N., Burke, W. and Ariga, J. (2018) 'Taking stock of Africa's second-generation agricultural input subsidy programs', *Food Policy* 75: 1–14 <http://doi.org/10.1016/j.foodpol.2018.01.003>.

Klein, N. (2007) *The Shock Doctrine: The Rise of Disaster Capitalism*, Penguin Books, London.

Kumbamu, A. (2020) 'The philanthropic-state-corporate complex: imperial strategies of dispossession from the "Green Revolution" to the "Gene Revolution"', *Globalizations* 17(8): 1367–1385 <http://doi.org/10.1080/14 747731.2020.1727132>.

Lynam, J. (2011) 'Plant breeding in sub-Saharan Africa in an era of donor dependence', *IDS Bulletin* 42(4): 36–47 <http://dx.doi.org/10.1 111/j.1759-5436.2011.00234.x>.

Madeley, J. (2002) *Food for All: The Need for a New Agriculture,* David Philip, Cape Town, and Zed Books, London.

Mahmood, I., Imadi, S.R., Shazadi, K., Gul, A. and Hakeem, K.R. (2016) 'Effects of pesticides on environment', in K.R. Hakeem, M.S. Akhtar and S.A. Abdullah (eds), *Plant, Soil and Microbes, vol. 1: Implications in Crop Science*, Springer International Publishing <http://dx.doi.org/10.1007/978-3-319-27455-3_13>.

MarketScreener (2017) 'BASF SE (BAS)' [website] <https://www.marketscreener. com/quote/stock/BASF-SE-6443227/company/>.

Mayer, A. and Runyon, L. (2016) 'Seeds, pesticides, fertilizer: how big companies harnessed the "holy trinity" of modern agriculture', *KCUR*, 31 October, <https://www.kcur.org/agriculture/2016-10-31/seeds-pesticides-fertilizer-how-big-companies-harnessed-the-holy-trinity-of-modern-agriculture>.

Mburu, F. (1986) 'The African social periphery', *Social Science and Medicine* 22(7): 785–790 <http://doi.org/10.1016/0277-9536(86)90232-7>.

Moore, J. (2015) *Capitalism in the Web of Life: Ecology and the Accumulation of Capital*, Verso, London and New York.

Moore, J. (2017) 'The Capitalocene, Part I: On the nature and origins of our ecological crisis', *The Journal of Peasant Studies* 44(3): 594–630 <http:// dx.doi.org/10.1080/03066150.2016.1235036>.

Morvaridi, B. (2012) 'Capitalist philanthropy and the new green revolution for food security', *International Journal of the Sociology of Agriculture and Food* 19(2): 243–256 <http://doi.org/10.48416/ijsaf.v19i2.228>.

MultiWatch (2016) 'March against Syngenta: Monsanto's Swiss twin unmasked' <https://multiwatch.ch/content/uploads/2017/12/March-on-Syngenta_ebook_def_klein.pdf>.

Otsuka, K. and Muraoka, R. (2017) 'A green revolution for sub-Saharan Africa: past failures and future prospects', *Journal of African Economies* 26(suppl_1): i73–i98 <http://doi.org/10.1093/jae/ejx010>.

Ozden, K. and Enwere, C. (2012) 'Urbanization and its political challenges in developing countries', *Eurasian Journal of Business and Economics* 5(10): 99–120 <http://www.researchgate.net/profile/Kemal-Ozden/publication/315718570_Urbanization_And_Its_Political_Challenges_In_Developing_Countries/links/5c0e4565299bf139c74dd5ce/Urbanization-And-Its-Political-Challenges-In-Developing-Countries.pdf?origin=publication_detail>.

Roumeliotis, G. and Stone, M. (2015) 'Dow, DuPont eye big tax savings in rare merger of equals', Reuters, 15 December <www.reuters.com/article/us-dow-m-a-tax-idUSKBN0TY01K20151215>.

Seed Co (2019) *2019 Annual Report*, Seed Co International Ltd, Gaborone <https://seedcogroup.com/wp-content/uploads/2022/07/Seed-Co-International-AR-Final-Version-July-2019_compressed_2.pdf>.

Smalley, R. (2013) *Plantations, Contract Farming and Commercial Farming Areas in Africa: A Comparative Review*, Working Paper 055, Future Agricultures <www.future-agricultures.org/wp-content/uploads/2013/03/FAC_Working_Paper_055.pdf>.

Sperling, L., Gallagher, P., McGuire, S. and March, J. (2021) 'Tailoring legume seed markets for smallholder farmers in Africa', *International Journal of Agricultural Sustainability* 19(1): 71–90 <http://doi.org/10.1080/14735903.2020.1822640>.

Stephan, H., Power, M., Hervey, A.F. and Fonseca, R.S. (2006) *The Scramble for Africa in the 21st Century: A View from the South*, Renaissance Press, Cape Town.

TASAI (2018) 'Industry competitiveness' <https://www.tasai.org/en/dashboard/data-summary/>.

Tilzey, M. (2006) 'Neoliberalism, the WTO and new modes of agri-environmental governance in the European Union, the USA and Australia', *International Journal of the Sociology of Food and Agriculture* 14(1): 1–28 <http://doi.org/10.48416/ijsaf.v14i.303>.

Tsatsakis, A., Nawaz, M.A., Tutelyan, V., Golokhvast, K., Kalantzi, O.-I., Chung, D.H., Kang, S.J., Coleman, M., Tyshko, N., Yang, S.H. and Chung, G. (2017) 'Impact on environment, ecosystem, diversity and health from culturing and using GMOs as feed and food', *Food and Chemical Toxicology* 107(A): 108–121 <http://doi.org/10.1016/j.fct.2017.06.033>.

Wager, C. (2016). 'A closer look at the Bayer-Monsanto merger and the seed licensing "cartel"', *Food & Power Newsletter*, 25 May <www.foodandpower.net/latest/2016/05/26/food-power-newsletter-bayer-monsanto-merger-seed-licensing-cartel>.

Werner, K. (2016) 'Bei Bayer und Monsanto reden auf beiden Seiten dieselben Investoren mit', *Süddeutsche Zeitung*, 21 September <www.sueddeutsche.de/wirtschaft/monsanto-und-bayer-beibayer-und-monsanto-reden-auf-beiden-seiten-dieselben-investoren-mit-1.3170377>.

Wibbelman, M., Schmutz, U., Wright, J., Udall, D., Rayns, F., Kneafsey, M., Trenchard, L., Bennett, J. and Lennartsson, M. (2013) 'Mainstreaming agroecology: implications for global food and farming systems', Discussion Paper, Centre for Agroecology and Food Security, Coventry <http://dx.doi.org/10.13140/RG.2.1.1047.5929>.

Zaad Holdings (no date) 'Champion crops together' <https://seedmarketing.co.za/>

CHAPTER 9

Corporate capture of agricultural and food policy in South Africa

David Fig

Corporate capture

The Zuma regime in South Africa (2009–18) became well known for sanctioning 'state capture' (Chipkin and Swilling, 2018). This involved extensive corruption of state appointment and procurement processes by elements in the private sector, the fraudulent diversion of state resources, and illicit control over decision-making instruments, especially with respect to cabinet, provincial, and local government, and numerous state-owned enterprises. One of the supposedly corrective actions of President Ramaphosa was to appoint the Judicial Commission of Inquiry into Allegations of State Capture, chaired by Deputy Chief Justice Raymond Zondo (Zondo, 2021). The Zondo Commission has set about trying to investigate specific instances of corporate corruption among individuals, offices, and projects. However, its mandate does not include the systematic appraisal of government policies which, for many years, have privileged the interests of larger corporations (both local and transnational) and aggravated major socio-economic inequalities, often across racial lines. Lobbying and influence-peddling by corporations are extensive and, while often shading easily into corrupt practices, not always regarded as illegal. Corporations have fought actively in many instances to evade, frustrate, or water down regulation.

This chapter will argue that, notwithstanding attempts to uncover and reverse recent instances of 'state capture', the broader issues of long-term corporate capture of national policy are not likely to come under official scrutiny for the time being. While short-term attention may be paid to the crises of corruption, pandemics, and governability, long-term policy needs (a just transition to a post-fossil-fuel economy, restoration of the power and effectiveness of state entities, national food sovereignty, and a check on rampant extractivism) recede into a policy mist.

Whether during the epochs of hunter-gatherer societies and more settled agrarian communities, or with the advent of colonial hegemony, food acquisition and production has generally been an important factor in the success or otherwise of the dominant mode of production. The initial logic behind colonial rule was for the Cape to provide the food needs of passing vessels belonging to a private Dutch mercantile corporation, the VOC (United East Indies Company), which gradually took control and colonized the Cape of Good Hope. Just over two centuries later, with the triumph of racial extractive capital in Southern Africa, the logic of food production had changed to the provision of an internal market, which included the need to reproduce an emerging semi-proletarian workforce.

In more recent times, there has been a great degree of continuity in the structure and shape of the South African agrifood chain. Neoliberal policies were introduced during the 1980s and have essentially continued to be dominant (Bernstein, 2013: 23). These policies have privileged the local large-scale private sector at the expense of small farmers, who have been deeply disadvantaged since as far back as the 1870s. The segregationist Union of South Africa (1910–61) and the apartheid republic which replaced it (1961–94) saw the intensification of restrictive and racially exclusive legislation that marginalized access to land and markets and thus rendered destitute most small and medium black agricultural producers.

Land reform under democracy has been a fraught process, especially for small-holders (Lahiff and Cousins, 2005). There has been insufficient state support for beneficiaries, and the constitution provides that land can only change hands between willing buyers and willing sellers (Cousins and Hall, 2017). This has led to the overwhelming majority of recent settlement schemes failing, and to renewed calls for the expropriation of agricultural land. South Africans are embroiled in heated debates about this expropriation (see, for example, Ngcukaitobi, 2021), but very little space is being devoted to how the country's scarce arable land could and should be used once it has been acquired.

This is an important part of the puzzle, given that the country's existing industrial agricultural system has essentially failed on a number of levels. These failures will be examined briefly, followed by an appraisal of the unequal structure of the agrarian economy. Finally some suggestions will be made for the correction of the problem.

Systemic failure of the current agrarian model

The chapter will argue that contamination, global warming, and dispro-portionate consumption of the country's fresh water supply have all been exacerbated by the increasing concentration and corporatization of the agrifood chain. National and multilateral policy spaces have been captured by the interests of large-scale national and transnational capital. Smallholders, including the beneficiaries of land reform, have been excluded from a fair share of any benefits that the system offers. Urgent steps need to be taken to reverse these policy failures so as to build popular food sovereignty.

Hunger

South Africa has a sophisticated agrifood chain, with significant commodity production, processing, and distribution capacity. Greenberg (2015 and in

Chapter 8 of this book) has shown that in almost every link in the chain, a few corporations dominate, and ownership is overlapping, interconnected, and networked. Private (e.g. banking and finance) and public (e.g. government pension funds) entities are also deeply implicated in ownership and control of the sector (Greenberg, 2016: 32, Figure 5). Its retail supermarket system is oligopolistic, with five dominant chains. Small retailers also play a significant role, but are not price-competitive.

It is sobering that, despite this extensive infrastructure, a quarter of South Africa's population go hungry every day, and half are at risk of hunger (Tsegay et al., 2014: 6, 9). Price inflation makes nutritious food increasingly unaffordable (Trading Economics, 2021), especially to the millions who are not in formal employment, particularly young people. This was clearly a factor in the way political discontent readily turned to looting and the destruction of shops and supermarket complexes in the most populated provinces of South Africa during July 2021 (*New Frame*, 2021).

Low levels of nutrition may result in negative impacts on intellectual development and physical growth, and higher infant and child mortality rates (Leathers and Foster, 2004). The *2020 Global Nutrition Report* (Mannar and Micha, 2020) notes that in South Africa 27.4 per cent of all children below the age of five suffer from stunted growth.

Health

Corporate monopolization of the food chain has resulted in a number of health problems for the population. Consumers, especially in poor communities, are driven towards purchasing foods that contain high levels of chemical additives, sugar, salt, saturated fats, bad cholesterol, and other substances that impact negatively on health and nutrition. This has resulted in epidemics of obesity, heart disease, diabetes, and other ailments. South Africa has also seen outbreaks of disease in food processing plants. The recent listeriosis scandal demonstrated how easily food safety could be compromised (WHO, 2018).

Inputs that contaminate

The country's commercial agriculture sector routinely relies on expensive and polluting pesticides, chemical fertilizers, and genetically modified (GM) seed. The laws regulating pesticides predate the apartheid era – see, for example, the Fertilizers, Farm Feeds, Agricultural Remedies and Stock Remedies Act (Union of South Africa, 1947) – and are no longer fit for purpose (Andrews, 2021; Shevel and Cramer, 2021). Weak regulation has resulted in contamination of farm workers and neighbouring fields and watercourses, especially from aerial spraying (London, 2003).

Ever since the warnings of Rachel Carson in the 1960s about the health and ecological dangers of the massive use of agricultural chemicals as pesticides (Carson, 1962), the world has become more intent on their regulation.

However, South Africa has lagged behind in many respects, and has continued to allow the use of pesticides banned in other jurisdictions.

Many pesticides persist in the environment far beyond the end of a planting season, affecting other plants, animals, and micro-organisms in the soil (Goering et al., 1993). Pesticide contamination of groundwater is difficult to reverse in the short term. The effectiveness of pesticides declines over time as the species affected develop forms of resistance (ibid.). Major chemical accidents at Seveso (Italy), Bhopal (India), and, recently, the UPL factory in Durban (South Africa) have resulted in human and environmental casualties (Homberger et al., 1979; Eckerman, 2005; Phillips, 2021).

At the 2006 African Union special summit of heads of state and government, the international chemical fertilizer lobby was able to effect the adoption of the Abuja Declaration on Fertilizers for an African Green Revolution (NEPAD-CAADP, 2011). Member states, including South Africa, resolved to increase fertilizer use from 8 kg to 50 kg per hectare by 2015. The South African government-backed New Partnership for Africa's Development (NEPAD) acts as the secretariat for the Abuja Declaration. South African consumption of fertilizer was measured at 72.8 kg/ha of arable land in 2018 (World Data Atlas, n.d.). The so-called Green Revolution, strongly promoted by the Bill & Melinda Gates Foundation, aims to encourage high-input agriculture as the solution to African hunger. The Gateses created the Alliance for a Green Revolution in Africa (AGRA) as an organization to embody the ambition of corporatizing African agriculture. A substantial number of agriculturalists and fisherfolk have resisted AGRA's vision, and severely criticized the growing corporate control over the food chain. Transnational food corporations have in recent years captured important multilateral institutions such as the Food and Agriculture Organization of the United Nations. A case in point was the United Nations Food Systems Summit of September 2021, with the AGRA president, Rwanda's Dr Agnes Kalibata, appointed as its chair. Many grassroots agricultural bodies, including the globally organized La Via Campesina, have noted the corporate capture of this initiative, and decided to withdraw from its deliberations (Food Systems 4 People, 2021).

South Africa is the only country in the world that permits its staple food, maize, to be grown from GM seed. More than 91 per cent of South Africa's maize is based on proprietary GM seed, using insect- and herbicide-resistant traits (Ala-Kokko et al., 2021). In addition, the Genetically Modified Organisms Act passed in 1997 (Republic of South Africa, 1997) has a number of shortcomings. Instead of a strict, impartial assessment of applications by the gene companies, the Act allows for self-regulated risk assessments to be submitted to the regulator based entirely on in-house tests conducted by the GMO-purveying corporates themselves. Critics of this loophole have called for more robust environmental impact assessment methods to be applied (African Centre for Biosafety, 2010). The Act opened the door to the import and release of GM seed and enabled GM seed experimentation and bulking in South Africa. Its implementation has been weak on the question of transparency in the decision-making process and on public participation. Furthermore, it leaves out the question of the labelling of GM products, which in practice, is left to other poorly regulated legislation (De Beer and Wynberg, 2018).

GMOs have been found to contaminate indigenous crops through 'the unwanted escape and spread of GMOs or genetic material from GMOs to non-GM plants, animals and foods' (CBAN, 2019: 1). Biowatch, a South African food sovereignty non-profit organization, has discovered that even when smallholder farmers apply agroecological practices, they still face the threat of contamination of their produce due to the escape of GM pollens or seeds into their environment (Iversen et al., 2014; Biowatch South Africa, 2017).

Because of the proprietary nature of GM seed, GM corporations can prevent traditional farmer practices such as seed saving and exchange, and oblige users to purchase new proprietary seed each season. This flies in the face of the principles of food and seed sovereignty, which respect farmer-led seed systems and allow farmers and consumers to control food outcomes, rather than leaving these to be controlled by capital and markets (Greenberg et al., 2021).

The use of proprietary seed goes along with strong recommendations from the GM corporations for the use of proprietary herbicides and pesticides like glyphosate. The ill-effects of this contaminant have resulted in litigation in jurisdictions including the United States. Dewayne 'Lee' Johnson, who suffered from non-Hodgkin lymphoma, sued Bayer for his exposure to Monsanto's glyphosate-based weedkiller when acting as a grounds manager for a school district outside San Francisco (Bayer had purchased Monsanto in 2016). The court awarded Johnson US$289 m, later reduced to $20.5 m on appeal (Gillam, 2021).The Johnson case set a precedent for a hundred thousand other victims, and, while not conceding liability, Bayer admitted it had set aside billions of dollars for settlements (Bega, 2021).

In 2002 the South African state was forced, for the first time, to provide the public with systematic information on GM permits after it was challenged in court by Biowatch (Wynberg and Fig, 2013).

But the power of the large corporations has intensified in the intervening years. In 2012 South Africa's Competition Appeal Court allowed the largest remaining local crop seed company, Pannar, to be purchased by DuPont subsidiary Pioneer Hi-Bred South Africa. This signalled the beginning of foreign monopoly control over local crop seed. This is now dominated by transnationals [Bayer]/Monsanto, DuPont, Dow and Syngenta.

The country's drive to adopt GMOs has resulted in some spectacular failures. One involved Monsanto attempting to persuade small-scale farmers on the Makhathini Flats, a floodplain on the Phongolo River in KwaZulu-Natal, to plant their proprietary GM cotton. The project was an attempt to convince the world that GM crops were suited to farmers like this. Monsanto flew representative Makhathini farmers around the world to advocate the corporation's position. But within only a few years the farmers found themselves deeply in debt and the GM cotton project was abandoned [Witt et al., 2006].

In the Eastern Cape province small-scale farmers were initially given free Monsanto GM and hybrid seed. Traditional farming practices were abandoned in favour of mechanical tilling and monocropping of maize. Called the Massive Food Production Programme, it failed to meet any of its key objectives over five years and swallowed R570 million in state funds [Tregurtha, 2009; Jacobson, 2013; Fischer and Hadju, 2015]. Productivity hardly improved and small-scale farmers were left with unpayable debts. (Fig, 2018)

Lion's share of fresh water consumption

Large-scale agriculture is also heavily reliant on irrigation: the commercial agricultural sector extracts 51 per cent to 63 per cent of the country's available surface water (Van Niekerk et al., 2018). Davies and Day, in their analysis of South Africa's water resources, identify modern irrigation-driven monocrop agriculture as 'probably the single most environmentally devastating development in the entire history of our species' (1998: 320).

Ignoring the climate emergency

As a water-scarce country, where drought conditions are periodic, South Africa is extremely vulnerable to rising temperatures linked to the production of greenhouse gases. McSweeney and Timperley (2018) claim that the climate change risks to agriculture include changing rain patterns, increased evaporation rates, higher temperatures, increased pests and diseases, shifting growing regions, and reduced yields. The continued promotion of the use of fossil fuels in agriculture, the strong commitment to coal, the abandonment of rail as the principal transportation medium for goods, and the tepid attitude to renewable energy are policies stubbornly maintained by the South African government that all promote more powerful corporate controls in the agrarian sector.

Structural support for industrial agriculture and the industrial food chain

Unlike most other countries on the continent, South Africa's agricultural sector is heavily skewed towards industrial farming. Its 40,122 commercial farmers produce most of the country's food (Statistics South Africa, 2020). Aliber and Hall (2010a: 6; 2010b: 14) suggest that there are approximately 2.3 million black small-scale farmers and 250,000 to 300,000 black commercial farmers in the country, while Kirsten (2011) claims there are a further 22,500 white small-scale farmers (Greenberg, 2015: 958).

Space limitations preclude a systematic analysis of the aggregation of the many instances in which the South African state has supported and safeguarded the interests of large-scale agrarian capital. Nevertheless, some examples will illustrate the point.

On the few occasions when lip service is paid to support for small-scale farmers, the usual formula is for the state to promote a minority of such farmers as potential candidates for fitting into the model of industrial agriculture. One of the favoured methods is to link small-scale farmers with mentors drawn from large-scale agriculture in their immediate vicinity. These mentors are charged with easing the small-scale farmers out of low-input subsistence production and into high-input, market-oriented monocrop production (Lahiff et al., 2012; Anseeuw et al., 2015). Wandile Sihlobo, an agricultural economist linked to the commercial agrifood sector, regards the emergence of a new class of black commercial farmers producing crops such as citrus, lucerne, sheep, pineapples, peppers, and blueberries, with the support of government and the co-ops, as proof of 'transformation happening in South Africa's agricultural sector' (2020: 10).

The co-ops referred to were originally formed by farmers at regional level to provide services to members. Such services might include the provision of farming equipment, insurance, vehicles, retail commodities, and grain sileage, using the members' joint purchasing power. When the commodity board system ended, the co-ops were commercialized and privatized. Since then, they have played a more prominent part in the agrifood value chain by expanding into new areas of agricultural investment (Greenberg, 2016).

Another result of neoliberal policy shifts was the state's practice of curtailing the provision of agricultural extension services. Instead of supporting farmers with independent advice on farming methods and inputs, the state paved the way for these services to be provided by corporations serving their own interests rather than the public good (Williams et al., 2008; Gelderblom et al., 2020).

Much has been made of the rising age profile of farmers, said to average 62 in South Africa (Sihlobo, 2020: 157).This points to the need to encourage younger entrants into the sector. So far this has been hard to incentivize, since agricultural wages are notoriously low and the rewards from subsistence farming are limited. The state could play more of a role in improving these conditions, but also in reinvigorating agricultural and ecological education within the school system and higher education. Training institutions such as the Owen Sitole College of Agriculture in KwaZulu-Natal are beginning to include modules on agroecology in their curriculum, but this is still only happening on an experimental basis (see Box G).

The state accommodates the interests of large-scale commercial farmers in their use of pollutants dangerous to the soil, air quality, water quality, and human and animal health. Aerial crop spraying is still allowed, despite the harm done to neighbouring communities and catchments, Little is done to protect small-scale farmers and farm workers exposed to pesticides and herbicides (Emmanuel, 1992; London, 1995, 2003; Rother et al., 2008). South Africa permits the use of certain poisons which are generally outlawed in the rest of the world. GMO producers are protected in a number of ways, through legislation which supports their interests, through the imbalance of power represented on the Executive Council for Genetically Modified Organisms (which acts as a regulator under the GMO Act), and through the refusal to make the labelling of GMO products compulsory. The continued toleration of the use of glyphosate-rich proprietary herbicides such as Roundup, regarded increasingly as a dangerous carcinogen (Kogevinas, 2019; Bega, 2021), allows these poisons to be used with impunity.

Public health concerns have begun to challenge certain corporate interests. In line with Mexico, Barbados, the UK, and other countries, the South African government, in 2018, implemented a tax on sugary soft drinks called the Health Promotion Levy. Proponents of the tax had advocated a rate of 20 per cent, but industrial lobbyists persuaded the state to charge only 10 per cent, and to exclude fruit juices and dairy-based drinks. These concessions have effectively watered down the full potential of the tax to slow down the incidence of obesity, heart disease, diabetes, and other ailments linked to excessive sugar intake. These concessions to the corporations seem permanent, since further lobbying to reopen the raising of the rate, despite early evidence of the successful outcomes of the tax, has so far proven unsuccessful (Hofman et al., 2021; Kruger et al., 2021; Stacey et al., 2021; Boachie et al., 2022).

Questioning current policy

Why are South African policymakers choosing to back large-scale farmers? The most cogent answer is that they have succumbed to pressures from transnational and some oligopolistic local corporations that dominate the local agrifood value chain. Such pressures have made farmers dependent on hybrid or GM proprietary seeds, herbicides, and fertilizers. Pressures from philanthropic bodies such as the Gates and Buffet foundations have promoted the industrial model that favours high inputs and large-scale private farming (Wise, 2020; GRAIN, 2021).

There is a growing attitude among politicians and planners that, with the increase of urbanization, there is little return for political parties in giving their support to small-scale farmers or a fundamental transformation of the rural economy.

For reasons iterated earlier, South Africa urgently needs to rethink its existing agricultural model. The current preference for large-scale, high-input farming enterprises fails to demonstrate confidence in the ability of small-scale family-based producers to provide more efficiently for the market. Employing agroecological methods – farming without GMOs, chemical pesticides, and artificial fertilizers – small-scale farmers can, with sufficient policy and practical state support, contribute significantly to food and nutritional security. This has been accomplished successfully elsewhere. For example, in the state of Santa Caterina in southern Brazil, the state government supported 60,000 small farmers, with the result that sales of their produce increased by 64 per cent after just one year (World Bank, 2017). In South Africa, it is also possible to make small-scale farms work efficiently.

Helping small-scale farmers requires a number of interventions. The first is practical support. South Africa used to provide extension services to farmers that consisted of independent advice. But budget cuts have reduced the quality of the service and opened the way for corporate agents to take on the role. For example, in the Hlabisa district, KwaZulu-Natal, the state and Monsanto have combined efforts to influence farmers to plant GM crops (Mahlase, 2016).

As part of the land debate, South Africans should be calling on the government to abandon its bias towards monopoly agribusiness. The first step would be to reverse the measures that favour international agribusiness interests. Second, biosafety regulations should be tightened, encouraging traditional practices of seed saving and exchange, reviving and building sustainable employment opportunities, and guaranteeing soil quality and food sovereignty. This would be a positive contribution to sustainable water usage and reducing carbon emissions. According to Lang and Heasman (2004: 294), '[s]ustainable food systems can both raise output and reduce inputs, while being more socially equitable, feeding people in need, providing jobs and being environmentally beneficial (not just benign)'.

The state needs to answer the call for land with practical agrarian reform measures that recognize the rights and needs of smallholder farmers. Often the cultivators are women, whose rights of inheritance are not fully secured under customary law. Such disparities need to be addressed. Protection of common grazing or cultivation areas needs to be secured. Public provision of basic infrastructure should be honoured. A broad agenda for agrarian reform, taking into account the promotion of smallholder interests, was developed by the Institute for Poverty, Land and Agrarian Studies, in a project undertaken to consider policy options in South Africa (Hall, 2009). This framework could easily be utilized as a basis for agrarian transformation. Lahiff and Cousins (2005) identify three areas of reform necessary for the expansion of smallholder farming: redistribution of land and other assets from large-scale to small-scale farmers; reform of agricultural markets; and support to new and existing smallholders.

In taking these steps, South Africa will have to loosen the grip of corporate domination on the agrifood chain, adopt fairer, safer, and more ecological models of food production and exchange, ensure that basic foods are more affordable to all, and initiate a form of land distribution that honours smallholder rights and sustains them with support.

References

African Centre for Biosafety (ACB) (2010) 'EIA regulations and GMOs in South Africa' ACB, Johannesburg <https://www.acbio.org.za/eia-regulations-and-gmos-south-africa>.

Ala-Kokko, K., Nalley, L.L., Shew, A.M., Tack, J.B., Chaminuka, P., Matlock, M.D. and D'Haese, M. (2021) 'Economic and ecosystem impacts of genetically modified maize in South Africa', *Global Food Security* 29 <https://doi.org/10.1016/j.gfs.2021.100544>.

Aliber, M. and Hall, R. (2010a) *The Case for Re-strategising Spending Priorities to Support Small-Scale Farmers in South Africa*, PLAAS Working Paper 17, Cape Town: Institute for Poverty, Land and Agrarian Studies, University of the Western Cape <http://hdl.handle.net/10566/4475>.

Aliber, M. and Hall, R. (2010b) *Development of Evidence-Based Policy around Small-Scale Farming*, Programme to Support Pro-Poor Policy Development, The Presidency, Pretoria <https://www.academia.edu/2128811/Development_of_Evidence_Based_Policy_Around_Small_Scale_Farming>.

Andrews, A. (2021) 'Submission by Unpoison on Draft Regulations under FFAR Act', 20 August, Unpoison, Cape Town <https://unpoison.org/wp-content/uploads/2021/08/2021_08_20-Unpoison-draft-regulations-comment-FFFARA-1.pdf>.

Anseeuw, W., Fréguin-Gresh, S. and Davis, N. (2015) 'Contract farming and strategic partnerships: A promising exit or smoke and mirrors?', in H. Cochet, W. Anseeuw and S. Fréguin-Gresh (eds), *South Africa's Agrarian Question*, pp. 296–313, HSRC Press, Cape Town.

Bega, S. (2021) 'A "cancer causing" herbicide has been found in South Africa's bread and flour', *Mail & Guardian*, 7 September <https://mg.co.za/environment/2021-09-07-a-cancer-causing-herbicide-has-been-found-in-south-africas-bread-and-flour/>.

Bernstein, H. (2013) 'Commercial agriculture in South Africa since 1994: "Natural, simply capitalism"', *Journal of Agrarian Change* 13(1): 23–46 <https://doi.org/10.1111/joac.12011>.

Biowatch South Africa (2017) 'GM testing of farmer seed', Biowatch internal brief, February, Biowatch, Durban.

Boachie, M.K., Theehla, E. and Hofman, K., (2022) 'New developments with the Health Promotion Levy in South Africa', *South African Medical Journal* 112(7): 454–455 <https://doi.org/10.7196/SAMJ2022v112I7.16579>.

Canadian Biotechnology Action Network (CBAN) (2019) *GM Contamination in Canada: The Failure to Contain Living Modified Organisms: Incidents and Impacts*, CBAN, Halifax, NS <https://cban.ca/wp-content/uploads/GM-contamination-in-canada-2019.pdf>.

Carson, R. (1962). *Silent Spring*, Fawcett, Greenwich, CT.

Chipkin, I. and Swilling, M. (eds) (2018) *Shadow State: The Politics of State Capture in South Africa*, Wits University Press, Johannesburg.

Cousins, B. and Hall, R. (2017) 'South Africa is still way behind the curve on transforming land ownership', *The Conversation Africa*, 13 November <https://theconversation.com/south-africa-is-still-way-behind-the-curve-on-transforming-land-ownership-87110>.

Davies, B. and Day, J. (1998) *Vanishing Waters*, University of Cape Town Press, Cape Town.

De Beer, T., and Wynberg, R. (2018) 'Developing and implementing policy for the mandatory labelling of genetically modified food in South Africa', *South African Journal of Science* 114(7/8) <https://doi.org/10.17159/sajs.2018/20170137>.

Eckerman, I. (2005) *The Bhopal Saga: Causes and Consequences of the World's Largest Industrial Disaster*, Universities Press, Hyderabad.

Emmanuel, K. (1992) *Poisoned Pay: Farmworkers and the South African Pesticide Industry*, Group for Environmental Monitoring and the Pesticides Trust, Johannesburg.

Fig, D. (2018) 'South Africa needs to reverse corporate capture of agricultural policy', *The Conversation*, 28 May <https://theconversation.com/south-africa-needs-to-reverse-corporate-capture-of-agricultural-policy-96661>.

Fischer, K. and Hadju, F. (2015) 'Does raising maize yields lead to poverty reduction? A case study of the Massive Food Production Programme in South Africa', *Land Use Policy* 46: 304–313 <https://doi.org/10.1016/j.landusepol.2015.03.015>.

Food Systems 4 People (2021) 'People's counter-mobilization to transform corporate food systems: NO to corporate food systems, YES to food sovereignty', Civil Society and Indigenous Peoples' Mechanism for Relations with the United Nations Committee on World Food Security, Rome <https://www.csm4cfs.org/wp-content/uploads/2021/09/Declaration-EN-2.pdf>.

Gelderblom, C., Oettlé, N., Clifford-Holmes, J., Malgas, R., Wilson, N. and Polonsky, S. (2020) *Transformative Cross-Sectoral Extension Services Dialogue: 2 & 3 March 2020 Synthesis Report*, Department of Environment, Forestry and Fisheries, WWF, South African National Biodiversity Institute (SANBI) and SA-EU Strategic Partnership, Cape Town <https://wwfafrica.awsassets.panda.org/downloads/transformative_cross_sectoral_extension_workshop_final_synthesis_report_low_res.pdf>.

Gillam, C. (2021) *The Monsanto Papers: Deadly Secrets, Corporate Corruption and One Man's Search for Justice*, Island Press, Washington, DC.

Goering, P., Norberg-Hodge, H. and Page, J. (1993) *From the Ground Up: Rethinking Industrial Agriculture*, Zed Books, London.

GRAIN (2021) 'How the Gates Foundation is driving the food system, in the wrong direction', 17 June <https://grain.org/en/article/6690-how-the-gates-foundation-is-driving-the-food-system-in-the-wrong-direction>.

Greenberg, S. (2015) 'Agrarian reform and South Africa's agro-food system', *Journal of Peasant Studies* 42(5): 957–979 <https://doi.org/10.1080/03066150.2014.993620>.

Greenberg, S. (2016) *Corporate power in the agro-food system and the consumer food environment in South Africa*, PLAAS Working Paper 32, Cape Town: Institute for Poverty, Land and Agrarian Studies, and Centre for Excellence on Food Security, University of the Western Cape <http://hdl.handle.net/10566/4522>.

Greenberg, S., Pelser, D. and Ranqhai, T. (2021) *Farmer-led seed systems: securing food sovereignty in the face of looming ecological and social crises*, Biowatch Briefing, September, Biowatch South Africa, Durban <https://biowatch.org.za/download/farmer-led-seed-systems/>.

Hall, R. (ed.) (2009) *Another Countryside? Policy Options for Land and Agrarian Reform in South Africa*, Institute for Poverty, Land and Agrarian Studies (PLAAS), University of the Western Cape, Cape Town <http://repository.uwc.ac.za/xmlui/handle/10566/4572>.

Hofman, K.J., Stacey, N., Swart, E.C., Popkin, B.M. and Ng, S.W. (2021) 'South Africa's Health Promotion Levy: excise tax findings and equity potential', *Public Health* 22(9) <https://doi.org/10.1111/obr.13301>.

Homberger, E., Reggiani, G., Sambeth, J., and Wipf, H.K. (1979) 'The Seveso accident: its nature, extent and consequences', *Annals of Occupational Hygiene* 22(4): 327–370 <https://doi.org/10.1093/annhyg/22.4.327>.

Iversen, M., Grønsberg, I.M., Van den Berg, J., Fischer, K., Aheto, D.W. and Bøhn, T. (2014) 'Detection of transgenes in local maize varieties of small-scale farmers in Eastern Cape, South Africa', *PloS One* 9(12): e. 116147 <https://doi.org/10.1371/journal.pone.0116147>.

Jacobson, K. (2013) *From Betterment to Bt maize: Agricultural Development and the Introduction of Genetically Modified Maize to South African Smallholders*, PhD thesis, Swedish University of Agricultural Sciences, Uppsala <https://pub.epsilon.slu.se/10406/1/Jacobson_k_130507.pdf>.

Kirsten, J. (2011) 'Most South African farmers are small-scale', *Farmer's Weekly*, 23 September: 38.

Kogevinas, M. (2019) 'Probable carcinogenicity of glyphosate', *British Medical Journal* 365: 1613 <https://doi.org/10.1136/bmj.l1613>.

Kruger, P., Karim, S.A., Tugendhaft, A. and Goldstein, S. (2021) 'An analysis of the adoption and implementation of a sugar-sweetened beverage tax in South Africa: a multiple streams approach', *Health Systems and Reform* 7(1) <https://doi.org/10.1080/23288604.2021.1969721>.

Lahiff, E. and Cousins, B. (2005). 'Smallholder agriculture and land reform in South Africa', *IDS Bulletin* 36(2): 127–131 <https://doi.org/10.1111/j.1759-5436.2005.tb00209.x>.

Lahiff, E., Davis, N. and Manenzhe, T. (2012) *Joint Ventures in Agriculture: Lessons from Land Reform Projects in South Africa*, International Institute for Environment and Development, London <https://pubs.iied.org/sites/default/files/pdfs/migrate/12569IIED.pdf>.

Lang, T. and Heasman, M. (2004) *Food Wars: The Global Battle for Mouths, Minds and Markets*, Earthscan, London.

Leathers, H.D. and Foster, P. (2004) *The World Food Problem: Tackling the Causes of Undernutrition in the Third World*, Lynne Rienner, Boulder, CO.

London, L. (1995) *An Investigation into the Neurological and Neurobehavioural Effects of Long-Term Agrichemical Exposures among Deciduous Fruit Farm Workers in the Western Cape, South Africa*, MD dissertation, Department of Community Health, University of Cape Town <https://open.uct.ac.za/handle/11427/26360>.

London, L. (2003) 'Human rights, environmental justice and the health of farm workers in South Africa', *International Journal for Occupational and Environmental Health* 9: 59–68 <https://www.researchgate.net/publication/10756934_Human_Rights_Environmental_Justice_and_the_Health_of_Farm_Workers_in_South_Africa>.

Mahlase, M.H. (2016) *Exploring the Uptake of Genetically Modified White Maize by Smallholder Farmers: The Case of Hlabisa, South Africa*, MSc dissertation, Faculty of Science, University of Cape Town <http://hdl.handle.net/11427/24452>.

Mannar, M.G.V. and Micha, R. (2020) *2020 Global Nutrition Report: Action on Equity to End Malnutrition*, Development Initiatives, Bristol <https://globalnutritionreport.org/documents/566/2020_Global_Nutrition_Report_2hrssKo.pdf>.

McSweeney, R. and Timperley, J. (2018) 'The carbon brief profile: South Africa', 15 October, Carbon Brief, London <https://www.carbonbrief.org/the-carbon-brief-profile-south-africa>.

NEPAD-CAADP (2011) *The Abuja Declaration on Fertilizer for an African Green Revolution: Status of Implementation at Regional and National Levels*, NEPAD, Johannesburg.

New Frame (2021) 'Durban food riots turn the wheel of history', 12 July, *New Frame*, Johannesburg <https://www.newframe.com/durban-food-riots-turn-the-wheel-of-history/>.

Ngcukaitobi, T. (2021) *Land Matters: South Africa's Failed Land Reforms and the Road Ahead*, Penguin Random House, Johannesburg.

Phillips, T. (2021) 'Criminal probe recommended against multinational UPL for chemical spill', *Mail & Guardian*, 8 October <https://mg.co.za/environment/2021-10-06-criminal-probe-recommended-against-multinational-upl-for-chemical-spill/>.

Republic of South Africa (1997) *Genetically Modified Organisms Act 15 of 1997*, Government Gazette 18029 <https://www.gov.za/sites/default/files/gcis_document/201409/act15of1997.pdf> .

Rother, H.A., Hall, R. and London, L. (2008) 'Pesticide use among emerging farmers in South Africa: contributing factors and stakeholder perspectives', *Development Southern Africa* 25(4): 399–424 <https://doi.org/10.1080/0376 8350802318464>.

Shevel, A. and Cramer, C. (2021) 'Poison on a plate: outdated agricultural chemical legislation means your avocado-topped pizza could be hazardous to your health', *Daily Maverick,* 26 January <https://www.dailymaverick.co.za/article/2021-01-26-poison-on-a-plate-outdated-agrichemical-legislation-means-your-avocado-topped-pizza-could-be-hazardous-to-your-health/>.

Sihlobo, W. (2020) *Finding Common Ground: Land, Equity and Agriculture*, Picador Africa, Johannesburg.

Stacey, N., Edoka, I., Hofman, K., Swart, E., Popkin, B. and Ng, S.W. (2021) 'Changes in beverage purchases following the announcement and implementation of South Africa's Health Promotion Levy', *The Lancet* 5(4): E200–E208 <https://doi.org/10.1016/S2542-5196(20)30304-1>.

Statistics South Africa (Stats SA) (2020) 'Stats SA releases Census of Commercial Agriculture 2017 Report' [media release], 24 March, Stats SA, Pretoria <http://www.statssa.gov.za/?p=13144>.

Trading Economics (2021) 'South Africa food inflation, October 2020 to July 2021' [website] <https://tradingeconomics.com/south-africa/food-inflation>.

Tregurtha, N. (2009) *Review of the Eastern Cape's Siyakhula/Massive Maize Project*, Trade and Industrial Policy Strategies, Pretoria <http://www.tips.org.za/files/u65/review_of_siyakhula_-_norma_tregurtha.pdf>.

Tsegay, Y.T., Rusare, M. and Mistry, R. (2014) *Hidden Hunger in South Africa: The Faces of Hunger and Malnutrition in a Food Secure Nation*, Oxfam International, Oxford <https://oxfamilibrary.openrepository.com/bitstream/10546/332126/1/rr-hidden-hunger-south-africa-131014-en.pdf>.

Union of South Africa (1947) *Fertilizers, Farm Feeds, Seeds and Remedies Act, 1947*, Government Gazette 3751 <https://www.gov.za/sites/default/files/gcis_document/201505/act-36-1947.pdf>

Van Niekerk, A., Jarmain, C., Goudriaan, R., Muller, S.J., Ferreira, F., Münch, Z., Pauw, T., Stephenson, G. and Gibson, L. (2018) *An Earth Observation Approach Towards Mapping Irrigation Areas and Quantifying Water Use by Irrigated Crops in South Africa*, WRC Report TT745/17, Water Research Commission, Pretoria <http://www.wrc.org.za/wp-content/uploads/mdocs/TT%20745%20 Final%20Report%20reprint%2025%2005%2018.pdf>.

Williams, B., Mayson, D., De Satge, R., Epstein, S. and Semwayo, T. (2008) *Extension and Smallholder Agriculture: Key Issues from a Review of the Literature*, Phuhlisani, Cape Town <http://www.phuhlisani.com/ oid%5Cdownloads%5CPhuhlisani%20extension%20reviewD1.pdf>.

Wise, T.A. (2020) *Africa's Choice: Africa's Green Revolution Has Failed, Time to Change Course*, policy brief, 28 July, Institute for Agriculture and Trade Policy, Minneapolis <https://www.iatp.org/sites/default/files/2020-07/2020_07_ AfricasChoice_PolicyBrief.pdf>.

Witt, H., Patel, R. and Schnurr, M. (2006) 'Can the poor help GM crops? Technology, representation and cotton in the Makhathini Flats, South Africa', *Review of African Political Economy* 109: 497–513 <https://doi. org/10.1080/03056240601000945>.

World Bank (2017) *Enhancing Small Farmers' Business Competitiveness in Santa Catarina, Brazil*, Results Briefs, 24 October, World Bank, Washington, DC <https://www.worldbank.org/en/results/2017/10/24/enhancing-small- farmers-competitiveness-santa-catarina-brazil>.

World Data Atlas (no date) 'South Africa: fertilizer consumption per unit of arable land', Knoema <https://www.knoema.com/atlas/South-Africa/ fertilizer-consumption>.

World Health Organization (WHO) (2018) 'Listeriosis outbreak in South Africa', 9 January, WHO South Africa <https://www.afro.who.int/news/ listeriosis-outbreak-south-africa>.

Wynberg, R. and Fig, D. (2013) *A Landmark Victory for Justice: Biowatch's Battle with the South African State and Monsanto*, Biowatch South Africa, Durban <https://biowatch.org.za/download/a-landmark-victory-for-justice/>.

Zondo, R. (2021) 'The Judicial Commission of Inquiry into Allegations of State Capture, Corruption and Fraud in the Public Sector including Organs of State' [website] <https://www.statecapture.org.za>.

CHAPTER 10

The slow and structural violence of agrochemicals: use, management, and regulation in South Africa

Morgan Lee

Introduction

Agrochemical dependence and a deepening agrarian crisis

Industrial agricultural models are fundamentally dependent on synthetic chemical inputs, which, as such, are firmly entrenched in our production systems and have a historically longstanding and significant role in crop productivity and protection (Handford et al., 2015; Shattuck, 2021). Agriculture is more dependent on agrochemicals than ever before, due to structural transformation in agricultural industries, such as decreased innovation, increased regulatory costs, industry consolidation, and a shift to generic chemical products (Shattuck, 2021: 232). It is difficult to capture detailed data about global agrochemical use as the data are usually neither reliable nor complete, because many countries do not have the capacity to adequately monitor and record import, trade, and use (Isgren and Andersson, 2021; Shattuck, 2021). Furthermore, patterns of agrochemical use change over time as chemicals are restricted or as new chemicals are approved (Ward et al., 2000). It is estimated that across the globe 3.5 billion kg of agrochemicals are sprayed every year, amounting to a total global market value of US$215.18 bn in 2016, which is expected to reach $272 bn by 2028 (Shattuck, 2021; Grand View Research, 2022). The promotion of input-intensive forms of agriculture, along with market liberalization and privatization, has made agrochemicals more accessible and affordable, and has created a global 'pesticide complex' (Isgren and Andersson, 2021; Shattuck, 2021). The 'chemical nature of agrarian capitalism' has created a multipolar pesticide complex, where commodity chains and environmental impacts are less visible, using market forces to manipulate economies of scale and reorder labour and ecological relations (Shattuck, 2021: 232). The standardization of chemical inputs has created a 'technology treadmill' which has forced farmers to adopt hybrid varieties and agrochemicals in order to stay competitive and profitable. This 'treadmill' has made agrochemicals ecologically and economically 'necessary' and allowed the expansion of commodity markets at profits and scales that would otherwise not have been possible (Shattuck, 2021). Despite the touted benefits and success of agrochemicals, their use is no longer proportional to production: a 1.8 per cent increase in agrochemical use is required to produce a 1 per cent increase in crop output per hectare (Shattuck, 2021). There is a deepening agrarian crisis unfolding which requires the urgent re-evaluation of the structure, practices, and regulation of the industrial agriculture model.

The increasing use of, and dependence upon, agrochemicals is especially concerning in light of poor regulatory practices. Methods of regulating agrochemicals vary between countries, especially between high-income and low-income countries, and are driven by culture, politics, economy, science, health, safety, food security and sustainability, trade, and pest management (Handford et al., 2015). This chapter aims to highlight key flaws in the South African agrochemical regulatory arena and illustrate how these flaws result in harm which is significantly underestimated. Two theoretical framings will be used: slow violence and structural violence. Human exposure to glyphosate and its persistence in the environment will be explored through the lens of slow violence: harm that occurs gradually and attritionally while largely out of sight. Poor environmental risk assessment (ERA) and inadequate regulation will be explored through the lens of structural violence: harm that is concealed, ingrained, and institutionalized beyond recognition. The scientific debate, misinformation, and regulatory inaction around the use of glyphosate are a prime example of how unaddressed regulatory flaws constitute structural violence and perpetuate slow violence. As such, glyphosate will be used as a lens for analysis. Addressing regulatory failure and acknowledging the harm it causes is especially important in Africa. Given Africa's largely absent legislation, poor capacity for testing and monitoring, and important biodiversity, the agrochemical situation across the continent is worrying. It is crucial that regulatory bodies be addressed and improved because Africa is a highly sought-after market for both genetically modified crops and the global oversupply of glyphosate (Watts et al., 2016). The agrochemical industry has prepared for Africa as the 'next pesticide frontier', and it is critical that African countries have legislation in place to prevent this (ACB, 2019).

The case of glyphosate-based herbicides

The unfolding agrarian crisis has triggered heated international debates about the use and regulation of agrochemicals. One chemical that has received most of the spotlight is glyphosate. Glyphosate (N-phosphonomethylglycine) is an active ingredient found in the most commonly used herbicides worldwide and has become the most heavily used chemical weedkiller, by volume, in human history (Murphy and Rowlands, 2016; ACB, 2019). First patented as a drain cleaner due to its chelating properties, glyphosate was patented by Monsanto in 1970 as a non-selective herbicide – the end-use product called Roundup (Benbrook, 2016; Murphy and Rowlands, 2016). The herbicidal properties of glyphosate result from metabolic poisoning as glyphosate shuts down the enzyme involved in the shikimate pathway (Murphy and Rowlands, 2016). The shikimate pathway ensures the synthesis of aromatic amino acids, and by shutting this process down glyphosate disrupts the plant's development (Defarge et al., 2016). Furthermore, the chelating properties of glyphosate bind vital nutrients (such as iron, manganese, zinc, and boron) in the soil, thereby preventing their uptake by plants (Huber, 2007). In 2010, Monsanto also patented glyphosate as an antibiotic, to be used against a wide variety of soil microorganisms (Murphy and Rowlands, 2016; Mesnage and Antoniou, 2017). However, glyphosate by itself is only slightly toxic to plants. The potency and toxicity of glyphosate formulations result from inert adjuvants and surfactants added to increase the efficiency of glyphosate (Antoniou et al., 2012).

The global glyphosate debate: corporate denial and misinformation

For the last two decades, glyphosate has been the world's most utilized herbicide due to its efficacy and the perception that it is one of the least chronically toxic chemicals to mammals (Benbrook, 2016). However, as glyphosate use has increased, debates about its impact have sparked political contention and lawsuits (ACB, 2019). Glyphosate's appealing combination of safety and effectiveness was thrown into dispute in 2015 when the International Agency for Research on Cancer (IARC) declared glyphosate a 'probable carcinogen to humans (group 2A)' (IARC, 2015). Around this time, many different studies emerged with findings that glyphosate's other environmental safety claims may be false too. Monsanto, since acquired by Bayer, denied the IARC's findings and still promotes glyphosate's safety, supported by the US Environmental Protection Agency's (US EPA) declaration that glyphosate is 'unlikely to be carcinogenic' (US EPA, 2020). Benbrook (2019) explains three key reasons why the findings of the IARC and the US EPA are at odds. First, the US EPA made use of unpublished industry-sponsored data, while the IARC made use of peer-reviewed data. Second, the US EPA assessed glyphosate on its own, while the IARC assessed glyphosate in combination with its adjuvants and surfactants. Third, the US EPA defined exposure very narrowly, while the IARC defined exposure more broadly along a spectrum. The polarized findings have produced contentious debates about glyphosate, generating misinformation that has contributed to the insufficient regulation of glyphosate-based herbicides.

Glyphosate use in Africa: the 'next pesticide frontier'

Glyphosate is one of the top five agrochemicals used in Africa, largely by cotton-producing countries (Mali, Burkina Faso, and Benin) and countries planting genetically modified (GM) herbicide-tolerant (HT) crops (ACB, 2019). Glyphosate kills plants by blocking the EPSPS (5-enolpyruvyl-shikimate-3-phosphate synthase) enzyme of the shikimate pathway

(ISAAA, 2020), and biotechnologists have created two strategies, using soil bacterium genes, to prevent this in GM HT crops. The first strategy is to incorporate a soil bacterium gene which produces a glyphosate-tolerant form of EPSPS (ISAAA, 2020). The second strategy is to incorporate a different soil bacterium gene which produces an enzyme that degrades glyphosate (ISAAA, 2020). Where previous use of glyphosate would kill each plant it came into contact with, genetic modification has made it possible to spray glyphosate-tolerant crops without damaging them, while still killing the weeds (Clapp, 2021). GM crops engineered to be resistant to glyphosate made it possible to use glyphosate as a broad-spectrum, post-emergence herbicide with an extended application time period (Benbrook, 2016). Glyphosate use with GM crops changed not only the way glyphosate is used in agricultural systems, but also the amount that is used, as it could be sprayed throughout the growing season (Clapp, 2021). However, determining the actual quantities of glyphosate used in Africa is difficult, because regulations are frequently circumnavigated, data on import volumes are hard to acquire, and country borders are porous (ACB, 2019). The position of African countries on glyphosate is unclear as there has been little response to the declaration by the IARC of glyphosate's carcinogenic potential and no permanent product bans have been made. This is largely due to the belief that glyphosate is critical to agriculture-based African economies, due to its efficacy and ease of use (ACB, 2019).

South Africa is the largest consumer of agrochemicals on the African continent and accounts for about 2 per cent of global agrochemical use (Handford et al., 2015; ACB, 2019). With 4,500 different agrochemical products registered in the country, South Africa is one of the most important African markets and a hub for international agribusiness (Clausing et al., 2020). Investment and agribusiness are growing continuously, and between 2009 and 2013 agrochemical expenditure rose by 55 per cent (UNEP, 2013). While many high-income countries have embarked on processes and policies to reduce agrochemical use, the use of agrochemicals continues to expand in South Africa, with little regard given to updating regulatory frameworks. For example, macroeconomic policies such as the 1994 Reconstruction and Development Programme, still encourage agrochemical use among smallholder farmers (London and Rother, 2000). South Africa has 96 registered glyphosate formulations and is the biggest glyphosate consumer on the continent, with 23 million litres sold at an estimated value of R641 m (about $33 m) in 2012 (Gouse, 2014). Glyphosate, now the most used herbicide in South Africa, was first introduced in the 1990s in tandem with the introduction of GM HT crops (Gouse, 2014; Lifestyle Reporter, 2019). Consequently, the use of glyphosate in South Africa is highly correlated to production of GM HT maize and soybeans – crops which form part of the staple diet of most citizens (Lifestyle Reporter, 2019). In the 2012–13 agricultural season, maize, wheat, and soybean farmers purchased 65 per cent of all glyphosate sold in South Africa (Gouse, 2014).

Chemical geographies of violence

The word 'violence' evokes thoughts of explosive and spectacular harm. A visceral and sensational event is expected, with violence occurring in a gory and dramatic fashion. With this notion of violence in mind, one tends to overlook harm that has occurred over time, harm that is not obvious and does not have a specific actor or a specific victim. Nixon (2011) urges us to look beyond 'spectacular' violence to a different kind of violence: a violence that is incremental and accretive, otherwise known as slow violence. Nixon's conceptualization of 'slow violence' refers to harm that is accumulative and occurs largely 'out of sight'. 'Slow violence' is inflicted and concealed by the passage of time and refers to harm that most would not consider violence at all (Nixon, 2011). To understand how slow violence is manifested, we must look to another form of violence that is entrenched within institutional structures and appears 'natural' or unalterable: structural violence (Davies, 2019). Structural violence was conceptualized by Johan Galtung in 1969 and describes social harms that are concealed, ingrained, and institutionalized beyond recognition (Davies, 2019). It is this embeddedness and fixity that gives structural violence its 'silent potency' (Davies, 2019: 5). Harm from structural violence is gradual and works slowly to erode human and environmental health. Davies (2019) argues that conceptualizations of slow and structural violence are irrevocably interlinked: slow violence is an inherently structural concept and both forms of violence extend beyond direct and immediate effects. Both structural and slow violence refer to systemic harm that is camouflaged within routine functions of society itself: a normalization of suffering (Davies, 2019). Davies posits that to analyse slow violence without attending to its structural foundations is 'an impoverishment of the concept' as both can be mutually reinforcing (2019: 6). The failure to adequately regulate the use of agrochemicals, which gives rise to environmental and social harms, constitutes a form of structural and slow violence. Slow violence is analysed in terms of human exposure to agrochemicals and environmental pollution: harm that is largely invisible and accumulative over large timeframes. The foundations of slow violence are analysed in this chapter as structural violence through poor ERA and inadequate regulations: harm that has been built into our governmental institutions.

Three routes of human exposure to agrochemicals

Human and environmental harm from agrochemicals is complicated by the issue of time and the uncertainty it produces. In toxicology studies, time is an important factor and determines the extent of damage a toxic substance can cause: longer exposure generally equates to more harm (Davies, 2018). However, when it comes to the everyday reality of agrochemical use, which is less controllable and occurs beyond clinical conditions, the issue of time serves to create serious ambiguity and 'toxic uncertainty' (Davies, 2018). Continuous application of agrochemicals over time gives the products the ability to 'defer their harmful consequences across time and space, putting distance and uncertainty between a toxic hazard and the people [and environment] it affects' (Davies, 2018: 1538). As such, long-term chronic ill-health, or poor soil health, is not attributed to agrochemical use. Studies of agrochemical harm on both human and environmental health do not consider this temporal longevity.

The proliferation of agrochemicals has created three routes of human exposure, each of which is looked at here, using glyphosate as a lens. Each route, with its associated harms, constitutes a form of violence. The first route of exposure is direct exposure. Direct exposure to glyphosate refers to the occupational exposure of farmers and farm workers, the frontline victims (ACB, 2019). The African agricultural context poses a number of challenges to safe agrochemical use (London and Rother, 2000; ACB, 2019). There is a lack of personal protective equipment and washing facilities, as well as inadequate control over the workplace. Farm workers often have little access to health care services that can diagnose and treat agrochemical exposure, and language barriers and low literacy levels can prevent adequate understanding of agrochemical labels and instructions. Agricultural extension services are often understaffed and undertrained and there are practical difficulties which prevent safe use practices (such as the proper disposal of agrochemical storage containers). For these reasons, farmers and farm workers in Africa are especially vulnerable to agrochemical exposure (ACB, 2019). The slow violence of exposure, due to poisoning events or a lack of personal protective equipment, is complicated by a paucity of data for individuals who are exposed to glyphosate occupationally (Gillezeau et al., 2019). It is therefore not possible to determine causality between exposure to glyphosate and longer-term illness, such as non-Hodgkin lymphoma. Long-term occupational exposure to glyphosate is a form of slow violence, but so too is the lack of adequate monitoring data in these working conditions (the structural foundation of exposure's slow violence).

The second route of human exposure to glyphosate is indirect exposure. Indirect exposure occurs when glyphosate-based herbicides drift off-farm and into communities that live close to treated areas (ACB, 2019). Glyphosate drift off-farm can occur for four reasons: problems with application equipment, wind conditions, human error, and negligence (Benbrook, 2019). For example, in Colombia, DNA damage as a result of spray drift has been recorded as far as 10 km away from the spray site, and in Argentina, communities living near fields that are sprayed with glyphosate have higher rates of reproductive disorders and congenital abnormality (Leahy, 2007; Avila-Vazquez et al., 2018). Spray drift is an ongoing phenomenon that largely affects vulnerable and marginalized groups, such as rural African farming communities who live on or in close proximity to farms, and results in a lower quality of life (London and Rother, 2000). The health consequences of agrochemical use are generally only recognized in the form of acute toxicity; long-term chronic ill-health is largely ignored as it is harder to determine (Rother et al., 2008). This is exacerbated by a lack of effective monitoring and reporting systems, particularly in the Global South (London and Rother, 2000). The absence of evidence of agrochemical impacts is viewed by policymakers as evidence of an absence of problems, and therefore regulatory structures are unable to acknowledge or redress harm from agrochemicals (London and Rother, 2000). A gradual decline in quality of life and human well-being, largely undocumented and dismissed by regulators, is slow violence.

The third route of glyphosate exposure is through residues in food. To control the extent of residue ingestion, maximum residue levels (MRLs) are statistically derived from field trials, and acceptable daily intake (ADI) levels are determined by toxicological data (Handford et al., 2015). MRLs and ADI vary per crop and per country. For example, the ADI for glyphosate in the US is 1.75 mg per kg of body weight per day, in the EU it is 0.3 mg, and in South Africa it is 1 mg (ACB, 2019). Glyphosate residues and their toxicity are widely contested. There is an increasing body of literature that points to the adverse environmental and social effects of glyphosate at ultra-low levels of 0.1 parts per billion (Mesnage et al., 2015a; Benbrook, 2016). Industry-sponsored studies have found the opposite, arguing that exposure below set thresholds poses no harm (Seneff, 2021). Many studies have linked glyphosate exposure to autism, cancer, birth defects, infertility, Alzheimer's disease, coeliac disease, colitis, heart disease, inflammatory bowel syndrome, Parkinson's disease, and liver and kidney disease (Samsel and Seneff, 2015; Kubsad et al., 2019). The findings of the IARC on glyphosate's carcinogenic potential were based on the relationship between glyphosate exposure and higher chances of developing non-Hodgkin lymphoma (IARC, 2015). A study by Mesnage et al. (2021) showed that glyphosate is able to affect the gut microbiome through the same mechanism with which it acts as a weedkiller. This is important because gut bacteria imbalances have been increasingly linked to a wide array of diseases, such as cancer, type 2 diabetes, obesity, depression, and Alzheimer's disease (Robinson, 2021). What is important to note is that some of these adverse human effects have been caused by glyphosate consumption within acceptable MRLs and ADI levels (Mesnage et al., 2015b). Data such as these challenge regulatory thresholds and the mechanisms used to set them. Given that causality between harm from accumulative residue exposure and chronic long-term ill-health is difficult to determine, it is often overlooked and under-regulated. In this way, it constitutes a form of slow violence.

Pseudo-persistence: glyphosate's presence in the environment

Environmental slow violence occurs via direct damage to the environment that threatens our own survival (Lee, 2016). Environmental harms are similarly complicated by the issue of time and the disconnect that occurs between the original action and its consequences. As such, regulatory inaction is excused since sources of environmental harm are temporally disconnected as well as 'dispersed and entangled in a complex assemblage of corporate power, state authority, local regulations, and capitalist structures of accumulation' (Davies,

2018: 1539). Harm from environmental slow violence is most often borne by vulnerable and marginalized groups who have been traditionally excluded from decision-making about environmental issues (London and Rother, 2000). The inaction of agrochemical regulation becomes 'violent' in the harm and slow violence it facilitates. The use of glyphosate and its contested environmental consequences are similarly entangled in this complicated nexus of science and uncertainty, action and inaction (Davies, 2018).

Environmental criticisms about glyphosate's use are premised on the idea that the increase in glyphosate use results in increased environmental loads and higher human exposure (Benbrook, 2016). The use of glyphosate inflicts slow environmental violence in three ways: weed resistance, soil and water contamination, and biodiversity loss. Glyphosate-resistant weeds have developed slowly, but as of 2018 there were 38 resistant species (Heap and Duke, 2018). The development of glyphosate resistant species was largely the result of GM HT crop adoption and persistent glyphosate application rather than using a variety of strategies to control weeds (Duke et al., 2018). Resistant weed species reduce the cost and efficacy advantages of glyphosate and GM HT crops but also induce environmental harm because farmers are forced to incorporate more toxic and ecologically harmful herbicides into their weed management regimes, for example paraquat in maize crops (Duke et al., 2018). Furthermore, glyphosate resistant weeds have encouraged biotechnology firms to genetically modify crops to be resistant to more than just glyphosate. For example, some soybeans now contain transgenes for both dicamba (a highly volatile broad-spectrum herbicide that can damage non-target species through spray drift) and glyphosate resistance (Duke et al., 2018). Environmental violence due to the overuse of glyphosate manifests as glyphosate resistant weed emergence that, in turn, results in intensified glyphosate use as well as the use of older and more harmful herbicides.

Environmental violence due to glyphosate also manifests as poor soil health. Glyphosate has a low soil persistence, and repeated applications are required to control weeds (ACB, 2019). This quality was thought to make glyphosate environmentally friendly. However, industrial agriculture is highly dependent on glyphosate and applies it frequently; as a result, glyphosate's presence in the soil has increased and soil organic matter has decreased (Virginia et al., 2018).

Chemicals are divided into groups based on their interactions with the environment. The first group is 'persistent chemicals' that, being very slow to degrade, create long-term environmental problems (Bernhardt et al., 2017). The second group is 'pseudo-persistent chemicals'. These chemicals are not technically persistent but are released at rates that exceed their degradation rates (Bernhardt et al., 2017). While technically glyphosate has low persistence, its excessive use in agricultural systems prevents its complete degradation in the soil as it is continuously applied. It is important to note that glyphosate's main metabolite, aminomethylphosphonic acid (AMPA), is categorized as persistent in soils. In this way, glyphosate has become a pseudo-persistent chemical with environmental consequences. A study conducted by Virginia et al. (2018) assessed AMPA and glyphosate presence in industrially farmed soil and found that 20 to 50 per cent of glyphosate remained in the soil for 60 days and that 1 mg of glyphosate accumulated per kg of soil after every five crop-spraying events. The persistence of glyphosate in the soil is important due to its antibiotic properties. As glyphosate accumulates in the soil, beneficial microbes and bacteria – such as rhizobia, which fix nitrogen in the soil – are killed, and the production of soil-borne pathogens is increased – such as *Fusarium* fungi, which are responsible for sudden-death syndrome in corn and soybeans (Vinje, 2013; Murphy and Rowlands, 2016). There is also concern that the chelating properties of glyphosate bind minerals in the soil, preventing their uptake by plants and leading to a reduction in growth, an increase in susceptibility to pathogens, and potentially even a reduction in the nutritional content of our food (Murphy and Rowlands, 2016). The chelating properties of glyphosate have also been used by the agrochemical industry as evidence that glyphosate cannot leach into groundwater. However, glyphosate is water-soluble and studies have shown that during heavy rainfall, glyphosate can wash out of the soil and contaminate water sources, persisting with a half-life of up to five months (Watts et al., 2016). Another manifestation of glyphosate's slow environmental violence is the reduction of overall biodiversity across farmland and surrounding areas. Glyphosate reduces the diversity of plants as well as pollinators. Research has revealed that bees exposed to glyphosate are more vulnerable to gut infections, disorientation, and slow larvae growth (Rogers, 2019). Continuous declines in soil health, water health, and biodiversity are particularly serious for rural African communities, who are reliant on natural resources for their livelihoods, and for the rich biodiversity of African landscapes.

The manifestations of glyphosate's environmental violence – glyphosate resistant weeds, soil health decline, water contamination, and biodiversity loss – are accumulative and are spatially and temporally dispersed. The resultant harm threatens our own survival as a species, and yet the structural foundations (ERA and agrochemical legislation) of this slow environmental violence fail to consider how glyphosate persists through multiple applications, how it moves through food webs, how it affects species interactions, how it shifts local species communities, or how it indirectly alters core ecosystem functions (Bernhardt et al., 2017). Consequently, the slow environmental violence of glyphosate persists.

Political structures of violence

When a chemical is so pervasive, so ubiquitous, so nearly impossible for even the most diligent person to avoid, it is especially incumbent on regulatory agencies and elected officials to ask tough questions, conduct rigorous investigations and hearings, and put the health and safety of its populace first. But in the case of glyphosate, this hasn't happened.

It's an abdication of responsibility and a disgrace to democracy. (Seneff 2021: 12)

Outdated environmental risk assessment

To better understand harm generated from slow violence, we need to understand the hidden and unrecognized structural violence present within our regulatory institutions (Lee, 2019). Structural violence has important implications as it is often the root cause of other forms of violence (Lee, 2019). The failure of governmental institutions to safeguard human and environmental health is a form of structural violence. This is particularly evident in Africa, where in many countries agrochemical legislation is missing or underdeveloped, plagued by issues such as narrow scope and lack of governmental enforcement capacity (Handford et al., 2015). As part of the failure to adequately assess and regulate agrochemicals, violent structures have been created, structures that fail to prevent harm. The ability to expose people to harm, not to 'make them die' but to 'let them die', was theorized by Mbembe as 'necropower' (2003). Harm that results from poor or missing agrochemical regulation is correctable and preventable through human agency, and it is therefore of critical importance that we attend to the violent failure of our regulatory arena (Lee, 2019).

The dimensions of the agrochemical regulatory arena are broad. Beyond MRLs and ADI, environmental risk assessment is used to determine whether the use of an agrochemical can be made safe for the receiving environment. ERA is one of the main regulatory procedures governing agrochemical use, and corporations often use ERA approval of their chemicals as a defence against accusations of adverse effects. However, just as MRLs are inadequate in governing human exposure to agrochemicals, ERA for agrochemicals is outdated and in dire need of an overhaul (Topping et al., 2020). ERA processes are problematic for numerous reasons, the primary one being their narrow scope of assessment. ERA is premised on the assumption that the risks of agrochemicals can be managed through single-product, single-crop assessments and that this will provide adequate environmental protection. This approach has been disproved as it does not account for the full extent of spatial and temporal exposure, given that a single product is not used on a single crop, but rather a mixture of chemicals is used in a process of sequential treatments (Topping et al., 2020). Whittingham et al. (see chapter 12) discuss how dominant models of Science-based risk assessment are largely incapable of assessing the relational consequences and implications of industrial agricultural technologies, especially in the diverse social and environmental context of the Global South. A politically driven Science-based approach to ERA, one that is tied to economic growth, is incapable of assessing more relational and 'non-scientific' (historic, social, economic, or political) concerns and thus struggles to safely or comprehensively assess the risk posed by agrochemicals (Whittingham et al., 2022). The scope of ERA is further restricted to the active ingredient of formulations, and not the product as a whole. For example, glyphosate is assessed for toxicity, but the end-use product Roundup is not assessed. This presents a challenge in determining the toxicity of a product, because its adjuvants and surfactants, which are often more toxic than the active ingredient itself and work to enhance its toxicity, are not analysed (Antoniou et al., 2012).

The narrow scope of ERA is complicated by the notion of 'confidential business information' (CBI). CBI permits agrochemical manufacturers to withhold information about the exact composition of their products. As such, the entire makeup of agrochemical products is not known and regulators cannot fully know their environmental and health consequences (ACB, 2019). By narrowing the scope of assessment and prioritizing corporate interests over human and environmental health, ERA processes fail to prevent harm and become a violent structure. Topping et al. (2020) outline how ERA, in its current form, is unable to account for the intensified stressors that our agricultural systems face (such as climate change, habitat destruction, and landscape homogeneity) and that its underlying assumptions do not hold in many conventional and intensive agricultural landscapes. The authors also explain the need to align agrochemical regulation with environmental reality and policy as 'regulatory ERA for pesticides has fallen out of step with scientific knowledge and societal demands for sustainable food production' (Topping et al., 2020: 360). If ERA systems are to prevent harm and become non-violent structures, the authorization of agrochemical use must escape the binary paradigm of 'safe' and 'unsafe' and, instead, create space for more alternative and relational approaches to risk assessment to emerge, where accepted risks and harm are transparently communicated to the public in order to facilitate sustainable agrochemical regulations and safer agrochemical use (Topping et al., 2020; (see chapter 12)).

South Africa's inadequate and inefficient regulatory capacity

In 2016, Prince Mangosuthu Buthelezi, president of the Inkatha Freedom Party (a political party in South Africa), released a press statement about GM crops and glyphosate in which he asked, 'Why is South Africa poisoning its people?' (Buthelezi, 2016). The answer to this question requires a deeper look into South Africa's current agrochemical legislation and the structural violence that pervades it. South Africa appears to conform to international agrochemical standards, given that residues in food are regulated, the International Labour Organization's standards for handling dangerous chemicals are followed, and the infrastructure of the chemical registration process is much more advanced than in other countries in Africa (London and Rother, 2000). Despite this, there are shortcomings in South Africa's policy framework for agrochemical safety, outlined by London and Rother (2000), that allow the slow violence of chronic exposure and poisoning to persist. These shortcomings illustrate the 'violent inaction' of regulatory structures that may not be 'making people die', but are most certainly 'letting people die' (Mbembe, 2003; Davies, 2018).

'Outdated, fragmented, and ineffective'. Legislation on agrochemicals in South Africa is fragmented, outdated, and ineffective (London and Rother, 2000). There are 14 different laws pertaining to agrochemicals which are administered by seven different government departments. As a result, there are overlaps, duplications, and gaps throughout the framework. The function of registering agrochemicals is the responsibility of the Department of Agriculture, Land Reform and Rural Development (DALRRD), but many other government departments are involved, in particular those responsible for labour and employment, water affairs, health, trade and industry, finance, science and technology, and transport (Cowen, 2019). The Act governing agrochemicals is the Fertilizers, Farm Feeds, Agricultural Remedies and Stock Remedies Act, which was passed by parliament in 1947 (Act 36 of 1947). Despite some minor amendments, the Act has remained largely unchanged for more than 70 years and registered agrochemical products have not been re-evaluated. The failure to update regulatory structures according to today's more stringent standards of risk assessment allows for harm to occur, since the regulations fail to minimize adverse effects on human health and the environment. This is the first instance of structural violence within our regulatory system. The urgent need to review Act 36 of 1947 was acknowledged by the Pesticide Management Policy of 2010 (DAFF, 2010), which made commitments to systematic revision and outlined numerous concerns relating to the Act, the most notable of which were: the Act does not adequately address constitutional requirements in relation to the Bill of Rights, that is, the rights to an environment which is not harmful to health, to access to information, to openness, transparency, and participation in decision-making, and also to just administrative action; there is no requirement for the review of registered agrochemicals or the re-evaluation of old chemicals; the establishment of agrochemical use surveillance and monitoring systems to gather information on common conditions of agrochemical use and their impact on human health and the environment has been poor; the Act does not adequately protect non-target areas (e.g. residential areas, schools, and hospitals) from exposure to spraying activities; and the Act does not adequately encourage registration that favours lower-risk products and reduced reliance on agrochemicals overall.

Despite its commitment to many laudable objectives, the 2010 Pesticide Management Policy did not come to fruition and few significant changes were made. Mbembe's (2003) 'necropower' is evident in the state's failure to act upon its acknowledgement that there are serious flaws in the system that may be causing social and environmental harm. There are many international frameworks available that, while not without flaws, provide assistance with designing adequate legislation regarding the registration, data requirements for registration, labelling, and monitoring of agrochemicals (Cowen, 2019). For example, the Food and Agriculture Organization of the United Nations' International Code of Conduct on Pesticide Management provides a reference framework for government authorities, industry, and stakeholders in general to guide processes of legislation review, update, or design (FAO and WHO, 2014). The failure of DALRRD to take advantage of assistance such as this starkly illustrates its 'violent inaction' and breaches the constitutional rights of South African citizens.

Conflicts of interest: control over agricultural regulation and promotion. The second shortcoming of South Africa's agrochemical regulation lies in DALRRD's responsibility for the primary legislation controlling agrochemicals and their registration, as well as for the promotion of agricultural production (London and Rother, 2000). Responsibility for both regulation and promotion begets a serious conflict of interest that has created a 'pesticide culture' of non-interference, with an ambivalent relationship between government and industry that continues to prevent the implementation of meaningful agrochemical control measures. Without meaningful measures of agrochemical control, harms such as spray drift and water contamination are allowed to exist within the system as forms of slow violence.

DALRRD's prioritization of agricultural trade and promotion over social health and environmental safety can be seen in the legislation that governs export-driven farming sectors in South Africa. Foreign market preferences have resulted in progressive agrochemical policies that are far ahead of internal South African regulation, favouring the health of consumers in the Global North. For example, the MRLs for produce bound for international markets are tightly regulated by DALRRD, the South African Bureau of Standards, the Perishable Products Export Control Board, industry working groups, and agrochemical companies (Sirinathsinghji, 2017). MRL regulation for domestic markets does not exist, and produce that violates international export standards is often sold domestically, disregarding the health of South African consumers. By failing to act on these inconsistencies and by jeopardizing social and environmental health for economic interests, the government is facilitating the persistence of violent structures which accommodate harm.

Absence of public participation and the 'wall of silence'. Poor provision for public participation is the third shortcoming of South Africa's agrochemical regulations, along with limited access to registration and appeal processes (London and Rother, 2000). This is largely due to the influence of powerful stakeholders over government departments and the concept of CBI. Agrochemical policy is seen as a matter to be decided solely between industry and government. This is evident in the absence of a public register that lists available products and active ingredients. DALRRD is mandated by Act 36 of 1947 to facilitate such a database and ensure its accessibility to the public. Instead, CropLife operates the Agri-Intel database and charges a substantial fee while reserving right of access (Clausing et al., 2020). There is a lack of transparency in South African agrochemical regulation, and this has created a 'wall of silence' that protects the agrochemical industry (Clausing et al., 2020). By failing to facilitate public participation, DALRRD is in contravention of the

National Environmental Management Act (Act 107 of 1998), which requires public participation in the development of environmental management plans, and is flouting the public's constitutional right to participation and collaboration (Cowen, 2019). The 'wall of silence' surrounding agrochemicals and their legislation restricts meaningful protection of environmental and social health and, in doing so, cements structural violence within the regulatory system.

Double standards of agrochemical use. The fourth shortcoming is the failure to adequately apply the precautionary principle. South Africa's agrochemical approval process is relatively lax and permits the registration and sale of chemicals that have been banned in other countries. In 2019, at least 17 active ingredients listed in the Rotterdam Convention annex that are banned in the EU were exported to South Africa – these included paraquat, aldicarb, pentachlorophenol, and methyl parathion (Clausing et al., 2020). For example, the wine-growing regions of the Western Cape make widespread use of Bayer's insecticide Tempo SC, which contains (beta-)cyfluthrin as its active ingredient – (beta-)cyfluthrin is highly hazardous, classified by the World Health Organization (WHO) as class 1B (Clausing et al., 2020). Regulatory inaction against known substances of harm illustrates structural apathy towards social and environmental health and a particular disregard for the health of citizens of the Global South. The structural violence of regulatory inaction is of particular concern because many neighbouring African countries make their decisions in line with agrochemical approval decisions made in South Africa.

Conclusion

To degrade and damage the natural environment is an act of violence, not only against our ecosystems, but also against those who are dependent on them. In turn, the failure to prevent harm to human health and the environment must also be seen as an act of violence. Agrochemical legislation and regulation in the Global South demonstrate a chronic lack of concern for human and environmental well-being. In failing to adequately protect people and the environment, agrochemical legislation perpetuates instances of slow violence, such as spray drift and residue contamination. Given these failures, the agrochemical system becomes violent in its inaction and apathy towards harm. To address structural failure and the slow violence it generates, we need to recognize that the use of agrochemicals is not only a technical issue, but also a political one. We need to acknowledge that agrochemicals are transformative agents, spatially and temporally, and treat them as such. Our capitalist agricultural assemblage is plagued by chemical ubiquity, and concerns about agrochemicals are not new, but to address these concerns requires new, more relational ways of thinking about human–nature assemblages. We must shift our focus away from individual agrochemicals as inputs for the agricultural commodity chain and instead look at the global pesticide complex itself. In doing so, we must recognize the structural violence of poor regulation and address the slow violence that inaction generates. Only by correcting failing regulatory structures and increasing governmental capacity to reduce harm, can we begin to address the numerous concerns surrounding the issue of agrochemicals.

References

African Centre for Biodiversity (ACB) (2019) *Africa must ban glyphosate now*, ACB, Johannesburg.

Antoniou, M., Habib, M.E.M., Howard, C.V., Jennings, R.C., Leifert, C., Nodari, R.O., Robinson, C.J. and Fagan, J. (2012) 'Teratogenic effects of glyphosate-based herbicides: divergence of regulatory decisions from scientific evidence', *Journal of Environmental & Analytical Toxicology* S: 4(006): 2161–0525 <https://www.hilarispublisher.com/open-access/terato-genic-effects-of-glyphosate-based-herbicides-divergence-of-regulatory-decisions-from-scientific-evidence-2161-0525.S4-006.pdf>.

Avila-Vazquez, M., Difilippo, F.S., MacLean, B., Maturano, E. and Etchegoyen, A. (2018) 'Environmental exposure to glyphosate and reproductive health impacts in agricultural population of Argentina', *Journal of Environmental Protection* 9(3): 241–253 <http://doi.org/10.4236/jep.2018.93016>.

Benbrook, C.M. (2016) 'Trends in glyphosate herbicide use in the United States and globally', *Environmental Sciences Europe* 28(1): 1–15 <https://doi.org/10.1186/s12302-016-0070-0>.

Benbrook, C.M. (2019) 'How did the US EPA and IARC reach diametri-cally opposed conclusions on the genotoxicity of glyphosate-based herbicides?', *Environmental Sciences Europe* 31(1): 1–16 <https://doi.org/10.1186/s12302-018-0184-7>.

Bernhardt, E.S., Rosi, E.J. and Gessne, M.O. (2017) 'Synthetic chemicals as agents of global change', *Frontiers in Ecology and the Environment* 15(2): 84–90 <https://doi.org/10.1002/fee.1450>.

Buthelezi, M. (2016) 'Why is South Africa poisoning its people?' [press statement], 17 November, Inkatha Freedom Party <https://www.ifp.org.za/newsroom/south-africa-poisoning-people/>.

Clapp, J. (2021) 'Explaining growing glyphosate use: the political economy of herbicide-dependent agriculture', *Global Environmental Change* 67 <https://doi.org/10.1016/j.gloenvcha.2021.102239>.

Clausing, P., Luig, L., Urhahn, J. and Beushausen, W. (2020) *Double Standards and Hazardous Pesticides from Bayer and BASF: A Glimpse behind the Scenes of the International Trade in Pesticide Active Ingredients*, Rosa Luxemburg Stiftung Southern Africa, Johannesburg <https://www.rosalux.de/fileadmin/images/publikationen/Studien/Double_Standards_and_Hazardous_Pesticides_ENG_20210422.pdf>.

Cowen, S. (2019) *Chemical Remedies: Submissions on Law Reform*, The Real Thing Food Supplements, Cape Town <https://therealthing.co.za/images/TheRealThing/Documents/Chemical_Remedies_Submission_on_Law_Reform.pdf>.

Davies, T. (2018) 'Toxic space and time: slow violence, necropolitics, and petro-chemical pollution', *Annals of the American Association of Geographers* 108(6): 1537–1553 <https://doi.org/10.1080/24694452.2018.1470924>.

Davies, T. (2019) 'Slow violence and toxic geographies: "out of sight" to whom?', *Environment and Planning C: Politics and Space* 40(2): 409–427 <http://doi.org/10.1177/2399654419841063>.

Defarge, N., Takács, E., Lozano, V.L., Mesnage, R., Spiroux de Vendômois, J., Séralini, G.E. and Székács, A. (2016) 'Co-formulants in glyphosate-based herbicides disrupt aromatase activity in human cells below toxic levels', *International Journal of Environmental Research and Public Health* 13(3): 264 <https://doi.org/10.3390/ijerph13030264>.

Department of Agriculture, Forestry and Fisheries (DAFF) (2010) *Adoption of Pesticide Management Policy for South Africa*, Government Gazette 33899 <https://www.gov.za/sites/default/files/gcis_document/201409/338991120.pdf>.

Duke, S.O., Powles, S.B. and Sammons, R.D. (2018) 'Glyphosate – how it became a once in a hundred year herbicide and its future', *Outlooks on Pest Management* 29(6): 247–251 <https://doi.org/10.1564/v29_dec_03>.

Food and Agriculture Organization (FAO) and World Health Organization (WHO) (2014) *The International Code of Conduct on Pesticide Management*, FAO, Rome <https://www.fao.org/fileadmin/templates/agphome/documents/Pests_Pesticides/Code/Code_ENG_2017updated.pdf>.

Gillezeau, C., van Gerwen, M., Shaffer, R.M., Rana, I., Zhang, L., Sheppard, L. and Taioli, E. (2019) 'The evidence of human exposure to glyphosate: a review', *Environmental Health* 18(1): 1–14.

Gouse, M. (2014) *Assessing the Value of Glyphosate in the South African Agricultural Sector*, Department of Agricultural Economics, Extension and Rural Development, University of Pretoria <https://www.up.ac.za/media/shared/108/2015%20Working%20papers/Value-of-glyphosate-in-sa-agriculture-Mgouse.zp56221.pdf>.

Grand View Research (2022) 'Agrochemicals market size worth $272.0 billion by 2028 | CAGR: 4.1%' [press release] <https://www.grandviewresearch.com/press-release/global-agrochemicals-market>.

Handford, C.E., Elliott, C.T. and Campbell, K. (2015) 'A review of the global pesticide legislation and the scale of challenge in reaching the global harmonization of food safety standards', *Integrated Environmental Assessment and Management* 11(4): 525–536 <https://doi.org/10.1002/ieam.1635>.

Heap, I. and Duke, S.O. (2018) 'Overview of glyphosate resistant weeds worldwide', *Pest Management Science* 74(5): 1040–1049 <https://doi.org/10.1002/ps.4760>.

Huber, D.M. (2007) 'What about glyphosate-induced manganese deficiency', *Fluid Journal* 15(4): 20–22 <https://fluidfertilizer.org/wp-content/uploads/2016/05/58P20-22.pdf>.

International Agency for Research on Cancer (IARC) (2015) *IARC Monographs Volume 112: Evaluation of Five Organophosphate Insecticides and Herbicides*, IARC, Lyon, France <https://www.iarc.who.int/wp-content/uploads/2018/07/MonographVolume112-1.pdf>.

International Service for the Acquisition of Agri-biotech Applications (ISAAA) (2020) *Herbicide Tolerance Technology: Glyphosate & Glufosinate*, Global

Knowledge Center on Crop Biotechnology, ISAAA, Los Baños, Philippines <https://www.isaaa.org/resources/publications/pocketk/foldable/Pocket%20K10%20(English).pdf>.

Isgren, E. and Andersson, E. (2021) 'An environmental justice perspective on smallholder pesticide use in sub-Saharan Africa', *The Journal of Environment & Development* 30(1): 68–97 <https://doi.org/10.1177%2F1070496520974407>.

Kubsad, D., Nilsson, E.E., King, S.E., Sadler-Riggleman, I., Beck, D. and Skinner, M.K. (2019) 'Assessment of glyphosate induced epigenetic transgenerational inheritance of pathologies and sperm epimutations: generational toxicology', *Scientific Reports* 9(1): 1–17 <https://doi.org/10.1038/s41598-019-42860-0>.

Leahy, S. (2007) 'Colombia-Ecuador: studies find DNA damage from anti-coca herbicide', Inter Press Service, 16 June <https://www.ipsnews.net/2007/06/colombia-ecuador-studies-find-dna-damage-from-anti-coca-herbicide/>.

Lee, B.X. (2016) 'Causes and cures VIII: environmental violence', *Aggression and Violent Behavior* 30: 105–109 <https://doi.org/10.1016/j.avb.2016.07.004>.

Lee, B.X. (2019) *Violence: An Interdisciplinary Approach to Causes, Consequences, and Cures*, John Wiley & Sons, Hoboken, NJ.

Lifestyle Reporter (2019) 'Why do we still use glyphosate?', *IOL*, 6 June <https://www.iol.co.za/lifestyle/health/why-do-we-still-use-glyphosate-25192574>.

London, L. and Rother, H.-A. (2000) 'People, pesticides, and the environment: who bears the brunt of backward policy in South Africa?', *New Solutions: A Journal of Environmental and Occupational Health Policy* 10(4): 339–350 <https://doi.org/10.2190%2FHAGW-QU9E-4H86-AW6W>.

Mbembe, A. (2003) 'Necropolitics', *Public Culture* 15(1): 11–40 <https://doi.org/10.1215/08992363-15-1-11>.

Mesnage, R. and Antoniou, M.N. (2017) 'Facts and fallacies in the debate on glyphosate toxicity', *Frontiers in Public Health* 5: 316 <https://doi.org/10.3389%2Ffpubh.2017.00316>.

Mesnage, R., Arno, M., Costanzo, M., Malatesta, M., Séralini, G.E. and Antoniou, M.N. (2015a) 'Transcriptome profile analysis reflects rat liver and kidney damage following chronic ultra-low dose Roundup exposure', *Environmental Health* 14(1): 1–14 <https://doi.org/10.1186/s12940-015-0056-1>.

Mesnage, R., Defarge, N., De Vendômois, J.S. and Séralini, G.E. (2015b) 'Potential toxic effects of glyphosate and its commercial formulations below regulatory limits', *Food and Chemical Toxicology* 84: 133–153 <https://doi.org/10.1016/j.fct.2015.08.012>.

Mesnage, R., Teixeira, M., Mandrioli, D., Falcioni, L., Ducarmon, Q.R., Zwittink, R.D., Mazzacuva, F., Caldwell, A., Halket, J., Amiel, C. and Panoff, J.M. (2021) 'Use of shotgun metagenomics and metabolomics to evaluate the impact of glyphosate or Roundup MON 52276 on the gut microbiota and serum metabolome of Sprague-Dawley rats', *Environmental Health Perspectives* 129(1): 17005 <https://doi.org/10.1289/ehp6990>.

Murphy, D. and Rowlands, H. (2016) *Glyphosate: Unsafe on Any Plate*, Food Democracy Now and The Detox Project <https://usrtk.org/wp-content/uploads/2016/11/FDN_Glyphosate_FoodTesting_Report_p2016-3.pdf>.

Nixon, R. (2011) *Slow Violence and the Environmentalism of the Poor*, Harvard University Press, Cambridge, MA, and London.

Robinson, C. (2021) 'Glyphosate and Roundup disturb gut microbiome and blood biochemistry at doses that regulators claim to be safe', GMWatch <https://gmwatch.org/en/news/latest-news/19677-glyphosate-and-roundup-disturb-gut-microbiome-and-blood-biochemistry-at-doses-that-regulators-claim-to-be-safe>.

Rogers, G. (2019) 'Controversial pesticide under spotlight as bees, butterflies wane', *HeraldLIVE*, 27 November <https://www.heraldlive.co.za/agrilive/2019-11-27-controversial-pesticide-under-spotlight-as-bees-butterflies-wane/>.

Rother, H.A., Hall, R. and London, L. (2008) 'Pesticide use among emerging farmers in South Africa: contributing factors and stakeholder perspectives', *Development Southern Africa* 25(4): 399–424 <https://doi.org/10.1080/03768350802318464>.

Samsel, A. and Seneff, S. (2015) 'Glyphosate, pathways to modern diseases III: manganese, neurological diseases, and associated pathologies', *Surgical Neurology International* 6: 45 <https://doi.org/10.4103/2152-7806.153876>.

Seneff, S. (2021) *Toxic Legacy: How the Weedkiller Glyphosate Is Destroying Our Health and the Environment*, Chelsea Green Publishing, White River Junction, VT.

Shattuck, A. (2021) 'Generic, growing, green? The changing political economy of the global pesticide complex', *The Journal of Peasant Studies* 48(2): 231–253 <https://doi.org/10.1080/03066150.2020.1839053>.

Sirinathsinghji, E. (2017) *No Safe Limits for Toxic Pesticides in Our Foods: Comments on Draft Regulations for MRLs*, African Centre for Biodiversity, Johannesburg <https://acbio.org.za/gm-biosafety/no-safe-limits-toxic-pesticides-our-foods/>.

Topping, C.J., Aldrich, A. and Berny, P. (2020) 'Overhaul environmental risk assessment for pesticides', *Science* 367(6476): 360–363 <https://doi.org/10.1126/science.aay1144>.

United Nations Environment Programme (UNEP) (2013) *Global Chemicals Outlook: Towards Sound Management of Chemicals*, UNEP, Nairobi <https://sustainabledevelopment.un.org/content/documents/1966Global%20Chemical.pdf>.

United States Environmental Protection Agency (US EPA) (2020) *Glyphosate Interim Registration Review Decision Case Number 0178* <https://www.epa.gov/sites/production/files/2020-01/documents/glyphosate-interim-reg-review-decision-case-num-0178.pdf>.

Vinje, E. (2013) 'Is Monsanto's Roundup killing our soil?' Planet Natural, 20 September <https://www.planetnatural.com/roundup-killing-soil>.

Virginia, A., Zamora, M., Barbera, A., Castro-Franco, M., Domenech, M., De Geronimo, E. and Costa, J.L. (2018) 'Industrial agriculture and agroecological transition systems: a comparative analysis of productivity results, organic matter and glyphosate in soil', *Agricultural Systems* 167: 103–112 <https://doi.org/10.1016/j.agsy.2018.09.005>.

Ward, M.H., Nuckols, J.R., Weigel, S.J., Maxwell, S.K., Cantor, K.P. and Miller, R.S. (2000) 'Identifying populations potentially exposed to agricultural pesticides using remote sensing and a geographic information system', *Environmental Health Perspectives* 108(1): 5–12 <https://doi.org/10.1289%2Fehp.001085>.

Watts, M., Clausing, P., Lyssimachou, A., Schütte, G., Guadagnini, R. and Marquez, E. (2016) *Glyphosate*, Pesticide Action Network Asia Pacific, Penang, Malaysia <http://pan-international.org/wp-content/uploads/Glyphosate-monograph.pdf>.

Whittingham, J., Marshak, M. and Swanby, H. (2024) 'Unsettling modernist scientific ontologies in the regulation of genetically modified crops in South Africa', in R. Wynberg (ed.), *African Perspectives on Agroecology: Why farmer-led seed and knowledge systems matter*, pp. 237–271, Practical Action Publishing, Rugby.

PART 3

Ways of seeing and knowing

CHAPTER 11

'Wild wayward free gifts':[1] a gendered view on agroecology and agricultural transitions

Vanessa Farr

Introduction

I acknowledge that the land from which I write is the ancestral territory of the San and Khoe peoples, who helped bring forth human life on Earth 77,000 years ago. The descendants of these people continue to live and work here today.

I acknowledge that the great standard of living enjoyed by many in this area, me included, is directly related to their resources and friendship, and that their contemporary existence is made precarious by their ongoing exclusion from that which brings and safeguards a good life.

I recognize and share their continued struggles for justice, and for life, waters, and lands.

By beginning this chapter with a Land Acknowledgement, a practice that I am grateful to have learnt from Indigenous teachers of Turtle Island (the Indigenous name for North America), I position my dissenting feminist self, descendant of colonial settlers in South Africa on my mother's side, in a respectful, responsible, reverent, and reciprocal relationship (Wall Kimmerer, 2013; Xiiem et al., 2019) with the land, water, struggles, stories, and knowledge practices of generations of peoples of Africa and other continents, whose theories of land and place deeply shape this writing.

As I write, I overlook Zeekoevlei, the largest of Cape Town's abundant shallow lakes, part of the city's extensive system of wetlands. In the past few months, after the city's release from the drought of 2017–2019 that threatened to turn it into the first city in the world to run out of water, this freshwater body has been repeatedly assaulted by flows of effluent that stream into it through two concrete channels built, in the style of colonial efforts to dominate and reshape nature, to 'manage' the flows of the Lotus River some time back in the apartheid era. Alongside it, both under and above the ground, runs the ongoing brutality of 'apartheid in the pipes',[2] an ageing infrastructure initially created to carry waste from dominant-class communities to a wastewater treatment plant south of here, built to make use of the natural abundance of water in this area – the 'ecosystem services', as neoliberal politics calls them – of the confluence of rivers, lakes, and groundwater; and beyond them, False Bay.

After a decade of local struggle to recognize its vital importance for migrating water birds, Zeekoevlei was declared a Ramsar site in 2015.[3] By the end of 2021 it had become the latest of the city's water bodies to be so fouled by sewage that it was closed – permanently, it seems – to recreational use. The *vlei* is sick. In the period following the COVID-19 pandemic, the whole of this postcard-perfect tourist destination at the southern tip of Africa – which is also one of the world's most violent urban settlements – is sick.

Citizens of this city are reeling with the unfolding impacts of the novel coronavirus, and with the pre-existing and intergenerational trauma that is produced by, and reproduces, pandemics of violence, hunger, and poverty. Each day, desperate newcomers, many of them climate refugees from the bone-dry Eastern Cape province, set up flimsy structures among thousands of others like them. To the south of the *vlei*, one informal settlement is home to around a thousand people, who share two drip-flow taps between them.[4] In the communities to the east, some abutting the concrete canals that were once the banks of a river flowing through seasonal wetlands, extreme hunger and thirst, gun violence, rape and assault, and alcohol and drug use take their daily toll. These are the violent legacies of slavery and Indigenous subjugation, originated when colonial settlers forcibly resettled, on the sandy soils of the Cape Flats, both the people who used to farm, hunt, and fish in this area, and the dissenters relocated here from other Dutch colonies.

Fearful people. Fouled waters. Failed systems. This is what has come of centuries of efforts initiated by European men to physically and psychically dominate, de-Africanize, and 'civilize' this area and the life it sustains; to impose scarcity while extracting immeasurable wealth for themselves; and to control both the science and the institutions that produce acceptable knowledge about this place, and the stories that can be told about the actions of these men (Mellet, 2020). As I gaze over the *vlei*, it strikes me how ironic it is that the first legal effort of the men sent here to subdue this land and exert control over this environment was to issue an edict, 'Placcaat 12 of 1655' (Green, 2020: 44), banning activities that would foul the fresh waters that had first drawn them into setting up a waystation at this halfway house for moving the spoils of Dutch colonial plunder back to Europe.

It is not explained, in the conversations my neighbours have as we try to make sense of the sickness of the wetlands, how the waters, birds, geckos, chameleons, and spiders of this land became, like the humans that inhabit it, so separated from each other that they can barely function, let alone flourish. The elephant in the metaphorical room of our community WhatsApp exchanges is our white privilege, which has protected us, in the past, from the reality of the filth that comes with political negligence. Yet I find it difficult to talk in new ways about this fouled water, in the face of continued political propaganda that tells us how privileged we are to live in this, 'the best-run city in the land'. We are angry at being let down, but we are ill-prepared to face the reality that the 'city has made its own Anthropocene' and cannot offer any conditions of liveability without 'a paradigm shift in … water management' or a profound commitment 'to finding and forming an ecopolitics that gives life' to all (Green, 2020: 59, 231).

This is the contemporary disaster that has come from old Europe's belief that it can remake the world in its own image by containing and controlling all that it encounters, dividing nature and city, women and men, dominated and superior; and by attempting to halt and redirect 'flows of rock, water, and life' (Green, 2020: 59). Colonial intrusions do not work in favour of life; but here on the *vlei*, we struggle to decry the imported technologies and engineering that generations of white male settlers have imposed in their efforts to drain, tame, and tax these wetlands (Scott, 2017). We cannot believe that we are being exposed to the sight and smell of this failed hiding of human excrement, so repulsive to Victorian minds, that can no longer be kept from us in water-borne sewage systems carried in pipes buried out of sight in the ground. Like everyone else in the city, we are facing the reality of our own shit.

Stress and addiction expert Gabor Maté would say that the fragmentation of ourselves and our systems is a manifestation of our as yet unexamined collective trauma, and that we will remain frozen in toxicity, expressed as rage but also inaction and nostalgia (Maté, 2009), until we are ready to tell different stories about who we are – to one another and to this land and water. For now, in the face of all the evidence that this system cannot work, we want to keep living in the 'sanctioned ignorance' (Spivak, 1988: 86; Morris, 2010) in which we grew up, as transmitted to us through the stories of brave white men in the history books we studied at school.

It is not explained

What has this story got to do with women and food and farming systems on the African continent? I tell it because, like several other authors in this volume, I am concerned with countering the well-oiled machinery of forgetting, exclusion, and epistemicide (De Sousa Santos, 2010) designed to highlight and authenticate a singular vision of the past and present, and therefore to dictate an imagined future of sameness, of continuity in patterns of exclusion and dominance, whose intentions and pathways are also imagined as continuing forever, unquestioned, and along known lines. In telling it, I hope to bring into the light the contemporary suppressions and distortions necessary to maintain the 'traditions of domination' (Eisler and Fry, 2019), and the trauma, required by settler-colonial patriarchy and enforced by its primary tools, racism and capitalism, which I will explore in this chapter as the foundational causes of African women's distress as the food and farming crises forced onto this continent escalate. The counter-narrative I assemble draws on decades of feminist efforts to unearth women's experiences of the world and (re-)assert their 'role as eternal guardians of lands, waters, and stories' (Xiiem et al., 2019: 11).

Following the meaning-making process proposed by Jo-ann Archibald Q'um Q'um Xiiem, I draw on 'Indigenous storywork' to guide my thoughts.

First, I turn to Wangari Maathai, who remains, a decade after her death in September 2011, the best-known African ecofeminist, to learn how she came to be such an outspoken protector of women's rights to their ancestral waters and lands. Maathai begins her autobiography (Maathai, 2007) with a story, a brief account of the cosmology of her people, the Kikuyu of Kenya. 'God created the primordial parents, Gikuyu and Mumbi,' who had 10 daughters together but no sons. When the girls reached maturity, a divine intervention sent suitable men to Earth to pair with them, and in this way, the 10 matrilineal clans of the Kikuyu, all tracing themselves back to the original daughters, came about. Since then, however, 'many privileges, such as inheritance and ownership of land, livestock, and perennial crops, were gradually transferred to men'. While the Kikuyu still tell this origin story, celebrate their direct descent via their mothers from the primordial mother, Mumbi, and retain some aspects of their original matrilinear practices, Kikuyu culture has become patrilinear. Maathai wryly observes: 'It is not explained how women lost their rights and privileges' over time (Maathai, 2007: 4–5).

Her observation frames an account of loss that is familiar to other feminists who have asked and tried to find answers to questions like hers for decades. How is it that African women, despite their ongoing and crucial association with and knowledge of farming and food systems, have been made so dispossessed and marginal? Ariel Sallah asks succinctly: 'Could there be a connection between the growth of violent, undemocratically imposed, unjust and unfair economic policies and the intensification in brutality of crimes against women?' (Mies and Shiva, 2014: xiv). In the 1950s, feminist economist Ester Boserup proposed that, while European settler-colonial expansion impoverished everyone in the subjugated land, as a patriarchal project reliant on gendered economic and political logics and hierarchies, it affected Indigenous women both more deliberately and more severely than men. One reason for this skewed impact is the gendered assumptions made by early settler-colonialists who could not see or give credence to expressions of women's knowledge, authority, and autonomy, or admit the value of their multiple contributions to social, spiritual, cultural, and material sustenance (Turner and Fischer-Kowalski, 2010).

In her research on pre-contact social and economic systems, anthropologist Jean Comaroff (1985) follows a similar pathway. Turning a feminist eye on archival evidence of 19th-century British military-settler-colonial encounters with the Tswana of Southern Africa, she finds that at the time the first male explorers set out to survey and describe the continent, they encountered societies in which issues such as control over seeds and agriculture were centrally bound up with gendered divisions of labour. That much, at least, was familiar to them from Europe's own rigidly hierarchized labour practices (Schreiner, 1911). Yet it is what they made of these gendered spaces that counts. Such divisions, as Riane Eisler's work has explored for decades, do not necessarily imply 'domination-leaning' societies (Eisler and Fry, 2019); and indeed, there are multiple archival testimonies, in Africa and elsewhere, indicating that pre-contact social formations were based on what Eisler characterizes as partnership, with great equality in the sharing of resources between female and male, young and old. Perhaps the most powerful of these accounts simply acknowledge how healthy, well-nourished, and strong communities were at first contact (Comaroff, 1985; Maathai, 2007; Green, 2020).

It is this sight that clearly startled and disoriented European men, accustomed as they were to the filthy, unsanitary conditions of near famine on the continent they had left behind, where peasants had endured centuries of immiseration from war, famine, pogroms, forced displacement, enclosure, and other deprivations. They had come from a world dependent on division, whose 'whole motley fabric [was] kept together by fear and blood' (Thompson, cited in Taylor, 1984: ix).[5]

So began a long process of gendered sense-making of the spaces and societies these male soldiers and settlers encountered. The archives show how these outsider observers, whose worldview over-associated masculinity with the power, knowledge, and practices that counted, began to paint a socio-economic and cultural picture through which it was possible to reorganize the scenes they were viewing into patterns they could understand. Part of this process required unpacking a physical puzzle, because the labour and domains of Indigenous men were centred on cattle and the *kraal* (cattle-holding pen), which was at the heart of the community. In their own embodied experience in Europe, it was men who lived at the edges of settlements and made dangerous journeys to engage with the wilderness, while women were protected in the home-hearth-heart configuration at the centre.[6] Encountering an inversion of the settlement practices with which they were familiar, and which they considered natural, the settlers were both challenged and disquieted in their patriarchal beliefs about what constituted male vitality and force; and they would make use of this strangeness in two ways as they established their dominion over lands and bodies. Firstly, by reading the *kraal* as a male space in which all important decision-making took place, they would give it primacy by associating with it all events of public, political, and economic importance and, as settler wars began to proliferate, by breaching it as if it were a fortified castle. An important part of justifying their violent, militarized domination of African men would be to devise narratives in which the fierce warriors they encountered were redrawn as effeminate because of where they had physically located their labour (Comaroff, 1985; Green, 2020; Mellet, 2020).

Having solved the problem of men, cattle, and *kraals* in their efforts to neatly rank and categorize the lives and work habits of the people they encountered, Europeans then had to work out what to do with the strength and physical freedom of Indigenous women, who were the very antithesis of the 'parasitic' and effete 'kept' woman simultaneously idealized and despised in the Victorian gender order (Schreiner, 1911). In their farming and food-gathering practices, these women also moved antithetically to the

European gaze, conducting their business away from the settled heart of the community, in what looked to Europeans like the periphery, the fields and the bush, close to the dangers of the wild. This meant wildness, too, had to be gendered as it was tamed, a process achieved through the invocation of tropes equating women and nature, with which settler-colonials would have been familiar from birth. Over time, Southern African women, too, were likened to wildness and, especially, to wild plants: wayward beings beyond the limits of settlements, having nutritional or medicinal properties and therefore being of the body, and unpredictable in their effects on humans when ingested. Defining them as unruly, in turn, helped justify their forced domestication and control by male soldiers and settlers (Schreiner, 1911; Comaroff, 1985).[7]

From its first imposition on African societies, the hierarchical and extractive logic of European domination required women's contributions, especially in the reproductive economy, to be viewed as marginal; a perspective that remains crucial in rendering the broad range of women's caregiving, agricultural, and other food-gathering practices unimportant and unmeasurable. With the proud certainty of their confirmation bias in place, settler-colonial administrations would go on to create the elaborate legal and economic structures that made their interpretation of the worlds they encountered 'true'.

Yet a close reading, especially of the footnotes of the work of a dissenting proto-feminist like Olive Schreiner, indicates that the male perspectives frozen in the archives are wildly off the mark. Male settlers relied not on what they saw, but on what they already knew. They misread and underestimated the relative value and importance to Indigenous communities of crops and cattle, and of work done by women and men, because they could not allow themselves to grasp simple facts, including, for instance, that the shape of the settlements was practical, not ideological, and was an effective means of safeguarding cattle by tucking them away in a *kraal* – not an ontological claim about cattle and men being at the centre of the world. Over time, this initial misreading would undermine entire food systems and create ruinous ecological imbalances. It set in motion a series of '[e]xpulsions and extinctions' (Green, 2020: 113) by imposing imported values on tame and wild animals, and on agricultural food production systems. It motivated further settler-colonial expansion enabled by land enclosure and the forced relocation of communities to inferior soils. Eventually, its logic resulted in today's mass production of commodities that are moved into 'a global food system to feed workers forced into towns' (Green, 2020: 119).

Moreover, the earliest settler-colonial proto-capitalist policies of land privatization for cash crops were explicitly focused on managing Europe's surplus male population, a project crucial to advancing established and normative white male hierarchies. Following established European practice, their imposition required subterfuges including deluding dispossessed young men into going to the colonies to make their fortune, and then wasting them in continuous warfare (Schreiner, [1897] 2019, 1911).

The establishment of white-male-bodied supremacy also relied on the forced movement of 'inferior' Indigenous men into emerging urban centres to serve as labourers. While such men had little choice but to comply, the price of their coercion was offset through the introduction of a gendered legal system by means of which to recruit Indigenous men into European patriarchy. African men were redefined as the 'owners' of lands that had historically been cared for communally, while land and labour mechanisms for leaving women behind, which were already well-practised in Europe, were imposed (Taylor, 1984; Comaroff, 1985; Millar et al., 1996; Criado Perez, 2019). Women's relationship with the land was fundamentally altered when the conditions were created for them to become temporary cultivators of fields over which they had no security of tenure, and from which they were not expected to accrue the benefits of their inputs, either as workers or as interlocutors of the soils they worked (Millar et al., 1996). So powerful were the intersectional mechanisms of exploitation established in the earliest phases of European 'industrial settler campaigns' (Whyte 2017: 208) – the emphasis on the military nature of this conquest is important – that they make African women and soils vulnerable to this day, continuing to burden both with the many-layered effects of 'socially constructed scarcity' (Yapa, 1995: 321), as discussed by Swanby in Chapter 15 of this volume.

As if women mattered

Wangari Maathai's musings are also interesting because, beyond exposing the specificity of women's losses as the new world order took hold, she highlights what happened when Indigenous cosmologies that recognized women's uniqueness and power as progenitors were overwhelmed with the arrival of Western patriarchal coloniality. Does her comment, then, also raise the question of why contemporary Kikuyu appear uncurious about why coloniality methodically dispossessed women and normalized their subordinate, and landless, status? As its proponents moved around the world, inventing racism (Grosfoguel, 2013) and reinforcing classism, proto-capitalist European patriarchy positioned both the repression of women and their children, and the enclosure of land and water, as shared tasks with rewards that reciprocally recognized and encouraged male collusion (Eisler and Fry, 2019). Perhaps, then, Maathai is asking why some Indigenous men chose to comply.

For Ramón Grosfoguel (2013), this patriarchal colonial-settler work of subordinating women's productive and reproductive labour succeeded because it had already been practised for several hundred years in Europe (see also Taylor, 1984). While the conquest of the Americas created 'a new racial imaginary and new racial hierarchy' (Grosfoguel, 2013: 80), Indigenous women were subjected to a misogynistic derision of their epistemologies that forms a continuum with, or extrapolates from, the epistemicide of Indo-European women's knowledge from the 15th to the 18th centuries, achieved by burning alive those marked as witches. Following Silvia Federici

(2004), Grosfoguel characterizes this campaign of violence as 'a strategy to consolidate Christian-centric patriarchy and to destroy autonomous communal forms of land ownership' in Europe long before these methods of control were exported around the world (Grosfoguel, 2013: 85–86). He concludes that when men arrogated to themselves the right to burn women alive they were intentionally destroying a multigenerational tradition of oral knowledge transmission about food systems, land, and farming practices, a violence as weighty in its impact on narrowing knowledge as the burning of the ancient texts that were immolated at around the same time.[8] Many of this volume's contributors would agree, observing that localized knowledge systems are routinely sacrificed as a homogeneous agricultural world order is imposed.

These violent physical erasures of women and their knowledge resulted, by the 20th century, in the global subsumption of women and the totalitarian dominance of 'the ideologically constructed category of western technological man as the uniform measure of the worth of classes, cultures and genders' (Shiva, 1988: 4). The reality produced within this male-fixated worldview makes women disappear, so that planning, policymaking, economies, public infrastructure and institutions, and legal systems accommodate, understand, respond to, and advance the narrowed interests of men. As Caroline Criado Perez examines in *Invisible Women: Exposing Data Bias in a World Designed for Men* (2019), the 'gender data gap' resulting from the generalization of women's experiences has, even in 'a world increasingly reliant on and in thrall to data', led to a loss of accuracy and analytical acuity, with extreme implications for women. When 'men confuse their own point of view with the absolute truth' (Criado Perez, 2019: xii–xiii), women experience a further deepening of patriarchal colonial erasure, being extinguished from African agricultural landscapes as they are reshaped through 'the alignment of Science, politics, and economics' of corporatism and industrialization (See chapter 12).

'Agriculture (from tools to scientific research, to development initiatives) has been designed around the needs of men' (Criado Perez, 2019: 41); and in a continuum with settler-colonialism, decisions are made today between corporates, paid-for science, governments, and international entities designed by, and to serve, Western men. They create ever-deepening cycles of inter-related and incremental loss (Shiva, 1988), denying women equal access to credit, despite their smaller share of cash resources, and overwhelming them with top-down technological transfer. Women are traumatized, losing confidence, autonomy, and control over decision-making when their small-scale, subsistence, and sustainable farming and food processing practices are undermined and replaced with technologized agribusiness, which compromises their long-term productivity and health, and that of the land itself (Millar et al., 1996).

Toxin- and input-reliant monocropping sacrifices women and soil, impacting differently on men's and women's sexual and reproductive health, producing lasting and intergenerational suffering (Nixon, 2011; Tobi et al., 2018).

Broken food systems increase unpaid work, taking a toll on women's mental health that is rarely either measured or mentioned. Following Grosfoguel, I would name this, the ongoing devastation of the intelligence of women's and soils' networks of care, the fifth epistemicide.

The systematic inhibition of 'antipatriarchal thinking and profeminist activism' (Enloe, 2013: 121) enables large corporations to infiltrate governments and regional entities, coercively introducing biotechnological regimes, including genetically modified crops and their accompanying planting and harvesting systems. Male-centred techno-science requires not only that other ways of knowing, but the knowers themselves, be assimilated. In this way, Indigenous people have been perpetually marginalized from formal decision-making since colonial engulfment began, using tactics – familiar to feminists – that permit the subordination and relegation of 'anything associated with femininity' to the realm of that which does not matter, and can therefore be overlooked as inconsequential (Enloe, 2013: 11, 136). Refusing to collect, or accurately analyse, sex-disaggregated data, especially that which could make inequities visible and lead to more effective, life-sustaining interventions for all, allows patriarchy to subordinate and capture the generative power of both women and soils (Millar et al., 1996; Mies and Shiva, 2014).

Same old, same old ...

Haidee Swanby (Chapter 15 of this volume) recalls how Norman Borlaug, the 'father' of the Green Revolution, raged against Rachel Carson, dismissing as 'hysterical' her analysis in *Silent Spring* (1962) of the implications of unleashing into agricultural systems the toxic additives and mechanisms of control devised initially as tools of 'cold-war America's military-industrial complex' (Nixon, 2011: xi). Borlaug's contempt invokes a well-worn trope invented by male European doctors in the 19th century to provide '"evidence" of ... the instability of the female mind' (Devereux, 2014: 20) and pathologize women's resistance to patriarchal control. Detractors used it with alacrity against Carson, who was neither a conventional nor a conformist woman. And it was an effective means of undermining her and her argument – although the prescience of her analysis has only deepened with time (Nixon, 2011: 311).

Nixon's discussion of this dissenting outsider, suspicious about and resistant to male scientism, lays bare the partiality of patriarchal agricultural scientific work, and exposes the falsity of this brotherhood's belief that their singular perspective is a sign of expertise, disinterest, and neutrality. Carson was right and those who opposed her were wrong, but misogyny quashed her dissent and led directly to the toxicity of the world's soils, waters, and air today.

'Same old, same old: so much is reproduced by the requirement to follow,' Sara Ahmed might shrug; 'such and such white man becomes an originator of a concept, an idea as becoming seminal, by removing traces of those who were there before' (Ahmed, 2017). Before she was recognized with the Nobel Peace Prize, Maathai, too, was belittled, blocked, and attacked for her efforts to free Kenyan women from the patriarchal ecocide brought about by the impositions of the technological men of Western science and their local enablers, who call on and seemingly endlessly expand the vast wealth and influence available to them through their financial and institutional networks.[9] At the same time they subject women, who represent 60 per cent of the agricultural labour force, to poorly designed, inappropriate agricultural policies, aggressive agrotechnologies that are neither designed for nor affordable to women, a lack of access to credit or other material support, and inaccessible markets – not to mention armed, everyday, and intimate-partner violence, illness, and overwork as labourers and carers. Their successive policy documents and budgetary decisions fail women and make them vulnerable, while accelerating 'a specific vision of industrial agricultural development' (see chapter 12).

'The enclosure of life is taking place everywhere; the privileged center is increasingly narrow,' writes the feminist Colectiva XXK (2021: 10). Similarly, Grosfoguel notes that what is recognized as authoritative insight 'is based on the knowledge produced by a few men from five countries ... [and on their] socio-historical experience and world views'. While presented as such, the influence of these men is not evidence of the superiority, universal 'applicability' or 'transferability' of their ideas and arguments, but a sign of their 'provincialism' (Grosfoguel, 2013: 74).[10] What Africa invites is a broadly inclusive approach to food and farming, not a singular, white-Western, masculinist 'scientific rationalism' that justifies endless extraction and privatized profit (see chapter 12). The deployment of gendered tropes and stereotypes silences African women's dissent, with very real consequences for their right to health and well-being, and that of their children to a good life and future.

Seriously?!

I have drawn attention to shrillness and rage as the predominant affective tools used in patriarchal systems to control women who dissent against dominant men's efforts to unilaterally impose their worldview. Yet there are, as I acknowledge by citing them, men who reject this, recognizing that women's knowledge of Indigenous foodways is essential to maintain and renew nature's abundance and gifts of continuing life. This is the spirit of the interview Mvuselelo Ngcoya conducted with Fakazile Mthethwa, fondly known as Gogo Qho (Box A), shortly before her death. It is an astonishing, hopeful, uncompromising exchange with a woman who lived with full dedication to freedom. By eating foods she had grown herself, Gogo Qho politically dissented not only from settler-colonial patriarchy, but from the agrotechnological food system that tried to subsume and pollute the soils and waters, and the human and non-human bodies, of her ancestral lands.

Looked at not only as an experience of reclaiming farming and food practices, but as a testimony of facing and overcoming colonial trauma, the power of her testimony lies in Gogo Qho's embeddedness in interconnections, especially those she makes between struggling for a good life and self-healing. Experts in the healing of ancestral trauma (Duran and Duran, 1995; Duran, 2006) would celebrate her somatic reintegration, her regaining of gut health as well as gut knowledge. Gogo Qho's life journey ended only when she had come to terms with her ancestors, herself, and her community, overcoming fragmentation and refusing the colonially imposed disassociation of intellect from soul and soil.[11] I was struck by her insight into how seeds of every kind travel in two dimensions and directions, in both women and plants – forwards from their mothers and backwards to their grandmothers. Such phenomena embody quantum social change, through which '[a]wakening to our innate mattering brings us back to life', making 'way for the flourishing of all life and future lives to come' (Christina Bethell, in O'Brien, 2021: xi). The emerging science of epigenetics, which is unravelling the causes of the crippling burden of metabolic disorders in the world today, offers further scientific corroboration of her insight. While Gogo Qho's own grandmother passed on to her a legacy of well-being from rain-fed food grown in healthy soils, both the earliest colonial settlers and the shrill contemporary proponents of Big Agro are the descendants of women in Europe who experienced not only the devastation of their ancestral knowledge, but waves of severe malnourishment as a result of that continent's endless wars. Their offspring are unusually prone to inflammatory diseases like diabetes, exacerbated by eating and drinking the chemically treated food Gogo Qho despised and avoided (Van der Kolk, 2015; Tzika et al., 2018). Made ill themselves by the toxins of enclosure and violence, Europeans have, over centuries, relentlessly colluded with, subsidized, and advanced the global agrotech industry behind the poisoned cabbages and packaged *imbuya* Gogo Qho observes robbing those around her of both their rightful health and the political promise of freedom her grandmother's generation struggled to advance. Settler-colonialism *is* slow violence, to borrow Rob Nixon's useful term, and it was achieved through the – often forced – movement of sick and traumatized bodies around the globe. By contrast, Gogo Qho's life-force runs strongly from her grandmother to her, following a pathway and 'a tradition that is rooted in a female mythology', a healing line that draws from 'direct experience of the world, spirit, and psyche' (Duran, 2012: 6; Tzika et al., 2018). I take courage from her recounting of how she rediscovered and returned to ancient agroecological practices, reclaimed her ancestral land, relearned sacred secrets, defied the social, economic, and gendered expectations of her community, and healed herself from the toxic effects of colonized life. Gogo Qho reclaimed her agency by recovering women's ancient status as seed improvers and custodians, preservers and gift-givers, commoners, traders, and interpreters of the land's will, and reaffirmed these knowledges as central components of the privilege and responsibility attached to women's social, economic, and political activities today.

'One day my grandmother comes to me in a dream,' said Gogo Qho, and 'showed me that my health and life was in the soil.' I have been dreaming her dream with her ever since I read those words. They connect me directly to the founding cosmology of the first peoples of Turtle Island and to Indigenous psychologist Eduardo Duran's writing on the healing power of dreams, because once an individual 'has become aware/conscious of earth via the thinking and feeling function, the opportunity arises for a more transcendent understanding' from which renewal can flow (Duran, 2012: 14).

'My dear ones, the work is about to begin'[12]

Before this Earth begins, Skywoman, pregnant with her only child, a daughter, plunges towards a watery world through a hole made when the celestial tree in the land from which she falls is uprooted. Her fall sets in motion many world-building events, none of which would be possible if she had not established, as she fell, the first agreement of mutual care between humans and the natural world: geese, seeing that she cannot fly, help her descend safely to a new land mass made for her, because she cannot swim, by the back of a turtle who rises from the ocean to meet her. Her landing is softened by soil brought from the bottom of the sea by a muskrat, a tiny animal, but one capable of very deep diving. It is on this new land that Skywoman's daughter is born and matures, is impregnated by the wind, and dies, giving birth to twin sons who will go on to build the features of the natural world including, eventually, humans, who appear because 'the common intersections of the female, animals, the spirit world, and the mineral and plant world' make our lives possible (Watts, 2013: 21).

Watering them with her tears, Skywoman plants in her daughter's body the fistful of World Tree seeds caught in her outstretched hand as she tumbled earthwards, rebirthing her as Mother Earth and making possible the emergence of what '[s]cientists refer to ... as ecosystems or habitats', but Indigenous people think of as complex societies in which humans have to make choices about 'how they reside, interact and develop relationships with other non-humans' who are equally active (Watts, 2013: 23). Thus set in motion, the human and non-human worlds continue to interact in a continuous cycle of observation, communication, and social organization until colonization disrupts their primal relationship, replacing Indigenous knowledge systems with narrow, hegemonic scientific ideas, diminishing Indigenous people's agency, and instituting separation and a 'hierarchy of beings' centred particularly on degrading the feminine. Women are no longer regarded as sacred protectors of a thinking, living natural world, but become 'synonymous with disappointment and stupidity' (Watts, 2013: 25). By this process, both women and land are made available for violation, exploitation, 'acquisition and destruction' (Watts, 2013: 31) in an emerging capitalist system that will eventually achieve an almost totalitarian control over an increasingly monolithic global food system in which the Earth is virtually stripped of its (bio)diversity.

In ways unimaginable to the coalition of 'scientists, technocrats, business-people, and lawyers, who have all played roles in engineering a specific lens through which to see the world and define what is acceptable in it' (see chapter 12), and whose interests dominate and attempt to control both Africa's agricultural and its cultural landscapes, this account of how seeds came to Earth along with femaleness celebrates one of the oldest human relationships, and ontologies, in the world: that imagining women, seeds, fertility, and soil as coequals conjoined by life, thoughtfulness, intentionality, and activity; a recognition of the land itself as 'full of thought, desire, contemplation and will' (Watts, 2013: 21, 23).

In her rendition of this story, Robin Wall Kimmerer, ecofeminist botanist from the Citizen Potawatomi Nation, says that Skywoman is a reminder 'not just of where we came from, but also of how we can go forward' (Wall Kimmerer, 2013: 5). I think of this as I look over the *vlei*. I think of Skywoman and her daughter, and of Gogo Qho, and of all they gave to the world.

At the water's edge, a red-knobbed coot carries short stems from one thicket of reeds to another, drawing from the ancient knowledge of her ancestors, building this year's nest for this year's chicks. She herself hatched on these waters only a year or two ago, and the simple beauty of her work fills me with hope.

Notes

1. Bessie Head, in a letter to Randolph Vigne, cited by Victoria Margree (2004).
2. This term is used by activists in the African Water Commons Collective.
3. It is the 22nd such site in South Africa, and the most urban (Zeekoevlei, 2015).
4. The City of Cape Town controls the water use of poor people using devices that regulate its flow. Countering this inhumane policy is a major focus of the African Water Commons Collective.
5. From a letter written in 1825 by the Irish social reformer William Thompson to the women's rights activist Anna Wheeler.
6. Olive Schreiner's allegorical tale *Trooper Peter Halket of Mashonaland* ([1897] 2019) focuses on the gendered psychological trauma experienced by the 'surplus' unemployed young men forced to leave the comforts of home behind when they were sent to the colonial killing fields. Also see Schreiner's *Woman and Labour* (1911).
7. Several of Olive Schreiner's works deal with aspects of this process of 'taming' Southern African women through various violent means, including rape.
8. Grosfoguel traces four epistemicides practised simultaneously by male Europeans throughout 'the long 16th century': against the Muslims in Andalusia, which included the burning, between the 13th and 16th centuries, of about 750,000 irreplaceable written texts in ancient

libraries; in the Americas, with the burning of Indigenous knowledge-recording processes and systems (*códices*); against enslaved Africans, by the destruction, through dispersal, of their knowledges; and of women's study of nature, as discussed above.

9. Attacks on dissenters are, of course, not confined to women, as was seen after the release of the report criticizing the Bill & Melinda Gates Foundation's intervention, the Alliance for a Green Revolution in Africa (AGRA) (Wise, 2020a). However, while AGRA's response questioned the research credentials of Timothy A. Wise, who led the report team, he was not subjected to attacks about his mental health in the ways women are. See Mkindi et al. (2020) and Wise (2020b).

10. Four of these five countries are in Western Europe: Italy, France, Germany, and the UK; the USA is the fifth. Criado Perez also points out that the world has become more dangerous for women because white male Americans are so over-represented as creators and designers of everyday items that often simply do not fit women, or provide safety for them.

11. The power of overcoming ancestral trauma through bodily integration was a key theme of the 2021 Collective Trauma summit, hosted online by Thomas Hübl, and informs my analysis in this section (Inner Science, 2021)

12. These words, as recounted by Jo-ann Archibald Q'um Q'um Xiiem of the Salish/Stó:lō, begin both stories and world-changing political work (Xiiem et al., 2019: 1).

References

Ahmed, S. (2017) 'Institutional as usual', Feminist killjoys [blog], 24 October <https://feministkilljoys.com/2017/10/24/institutional-as-usual/>.

Carson, R. (1962) *Silent Spring*, Houghton Mifflin, New York.

Colectiva XXK (2021) *Together and Rebellious: Exploring Territories of Feminist Economics*, Sempreviva Organização Feminista, São Paulo, and Colectiva XXK, Bilbao <https://colectivaxxk.net/wp-content/uploads/2021/07/Together-and-Rebellious_V6.pdf>.

Comaroff, J. (1985) *Body of Power, Spirit of Resistance: The Culture and History of a South African People*, University of Chicago Press, Chicago.

Criado Perez, C. (2019) *Invisible Women: Exposing Data Bias in a World Designed for Men*, Random House, New York.

De Sousa Santos, B. (2010) *Epistemologias del sur*, Siglo XXI, Mexico City.

Devereux, C. (2014) 'Hysteria, feminism, and gender revisited: The case of the second wave', *ESC: English Studies in Canada* 40(1): 19–45 <http://doi.org/10.1353/esc.2014.0004>.

Duran, E. (2006) *Healing the Soul Wound: Counseling with American Indians and Other Native Peoples*, Teachers College Press, New York.

Duran, E. (2012) 'Medicine wheel, mandala, and Jung', in *Spring, 87: Native American Cultures and the Western Psyche: A Bridge Between*, pp. 125–153 <https://www.academia.edu/40675303/MEDICINE_WHEEL_MANDALA_AND_JUNG>.

Duran, E. and Duran, B. (1995) *Native American Postcolonial Psychology*, State University of New York Press, Albany, NY.

Eisler, R. and Fry, D.P. (2019) *Nurturing Our Humanity: How Domination and Partnership Shape Our Brains, Lives, and Future*, Oxford University Press, New York.

Enloe, C. (2013) *Seriously! Investigating Crashes and Crises as If Women Mattered*, University of California Press, Berkeley, CA.

Federici, S. (2004) *Caliban and the Witch*, Autonomedia, New York.

Green, L. (2020) *Rock | Water | Life: Ecology and Humanities for a Decolonial South Africa*, Duke University Press, Durham, NC.

Grosfoguel, R. (2013) 'The structure of knowledge in westernized universities: epistemic racism/sexism and the four genocides/epistemicides of the long 16th century', *Human Architecture: Journal of the Sociology of Self-Knowledge* 11(1): 73–90 <https://www.okcir.com/product/journal-article-the-structure-of-knowledge-in-westernized-universities-epistemic-racism-sexism-and-the-four-genocides-epistemicides-of-the-long-16th-century-by-ramon-grosfoguel>.

Inner Science (2021) 'Collective Trauma Summit 2021' [website], Inner Science LLC <https://thomashuebl.com/event/collective-trauma-summit-2021/>.

Maathai, W. (2007) *Unbowed: A Memoir*, Anchor Books, New York.

Margree, V. (2004) 'Wild flowers: Bessie Head on life, health and botany', *Paragraph* 27(3): 16–31 <https://www.jstor.org/stable/43151760>.

Maté, G. (2009) *In the Realm of Hungry Ghosts: Close Encounters with Addiction*. Vintage Canada, Toronto.

Mellet, P.T. (2020) *The Lie of 1652: A Decolonised History of Land*, Tafelberg, Cape Town.

Mies, M. and Shiva, V. (2014) *Ecofeminism*, Zed Books, London.

Millar, D., Ayariga, R. and Anamoh, B. (1996) '"Grandfather's way of doing": gender relations and the *yaba-itgo* system in Upper East Region, Ghana', in C. Reij, I. Scoones and C. Toulmin (eds), *Sustaining the Soil: Indigenous Soil and Water Conservation in Africa*, pp. 117–125, Earthscan, Abingdon <https://doi.org/10.4324/9781315070858>.

Mkindi, A.R., Maina, A., Urhahn, J., Koch, J., Bassermann, L., Goïta, M., Nketani, M., Herre, R., Tanzmann, S., Wise, T.A., Gordon, M. and Gilbert, R. (2020) *False Promises: The Alliance for a Green Revolution in Africa (AGRA)*, Rosa-Luxemburg-Stiftung <https://www.rosalux.de/en/publication/id/42635/false-promises-the-alliance-for-a-green-revolution-in-africa-agra>.

Morris, R.C. (ed.) (2010) *Can the Subaltern Speak? Reflections on the History of an Idea*, Columbia University Press, New York.

Nixon, R. (2011) *Slow Violence and the Environmentalism of the Poor*, Harvard University Press, Cambridge, MA.

O'Brien, K. (2021) *You Matter More Than You Think: Quantum Social Change for a Thriving World*, Change Press, Oslo.

Schreiner, O. ([1897] 2019) *Trooper Peter Halket of Mashonaland*, Good Press.

Schreiner, O. (1911) *Woman and Labour*, T. Fisher Unwin, London.

Scott, J.C. (2017) *Against the Grain: A Deep History of the Earliest States*, Yale University Press, New Haven, CT.

Shiva, V. (1988) *Staying Alive: Women, Ecology and Development*, Zed Books, London.

Spivak, G.C. (1988) 'Can the Subaltern Speak?', *Die Philosophin* 14(27): 42–58 <https://doi.org/10.5840/philosophin200314275>.

Swanby, H. (2024) 'A movement for life: African food sovereignty', in R. Wynberg (ed.), *African Perspectives on Agroecology: Why farmer-led seed and knowledge systems matter*, pp. 311–327, Practical Action Publishing, Rugby.

Taylor, B. (1984) *Eve and the New Jerusalem: Socialism and Feminism in the Nineteenth Century*, Virago Press, London.

Tobi, E.W., van den Heuvel, J., Zwaan, B.J., Lumey, L.H., Heijmans, B.T. and Uller, T. (2018) 'Selective survival of embryos can explain DNA methylation signatures of adverse prenatal environments', *Cell Reports* 25(10): 2660–2667 <https://doi.org/10.1016/j.celrep.2018.11.023>.

Turner, B.L. II and Fischer-Kowalski, M. (2010) 'Ester Boserup: an interdisciplinary visionary relevant for sustainability', *Proceedings of the National Academy of Sciences of the United States of America* 107(51): 21963–21965 <https://doi.org/10.1073/pnas.1013972108>.

Tzika, E., Dreker, T. and Imhof, A. (2018) 'Epigenetics and metabolism in health and disease', *Frontiers in Genetics* 9: article 361 <https://doi.org/10.3389/fgene.2018.00361>.

Van der Kolk, B.A. (2015) *The Body Keeps the Score: Brain, Mind, and Body in the Healing of Trauma*, Penguin Books, New York.

Wall Kimmerer, R. (2013) *Braiding Sweetgrass: Indigenous Wisdom, Scientific Knowledge, and the Teachings of Plants*, Milkweed Editions, Minneapolis, MN.

Watts, V. (2013) 'Indigenous place-thought and agency amongst humans and non-humans (First Woman and Sky Woman go on a European world tour!)', *Decolonization: Indigeneity, Education and Society* 2(1): 20–34 <https://jps.library.utoronto.ca/index.php/des/article/view/19145/16234>.

Whittingham, J., Marshak, M. and Swanby, H. (2023) 'Unsettling modernist scientific ontologies in the regulation of genetically modified crops in South Africa', in R. Wynberg (ed.), *African Perspectives on Agroecology: Why farmer-led seed and knowledge systems matter*, pp. 237–271, Practical Action Publishing, Rugby.

Whyte, K.P. (2017) 'Our ancestors' dystopia now: Indigenous conservation in the Anthropocene', in U. Heise, M. Niemann and J. Christensen (eds), *The Routledge Companion to the Environmental Humanities*, pp. 206–215, Routledge, Abingdon.

Wise, T.A. (2020a) *Failing Africa's Farmers: An Impact Assessment of the Alliance for a Green Revolution in Africa*, Working Paper No. 20–01, July, Global Development and Environment Institute, Tufts University, Medford, MA <https://sites.tufts.edu/gdae/files/2020/07/20-01_Wise_FailureToYield.pdf>.

Wise, T.A. (2020b) 'Response to Alliance for a Green Revolution in Africa (AGRA) statement on "False Promises" report', 10 August, Institute for Agriculture & Trade Policy <https://www.iatp.org/blog/202008/response-alliance-green-revolution-africa-agra-statement-false-promises-report>.

Xiiem, J.-A.A.Q.Q., Lee-Morgan, J.B.J. and De Santolo, J. (2019) *Decolonizing Research: Indigenous Storywork as Methodology*, Zed Books, London.

Yapa, L. (1995) 'Response: A hybrid by any other name …', *Economic Geography* 71(3): 319–321 <https://doi.org/10.2307/144315>.

Zeekoevlei (2015) 'False Bay Nature Reserve a Ramsar site!', Zeekoevlei: Community News, Info and Blogs, 3 February <http://www.zeekoevlei.co.za/2015/02/false-bay-nature-reserve-a-ramsar-site/>.

Box F My grandmother's farm: a story by Mugove Walter Nyika

The wild fruit trees which were my boyhood playground are not there any more. There is no shade in which to shelter, no sweet fruit to eat. What remains is just row after row of hybrid maize in neat lines. The soil beneath my feet is hard as cement. This is my grandmother's farm where I grew up – but it was different then.

My grandmother farmed in south-central Zimbabwe in the 1960s. She was a smallholder farmer using local seeds that she saved from each harvest, and traditional methods both to protect the seeds from pests and to grow the crops. She used manure from the cattle pen, termite mound soil, and leaf and crop residue litter to maintain the fertility of her soils. She intercropped legumes with her other crops. I remember watching her select the best seed from her harvest every year, and the many ways she had to keep it safe from pests. She would hang some of the seed above the fireplace and keep the rest in her sealed granary under a layer of *rapoko* grains.

I remember when the government extension officer came. 'You can now become Master Farmers,' he said. In order to achieve this status, farmers had to remove all trees from their arable land and plough it uniformly. Then they needed to plant maize in straight lines with uniform spacing and no other crops in between. They were encouraged to buy ox-drawn cultivators to clear the weeds in between those neat rows.

By the 1980s my grandmother had become a modern farmer. She was buying and using hybrid seeds, chemical fertilizers, and chemical pesticides. She was practising monoculture, growing mostly just maize. But with this transformation came massive deforestation, soil erosion, siltation, loss of soil fertility, soil compaction, dependency on external inputs, and malnutrition, especially among the children. Far from this Green Revolution solving Africa's problems, as we were taught it would, things seemed only to get worse.

Today Mugove Walter Nyika is a permaculturalist and proponent of ecovillages, running a regional NGO, ReSCOPE, transforming schoolyards into verdant food forests and teaching children that one can grow food without money and that we are what we eat and grow. He is also returning to his roots, the rural home where his grandmother farmed, to turn around the damage of the Green Revolution and has committed himself to building resilience in his community and across landscapes. Walter is part of a growing network of people around the continent who are committed to changing the mindsets and agricultural practices that created millions of farms like his grandmother's.

Source: SKI, Seed sovereignty writeshop, September 2016, unpublished.

CHAPTER 12

Unsettling modernist scientific ontologies in the regulation of genetically modified crops in South Africa

Jennifer Whittingham, Maya Marshak, and Haidee Swanby

Introduction

The hegemony of modernist Science as an ontological framework is reflected in the centrality of Science[1]-based risk assessment of genetically modified (GM) crops (Adenle et al., 2020). This method of appraisal has closely accompanied the introduction of GM crops, yet scholars and activists have drawn attention to the narrow framing of early risk assessment approaches and the ways in which this has enabled their efficient release into landscapes (Scoones, 2008; Herrero et al., 2015). Current assessments continue to focus on a limited range of risks, relating principally to human health and the environment, with narrow socio-economic concerns being considered more recently (Binimelis and Myhr, 2016). Attention has been drawn to the need to assess a wider range of concerns ranging from social, cultural, political, and economic, to eco-toxicological and social-ecological, which go far beyond current risk assessment frameworks (Herrero et al., 2015; Preston and Wickson, 2016). While broadening the dimensions of risk assessment is important, we articulate the need to interrogate the ontological underpinnings that inform and legitimize risk assessment and decision-making processes informed by it. Drawing on the work of decolonial theorists from the Global South we explain how modernist Science and politics have intertwined to legitimize risk assessment as a neutral regulatory tool while excluding other ways of conceptualizing GM crops and their place in agroecosystems. We seek to unravel how modernist scientific authority 'reflects and refracts a particular set of biases and assumptions' that legitimize GM crops in South African agroecosystems and risk assessment as the best means to assess them (Schnurr, 2019: 18). In doing this, we suggest the importance of including a diverse set of knowledge systems and ontologies that can inform more appropriate and equitable approaches to imagining and co-creating agroecological futures.

As the first African country to allow the cultivation of GM crops, South Africa sits at a unique intersection of national, regional, and international regulatory influences. This position not only encompasses divergent regulatory approaches but is also characterized by distinct and often conflicting knowledge and value systems. In South Africa, as in other parts of the Global South, GM seeds have come with promises of boosting yields (Brookes and Barfoot, 2017), ensuring food security (Muzhinji and Ntuli, 2020), battling climate uncertainty (Thomson, 2008), drought resilience (Edge et al., 2018), pest resistance (Thomson, 2008), poverty alleviation, and technological progress (Adenle et al., 2020). We explore how GM crops have been coerced into South Africa's landscape not only by geneticists dealing with microscopic base pairs, expressions, and traits, but by scientists, technocrats, businesspeople, and lawyers, who have all played roles in engineering a specific lens through which to see the world and define what is acceptable in it. This analysis hopes to illuminate the specific ways in which science, economy, and politics have come together in an effort to industrialize African agricultural systems (Boyd, 2003).

As GM seeds and the political and ideological machinery that supports them have been aggressively promoted, other ways of knowing and knowers themselves have been opposed and marginalized to the peripheries of formal decision-making processes. We suggest that the dominant Science-based approach to GM regulation has been legitimized through power imbalances, foreign interests, political gains, and the fervent pursuit of economic growth; forces that arise from a distinct ontology. We question whether dominant, institutionalized models of Science-based risk assessment are capable of assessing the relational consequences and implications of modern agricultural biotechnologies in the Global South, where so much of the world's biocultural diversity is located (Wynne, 2007).

A turn to ontology: unearthing the foundations of genetically modified landscapes

Much of the scholarship on the failings of the Science-based approach to risk assessment draws on the ways in which risk is a multifaceted and relational issue and how this institutionalized version of scientific rationality cannot accurately account for the multitude of variables and uncertainties that may arise when assessing the risks of planting GM crops (Wynne, 2007; Hilbeck and El-Kawy, 2015; Hilbeck et al., 2020). Current GM crop assessments tend to focus on a narrow range of risks, relating principally to human health or the environment, that have been proven through 'sound science'. In recent years scholars, activists, and policymakers have drawn attention to the need to not only widen the breadth and improve the complexity of scientific enquiry being undertaken in risk assessment but expand the range of risks that are accounted for in risk appraisal (Wickson et al., 2017; Hilbeck et al. 2020). This has resulted in the addition of socio-economic concerns to the risk assessment framework, though in a rather restricted way. While broadening the dimensions of risk assessment could incorporate additional matters of concern (Latour, 2004) and help to uncover some of the shortcomings of the Science-based approach, we suggest the need to move beyond this 'additive approach' and interrogate the ontological underpinnings that inform the regulatory decision-making process around GM crops.

In industrial-capitalist society, what we term 'science' enjoys a privileged status among the possible ways of establishing knowledge about the world and 'reality' (Kloppenburg, 1991). Since the 1960s, however, social scientists like Kuhn, Strathern, Haraway, and Latour have questioned the progressive accumulation of 'rational' modernist scientific knowledge and explored the idea that science itself is a 'culture' that can be studied. Their critiques have drawn into focus how science 'is not a transcendent mirror of reality' (Jasanoff, 2004: 3) but, rather, a set of partial truths that may be considered truthful by those who hold views that align with them (Sismondo, 2011). The field of science and technology studies (STS) has grappled with the cultural dimensions of science and the impact that scientific epistemologies and their technologies have on the world. Despite strong and institutionally legitimated narratives of universality and objectivity, modernist Science 'is a story that casts itself in universal and culturally neutral tones, [while] expressing ... a particular cultural and philosophical tradition' (Reddekop, 2014: 3). It is also important to recognize that science itself is not a universal category: certain methods and projects may be favoured in how they support certain truths while other methods may be ignored (Hilbeck et al., 2020).

STS scholars have shown how the institutionalization of scientific reason and rationality emerges through a particular ontology, one that traces to the Western European Enlightenment on the quest for modernity (Seth, 2009). Since the Scientific Revolution (16th and 17th centuries), social theorists have joined the public, industry, and policymakers in treating scientists as the holders of ultimate 'truth'. Due to modern Science's rational methodological foundation and normative characteristics, scientists and their institutions have been perceived as generating knowledge that – unlike other ways of knowing – leaves no trace of its genesis in a particular social context (Merton, 1973).

Postcolonial science and technology studies (PCSTS) have endeavoured to reframe STS in the context of the Global South, in which Science and technology have a particularly oppressive history (Harding, 2011). Scholars in this field have shown how Science has played a significant part in shaping colonial power and control over people and landscapes and disrupting socio-ecological interconnections and relational philosophies (Seth, 2009; Tilley, 2011). Decolonial theorists have illuminated how scientific knowledge production and its core principles of scientific objectivity, technical efficiency, and economic profitability (Latour, 2007) cannot be disentangled from its roots in serving colonial and capitalist expansion (Quijano, 2007; Moore, 2015). The growth and globalization of modernist Science as a legitimate measure of reality has gone hand in hand with the dismissal and erasure of diverse ways of knowing and being and has replaced them with a homogenized concep-tualization of reality underpinned by modernist dualisms, such as objective/subjective, mind/body, and nature/culture. Growing scholarship and activism from the Global South in recent decades has engaged with offsetting the Western domination of knowing and understanding the world. While this scholarship comes from diverse fields, the ontological turn is a useful umbrella concept for understanding the shift in theoretical lens (Holbraad and Pedersen, 2017).

Scientific rationalism has come to dictate the value system of the institutions that govern capitalist society by excluding, and to the detriment of, other knowledge systems and knowers. De Sousa Santos (2016: 20) argues that ways of knowing that have been marginalized by colonial and capitalist systems very often 'do not count as knowledge' and may be dismissed as 'not rigorous' and 'not monumental' and viewed as 'superstitions, opinions, subjectivities, common sense'. In this context, decolonial scholars work to de-centre modernist ontology by bringing into focus previously marginalized ontologies – or multiple ways of knowing and being in the world that exist at the same time (Escobar, 2016). This work destabilizes the 'one-world-world' (Law, 2015) that is underpinned by a dualist ontology, separating nature from culture. It is a position that assumes there is one, 'real' world, only knowable through the application of science and technology, while the rest of us, the non-scientists, only hold perspectives on that world. The ontological turn acknowledges that there are multiple 'worlds' rather than simply multiple perspectives on 'one world' (Rosenow, 2019: 82).

A shift to ontology rejects rationalist scientific claims to objectivity and its ability to remain value-free, outside and above the social and cultural sphere, troubling fundamental assumptions about what 'nature' is and how we should relate to it. When such claims are interrogated and when other forms of knowledge and ways of knowing are valued and legitimated, the ultimate truthfulness of scientific rationality is called into question. This acknowledgement is theoretically grounded in a call for the mobilization of a pluriversal approach to ontology that can better engage with multiple ways of knowing and being (Escobar, 2016, 2018; Blaser and De la Cadena, 2018). Such alternative ontologies that reflect a relational understanding of life are often referred to as cosmovisions (e.g. 'Muntu and Ubuntu in parts of Africa; the Pachamama or Mama Kiwe among south American Indigenous peoples; U.S. and Canadian American Indian cosmologies'; Escobar, 2016: 22–23). Work within the ontological turn shows how modern Science and rationality alone are insufficient for solving the myriad of complex social-ecological challenges that the world faces, a clear representation of De Sousa Santos's paradox: that we face modern problems for which there are no longer sufficient modern solutions (2014: 44). A shift to ontology highlights this nuance and puts forward that if modern science is just one way of knowing the world and, like all knowledge, is subject to social factors, values, and intentions, there is room for different values and different intentions.

We argue that by interrogating the ontological underpinnings of GM crops and their assessment in South Africa, discussions can move beyond deliberating what should or should not be included in a risk assessment to make it inclusive of more concerns, and rather begin to question how Science-based risk assessment has been institutionalized as the default appraisal approach and why it continues to dominate, despite its well-known failings. Rather than maintaining an 'additive approach', a shift towards ontology destabilizes the alignment of Science, politics, and economics that gives rise to and legitimizes risk assessment as the best approach and gives room for alternative, more relational approaches to emerge.

Engineering a hospitable landscape for genetically modified organisms in South Africa

South Africa has a highly industrialized food system that has been built upon its colonial and apartheid legacy (Greenberg, 2003). During the colonial and apartheid periods, racially skewed land and segregationist policies (e.g. the Glen Grey Act of 1894 and the Natives Land Act of 1913) actively supported the emergence of white commercial agriculture by, among other things, restricting and dismantling established traditional smallholder farming systems, denying access to formal markets, and forcing black South Africans to live in 'homelands', often located on unfavourable land (Freund, 1984). The white, capital-intensive production system, based on Green Revolution technology, and the marginalized subsistence agricultural systems of the homelands were historically characterized as 'modern' and 'traditional' respectively (Hall, 2004). An economic function of the homelands was to provide cheap labour for the industrialization and economic growth of white South Africa (Freund, 1984). This collusion of race and capitalist interests was a key feature of the apartheid regime (Hall, 2004). By the end of apartheid, South Africa's agricultural sector was dualistic and imbalanced. It was characterized by a concentrated agricultural production structure made up of state-supported capital and technology-intensive agriculture, in areas formerly restricted to white commercial agriculture, and smallholder farming, located primarily in former homeland areas. This equated to approximately 60,000 commercial farms utilizing approximately 85 per cent of South Africa's most favourable agricultural land while an estimated 1.3 million smallholders were confined to farm on the remaining 15 per cent (DALRRD, 2020: 5).

Shortly after the onset of democracy the South African government adopted a series of neoliberal policies that became integral to the trajectory of agricultural development (Bayley, 2000). Negotiations and trade-offs in the transition to democracy ensured continuity of the economic structures and landownership that were already consolidating corporate power in South Africa's agrifood system (Greenberg, 2013). Despite rhetoric prioritizing smallholder agriculture since the transition to democracy, a dualistic agricultural structure remains today, with large-scale commercial production supported as the mainstay of national food security (Greenberg 2013).

New agricultural policies focused on global competitiveness, export markets, and value chain integration and supported the involvement of public–private partnerships with multinational organizations, international development agencies, and philanthropic organizations. Genetically modified crops became available at a very particular moment for South Africa, when an entrenched Green Revolution production system that had previously been state-supported quickly became liberalized and catapulted into global trade. GM crops promised a competitive technological edge for commercial farmers at this crucial moment. Furthermore, a series of smallholder development programmes were implemented. These primarily focused on bringing smallholders into the commercial system, from which they had previously been excluded, through a focus on commercial crops such as maize, cotton, and sugar cane (Greenberg, 2003) and included the promotion of GM seed. GM seeds have been part of the 'basket of technologies' that have been aggressively promoted to smallholder farmers across South Africa. Subsidized fertilizers, pesticides, and seeds (both conventional hybrids and GM) were promoted as 'pro-poor' technology packages promising to increase yields, reduce pesticide use, and improve market access (Mayet, 2007). Another, often overlooked dimension of these development programmes is epistemological: smallholder development programmes often include a unidirectional transfer of knowledge and Science-based technologies that treat smallholders as passive recipients of 'modern' agricultural knowledge rather than active partici-pants in the creation and reproduction of contextual agricultural knowledge (Kloppenburg, 1991). Such programmes often fail to consult smallholders on their opinions and concerns surrounding the use of biotechnologies and the agricultural problems they intend to solve. There has been little recognition of the disruption and destruction of farmer knowledge brought about by regimes of modernity and the continued erasures that occur through the dominance of industrial agriculture.

In South Africa, GM maize was introduced in 1998, making South Africa the first country in Africa to allow its cultivation, and the first country globally to approve a GM staple crop that was planted by both large-scale commercial and smallholder farmers (Jacobson, 2013). When genetically modified organisms (GMOs) entered South Africa, they did so in a global climate that was highly attuned to their controversies. However, they also arrived into a new democracy enthusiastic for 'transformation towards a competitive outward-oriented economy' (DoF, n.d.: 1) where science and technology would play a major role in rebuilding a deeply fractured and unequal nation. While GMOs were only officially approved for commercial release in 1998, their approval was decades in the making. In 1979, the South African Committee for Genetic Experimentation (SAGENE) was established by the apartheid government, marking the start of the institutionalization of biotechnology regulation in the country (Morris, 1995). It comprised a group of South African scientists with the intention of leading the drive for biotechnology uptake in the country and ensuring compatibility with rapidly changing standards emerging from the United States (Schnurr, 2019). Before the Genetically Modified Organisms Act (RSA, 1997) came into operation in 1999, at a time when no GMO legislation existed, 178 permits for open field trials were granted by SAGENE (Mayet, 2007: 10). In 2002, it was acknowledged by the chair of the National Assembly's Portfolio Committee on Environmental Affairs and Tourism that 'the legislation that has allowed South Africa to pioneer GMOs was prepared by the apartheid government and rushed into law months after the election of the first democratic government' (Reuters, 2002). As a result, government and civil society were left in the dark about the legislation and ongoing GM activity in the country. This situation privileged the techno-scientific perspective and further institutionalized scientific rationalism as the foundation for development of South Africa's agricultural trajectory. The legacy of these pre-democratic structures and their ontological foundations passed through the democratic transition seamlessly and without consultation of more broadly affected parties. SAGENE, as both a scientific and political instrument, was conducive to transferring this logic through the transition to democracy into the new political dispensation.

This section has demonstrated how the introduction of GMOs into South African soils was not neutral but part of a wider ideological shift towards a globalized, industrial, and commercially oriented social-agricultural system. As Schnurr (2019) has argued, biotechnologies have become central to visions of development, which have been deeply shaped by political and economic agendas. As we have shown, this biotechnological vision was nurtured during apartheid and carried through into the new democracy. The history of agricultural Science in South Africa (as throughout much of the Global South) has been a site of epistemological and ontological violence that has produced a fractured agricultural landscape that often does not best serve those whom it claims to (Quijano, 2007; Tilley, 2011). Instead, local farmers, their communities, and their ontologies have been and continue to be undermined.

The capital 'S' in Science-based risk assessment

In this section we outline some of the shortcomings of a Science-based risk assessment and examine how it fails to account for relational concerns that cannot be neatly slotted into a risk framework. Harding's (2008) concept of 'Science' with a capital 'S' is useful in differentiating between 'Science' produced through the legitimizing apparatus of various institutions and in service of politics and capital, and 'science' as a unique tool of enquiry consisting of a vast and diverse set of scientific knowledge projects (Subramaniam and Willey, 2017).

The emphasis on the production of scientific evidence for demonstrating the risks of planting GM seeds is pervasive in their regulation: Biosafety South Africa (2021), the national biotechnology platform in South Africa, defines risk assessment as a 'structured approach to determine the chance of harm from activities with a particular GMO based on scientific evidence by consideration of what could go wrong and how this may occur'. Within this framing,

Science is widely presented as entirely objective and value-free (Harding, 2008) and does not recognize that some of the steps involved in the characterization of risk (e.g. hazard identification and mitigation) are not restricted to scientific methodology alone (Pavone et al., 2011). Value-laden assumptions, such as the significance given to the distribution of risks, what constitutes a benefit worth taking a risk for, and what level of risk is acceptable, are also incorporated into this process (Hilbeck et al., 2020).

In a risk assessment, only risks to human health and the environment that have been proven through 'sound science' are considered, leaving socio-economic, cultural, and political dimensions largely outside the frame of view (Binimelis and Myhr, 2016). The inclusion of socio-economic factors in risk assessment and what that might entail became a contentious issue at the negotiations of the Cartagena Protocol on Biosafety in 1999. High-income, industrialized nations contended that including socio-economic considerations would constitute unfair trade barriers, while low-income nations feared that the introduction of GMOs could displace local agricultural resources, cultures, and livelihoods (Khwaja, 2002). Agreement was finally reached on allowing for the limited and voluntary inclusion of social-economic considerations in risk assessment (UNEP, 2003). However, guidelines on how to interpret this were only developed in 2011. Although these guidelines were developed in consultation, there remains a lack of consensus regarding the definition, scope, and methodology of how such risks should be assessed (Hilbeck and El-Kawy, 2015). In South Africa, socio-economic studies are not mandatory, and the Executive Council for GMOs considers this on a case-by-case basis.

South Africa's GMOs Act of 1997 (RSA, 1997), as amended in 2006 (RSA, 2006), entrenches a separation between Science and society through separating Science-based risk 'assessments' that are managed by an expert scientific 'Advisory Committee' and policy-based risk 'management', governed by an 'Executive Council', which is the ultimate decision-making body that approves or rejects GMO applications. As risk assessment is commonly seen as the domain of pure science that should test hypotheses and make predictions based on evidence; anything 'non-scientific' that addresses social concerns is seen as lying outside risk assessment. This institutional separation between Science and society, between 'Science' and 'non-science', it has been argued, only serves the actors who created the system and favours a narrow conception of risk to human health and the environment over that to biodiversity, and social and economic concerns (Schnurr, 2019). Harding (2015: 90) points out that while modern sciences such as physics, chemistry, and genetics have become the most powerful knowledge systems around the globe and have realized significant achievements, this is only part of their history: 'There are often unrecognized problems lurking beneath this rosy picture of success.' While there is a definite place for statistical and science-based analysis in establishing risk, dependence on rational Science alone ignores the social construction of knowledge and how the entwinement of culture, politics, and scientific research may privilege economic considerations and be complicit with the agendas of dominant social groups (Latour and Woolgar, 1979; Stirling, 2007).

Social critiques of scientific knowledge production destabilize the triumphalism of modern Science and provide 'new theoretical resources for challenging that voice of decontextualized rationality which agricultural science has used to such dominating effect' (Kloppenburg, 1991: 520). A moment that challenged the hegemony of Science-based risk assessment and its embedded ontologies lies in a community objection to the extension of Monsanto field trials for a drought-tolerant GM maize variety in the Western Cape, South Africa. The objection was submitted in 2010 to the Department of Agriculture, Forestry and Fisheries by a coalition of civil society organizations, declaring that the challenges faced by smallholder farmers were induced by historic and ongoing structural violence, such as racism, land dispossession, lack of access to nutritious food, and lack of agency regarding decisions about producing, distributing, and consuming food. They argued that South Africa's adoption of Monsanto's industrial production model made the state complicit in the ongoing food insecurity of smallholder farmers, citing, for example, further concentration of land for industrial monocropping, job losses, the continued use of cheap, black labour in food production, and the loss of seed sovereignty and agricultural skills and knowledge (Food Sovereignty Campaign, 2010). The African Centre for Biosafety/Biodiversity (ACB)[2] (2011) stated that Monsanto had dismissed the objection by taking refuge in South Africa's Science-based biosafety regulations, writing that the objection made 'numerous unsubstantiated and ideological claims and allegations not specifically relevant to Monsanto's application for permit extension to conduct field trials with maize MON 87460' (Monsanto, 2010, in ACB, 2011).

This objection demonstrates the ways in which a Science-based risk assessment does not account for concerns that lie outside the realm of scientific rationalism. Not only can this approach not accommodate complex, historic, social, economic, or political concerns, it actively shuts them out and denies their relevance. This is where our concern with an additive approach lies. We propose that a Science-based approach – one that is politically driven and tied to a trajectory of economic growth – is incompatible with the assessment of what are often termed 'non-scientific' concerns or risks that materialize through more relational ways of thinking and being in the world. Rather, when the dominant ontology is interrogated, the Science-based risk approach and its decision-making framework struggle to comprehensively or safely assess GM crops.

Assessing relational risks with Science-based risk assessment: incompatible natures?

We have shown how the hegemony of modern Science has been integral to ripening the South African landscape for GMOs. While the transition to democracy in South Africa initiated a shift in political power, the hegemony of rational Science was seamlessly carried into a new era and remains largely uninterrogated. This sanctioned authority allows Science to permeate the regulatory infrastructure for GM crops, despite this approach having serious shortcomings. By examining how the regulation of GM crops in South Africa is deeply structured around modern scientific principles, it becomes clear how their expansion aids in reconstructing and preserving a specific vision of industrial agricultural development. This vision holds on to colonial power inequities rather than forging transformation, follows an industrial-capitalist mode of production and proliferation, and again pushes aside those who have alternative visions of the relationship between nature and society (Quijano, 2007; Mignolo, 2011).

GM seed technologies are often marketed as 'silver bullet' solutions to drought, hunger, poverty, and climate change and promoted as progressive and somehow inevitable techno-objects of scientific advancement. Through other lenses formed through a more relational ontology, GM technologies and their social-ecological complexities raise questions around rational claims of Science and the idea that humans can and should control complex ecological systems. South Africa's approach to GM crop regulation is also a product of this particular worldview or ontology and serves to legitimize GMOs as the solution to complex social-environmental problems. Through the lens of scientific rationalism, GM crops can appear to be the logical response to a host of global problems, and, at the same time, risk assessment and its decision-making framework can seem the logical means to accurately assess and manage their risks. However, as shown in the examples below, the current Science-based approach to assessment fails to account for the complex ways in which GMOs become entangled in real life social-ecological contexts.

The entwinement of science and certain agendas can distort the kind of Scientific assessment being carried out. In the context of GM risk assessment there is a danger, as Hilbeck et al (2020: 12) point out, of scientific methodologies being used that are 'not just unscientific but anti-scientific' and reductionist and thus fail to incorporate important dimensions of ecological complexity. An example lies in the risk assessment of herbicide-tolerant GM crops. While the claimed benefits of herbicide tolerance result from planting the GM crop *and* spraying it with herbicide, the risk assessment does not account for this. Rather, in South Africa, if the use of the herbicide has been previously authorized then the safety of the herbicide application with the GM crop is assumed and no assessment of the combinatorial effects of multiple or increasing herbicide applications over time is required (ACB, 2019; Hilbeck et al., 2020). In the current approach to GM crop risk assessment, the complexity of nature and its social-ecological constitution is drastically reduced, and the interdependence of humans and their ecological environments is largely ignored. The narrow, Science-based risk framework does not consider the real, lived experience of herbicide application, noting discrepancies in farmer applications, nor the accelerating evolution of herbicide-resistant weeds that result in much higher and frequent doses of herbicide (Hilbeck et al., 2020).

Separating and reducing the process of herbicide use and regulation into discrete components and analysing them in isolation from one another is a clear example of Cartesian reductionism and is emblematic of a mechanistic worldview (Merchant, 1990). This way of seeing and knowing about the world decontextualizes nature from its social, political, and ecological constituents, and nurtures a hierarchical and linear view of nature. Risk assessment is a method that arises through this mechanistic and hierarchical ontology, and operationalizing concerns that are not so easily broken down and separated becomes difficult within such a framework.

Similarly, the way in which the principle of the 'refuge' is calculated and implemented demonstrates how the presumed efficiency of objective Science cannot reliably assess the true nature of risk in a social-ecological system. A refuge is a portion of a farmer's field that is planted with non-GM seed adjacent to the GM plot with the aim of reducing insect resistance. Ideally, the farmer commits to planting a refuge in a legally binding technical agreement signed with the industry when buying seeds (Thomson, 2008). While this model recognizes the principle of the refuge in theory, it does not account for a multitude of on-the-ground realities in which insect resistance can develop (Van Rensburg, 2007; Kruger et al., 2009, 2012). In many cases, smallholder farmers do not plant a refuge either because they are unaware that they need to comply with certain requirements, or because doing so is too labour-intensive and time-consuming (Kruger et al., 2012; Mahlase, 2017). Kruger et al.'s (2012) survey data showed that no first-time technical agreements were signed between 1998 and 2006 although the farmers had been planting Bt maize since 1998. The impracticality of refuge design for smaller farms also complicates the management of refugia (Kruger et al., 2009), and compliance with refuge requirements cannot be easily implemented without significant changes to agricultural practices and landscapes (Bøhn et al., 2016). Not only have smallholders struggled to comply with refuge requirements, but the limited compliance by large-scale farmers has been linked to the development of insect resistance in South Africa (Van Rensburg, 2007; Kruger et al., 2009). From a relational ontological perspective, the concept of refugia assumes a reductionist and anthropocentric lens in its assertion that humans can control multi-species relationships through the application of Science and technology.

The studies above depict not only an incompatibility between the social-ecological complexity of smallholder practices and stringent, technical biosafety requirements, but also an acute disconnect and clashing of worldviews between a fiercely controllable agricultural system and a uniquely tangled mosaic of social-ecological relations that are present on and in the ground. As Herrero et al. (2015: 11321) remark, 'agricultural biotechnologies cannot be usefully assessed as isolated technological entities' that are separate from their biotic and social environments 'but need to be evaluated within the context of the broader social-ecological system that they embody and engender'. Yet a risk assessment is based on the reduction and exclusion of these factors and is interested only in the immutable components of a phenomenon. As a result, a Science-based risk approach to assessment loses connection with the variability of local systems (Kloppenburg, 1991). Despite these concerns, the values of scientific rationalism continue to cradle a risk-based approach and thus still mark the core of GM crop regulation and decision-making.

Conclusion

This chapter looks through an ontological lens to explore the ways in which GMOs have made their way into South Africa's agricultural landscapes. It has shown that the central role of scientific authority in the risk assessment and decision-making structures that underpin GM crop appraisal is not incidental, nor entirely impartial, as many would argue, but is rather actualized through the institutionalization of a modern scientific ontology. This process privileges its own set of values – of objectivity, profitability, and efficiency (Latour, 2007) – under a guise of neutrality while marginalizing other values, concerns, and affected parties. We shed light on some of the issues that arise when reductionist scientific principles are applied within a Science-based risk assessment and ask whether this framework is capable of successfully accounting for relational and complex risks. While we recognize the call by many to broaden the framing of risk to include more concerns, we have suggested that a turn to ontology may be a useful starting point to begin unstitching the tightly woven seam that holds modernist Science in its place in GM crop regulation.

An ontological lens can provide alternative ways of thinking, knowing, and being within social-ecological systems and start to de-centre modernist hegemony. It seeks to make space for 'science', as a vast and diverse set of knowledge projects beyond rational modernist 'Science' (Harding, 2008), and a multitude of pluriversal ontologies from the margins. In the context of South Africa and the Global South more widely, making this space is key. When the majority of the world's biocultural diversity is found in the Global South, a vastly different set of questions need to be asked of GMOs that are relevant to the diversity of social-ecological contexts and communities (Egziabher, 2003). By acknowledging multiple ways of being in the world and inviting different knowledge systems and ways of knowing to approach a problem or issue, 'the horizon of possibilities' is opened up and 'strong questions'[3] can be asked about the nature of these technologies and their place in the world – now and in future agroecological pathways (De Sousa Santos, 2014: 20).

Notes

1. Throughout the chapter, we distinguish between 'Science' (capital 'S') as state- and industry-supported scientific knowledge production and 'science', a unique epistemological tool that produces vast and diverse scientific knowledges (Harding, 2008).

2. In 2015, the African Centre for Biosafety changed its name to the African Centre for Biodiversity. In this chapter, 'ACB' refers to information published by the organization under either name.
3. Such questions address 'the societal and epistemological paradigm that has shaped the current horizon of possibilities within which we fashion our options' and inform what is possible (De Sousa Santos, 2014: 20).

References

Adenle, A., Steur, H., Hefferon, K. and Wessler, J. (2020) 'Two decades of GMOs: how modern agricultural biotechnology can help meet sustainable development goals', in A. Adenle, M.R. Chertow, E.H.M. Moors and D.J. Pannell (eds), *Science, Technology, and Innovation for Sustainable Development Goals: Insights from Agriculture, Health, Environment, and Energy,* pp. 401–422, Oxford University Press, New York.

African Centre for Biosafety (ACB) (2011) *Water Efficient Maize for Africa: Pushing GMO Crops Onto Africa,* African Centre for Biosafety, Melville <https://www.acbio.org.za/sites/default/files/2015/02/WEMA-Pushing-GMO-crops.pdf>.

ACB (2019) 'Objection against general release of three 2,4-D GM varieties', African Centre for Biodiversity, Johannesburg <https://www.acbio.org.za/objection-against-general-release-three-24-d-gm-maize-varieties>.

Bayley, B. (2000) *A Revolution in the Market: The Deregulation of South African Agriculture,* Oxford Policy Management, Oxford.

Binimelis, R. and Myhr, A.I. (2016) 'Inclusion and implementation of socio-economic considerations in GMO regulations: needs and recommendations', *Sustainability* 8(1): 62 <https://doi.org/10.3390/su8010062>.

Biosafety South Africa (2021) 'Risk analysis: a tool for decision making', Biosafety South Africa, Somerset West <http://biosafety.org.za/information/dig-deeper/sustainable-gm-product-development/risk-analysis-a-tool-for-decision-making>.

Blaser, M. and De la Cadena, M. (2018) 'Introduction. Pluriverse: proposals for a world of many worlds', in M. Blaser and M. de la Cadena (eds), *A World of Many Worlds,* pp. 1–22, Duke University Press, Durham, NC.

Bøhn, T., Aheto, D.W., Mwangala, F.S., Fischer, K., Bones, I.L., Simoloka, C., Mbeule, I., Schmidt, G. and Breckling, B. (2016) 'Pollen-mediated gene flow and seed exchange in small-scale Zambian maize farming: implications for biosafety assessment', *Scientific Reports* 6(1): 1–12 <https://doi.org/10.1038/srep34483>.

Boyd, W. (2003) 'Wonderful potencies? Deep structure and the problem of monopoly in agricultural biotechnology', in R.A. Schurman and D.D.T. Kelso (eds), *Engineering Trouble: Biotechnology and Its Discontents,* pp. 24–62, University of California Press, Berkeley, CA <https://papers.ssrn.com/sol3/papers.cfm?abstract_id=2080128>

Brookes, G. and Barfoot, P. (2017) 'Farm income and production impacts of using GM crop technology 1996–2015', *GM Crops & Food* 8(3): 156–193 <http://dx.doi.org/10.1080/21645698.2017.1317919>.

Department of Agriculture, Land Reform and Rural Development (DALRRD) (2020) *Abstract of Agricultural Statistics 2020*, Department of Agriculture,

Forestry and Fisheries, Pretoria <http://www.dalrrd.gov.za/phocadown-loadpap/Statistics_and_Economic_Analysis/Statistical_Information/Abstract%202022.pdf>.

Department of Finance (DoF) (no date) *Growth, Employment and Redistribution: A Macroeconomic Strategy*, Department of Finance, Pretoria <http://www.treasury.gov.za/publications/other/gear/chapters.pdf>.

De Sousa Santos, B. (2014) *Epistemologies of the South: Justice against Epistemicide*, 1st edn, Paradigm Publishers, Boulder, CO.

De Sousa Santos, B. (2016) *Epistemologies of the South: Justice against Epistemicide*, 2nd edn, Routledge, Oxford.

Edge, M., Oikeh, S.O., Kyetere, D., Mugo, S. and Mashingaidze, K. (2018) 'Water efficient maize for Africa: a public-private partnership in technology transfer to smallholder farmers in sub-Saharan Africa', in N. Kalaitzandonakes, E.G. Carayannis, E. Grigoroudis and S. Rozakis (eds), *From Agriscience to Agribusiness*, pp. 391–412, Springer, Cham.

Egziabher, T.B.G. (2003) 'The use of genetically modified crops in agriculture and food production, and their impacts on the environment: a developing world perspective', *Acta Agriculturae Scandinavica B* 53(Sup1): 9–13 <https://doi.org/10.1080/16519140310015148>.

Escobar, A. (2016) 'Thinking-feeling with the Earth: territorial struggles and the ontological dimension of the epistemologies of the South', *AIBR Revista de Antropología Iberoamericana* 11(1): 11–32 <https://doi.org/10.11156/aibr.110102e>.

Escobar, A. (2018) *Designs for the Pluriverse: Radical Interdependence, Autonomy, and the Making of Worlds*, Duke University Press, Durham, NC <https://doi.org/10.1215/9780822371816>.

Food Sovereignty Campaign (2010) *The Right to Agrarian Reform for Food Sovereignty Campaign's Objection to Monsanto's Application for a Time Extension of an Existing Permit for Activities with GMOs Drought Tolerant Maize in South Africa: Trial Release*, 1 June <https://acbio.org.za/wp-content/uploads/2022/03/ACB_objection_MON87460_Lutzville.pdf>.

Freund, B. (1984) 'Forced resettlement and the political economy of South Africa', *Review of African Political Economy* 11(29): 49–63 <https://doi.org/10.1080/03056248408703567>.

Greenberg, S. (2003) 'Land reform and transition in South Africa', *Transformation: Critical Perspectives on Southern Africa* 52: 42–67 <https://doi.org/10.1353/trn.2003.0030>.

Greenberg, S. (2013) *The Disjunctures of Land and Agricultural Reform in South Africa: Implications for the Agri-Food System*, Working Paper 26, PLAAS, UWC, Bellville <http://repository.uwc.ac.za/xmlui/handle/10566/4489>.

Hall, R. (2004) 'A political economy of land reform in South Africa', *Review of African Political Economy* 31(100): 213–227 <https://doi.org/10.1080/0305624042000262257>.

Harding, S. (2008) *Sciences from Below: Feminisms, Postcolonialities, and Modernities*, Duke University Press, Durham, NC.

Harding, S. (ed.) (2011) *The Postcolonial Science and Technology Studies Reader*, Duke University Press, Durham, NC.

Harding, S. (2015) *Objectivity and Diversity: Another Logic of Scientific Research*, University of Chicago Press, Chicago.

Herrero, A., Wickson, F. and Binimelis, R. (2015) 'Seeing GMOs from a systems perspective: the need for comparative cartographies of agri/cultures for sustainability assessment', *Sustainability* 7(8): 11321–11344 <https://doi.org/10.3390/SU70811321>.

Hilbeck, A. and El-Kawy, O. (2015) 'The Cartagena Protocol on Biosafety's negotiations: science-policy interface in GMO risk', *Journal of Health Education Research & Development* 3(1): 1000e120 <https://doi.org/10.3929/ethz-b-000338468>.

Hilbeck, A., Meyer, H., Wynne, B. and Millstone, E. (2020) 'GMO regulations and their interpretation: how EFSA's guidance on risk assessments of GMOs is bound to fail', *Environmental Sciences Europe* 32: 54 <https://doi.org/10.1186/s12302-020-00325-6>.

Holbraad, M. and Pedersen, M.A. (2017) *The Ontological Turn: An Anthropological Exposition*, Cambridge University Press, Cambridge.

Jacobson, K. (2013) *From Betterment to Bt Maize: Agricultural Development and the Introduction of Genetically Modified Maize to South African Smallholders*, PhD thesis, Swedish University of Agricultural Sciences, Uppsala <https://pub.epsilon.slu.se/10406/1/Jacobson_k_130507.pdf>.

Jasanoff, S. (2004) 'The idiom of co-production', in S. Jasanoff (ed.), *States of Knowledge: The Co-Production of Science and Social Order*, pp. 1–12, Routledge, New York.

Khwaja, R.H. (2002) 'Socio-economic considerations', in C. Bail, R. Falkner and H. Marquard (eds), *The Cartagena Protocol on Biosafety: Reconciling Trade in Biotechnology with Environment and Development?*, pp. 362–365, Earthscan, London.

Kloppenburg, J. (1991) 'Social theory and the de/reconstruction of agricultural science: a new agenda for rural sociology', *Sociologia Ruralis* 32(1): 519–548 <https://doi.org/10.1111/j.1549-0831.1991.tb00445.x>.

Kruger, M., Van Rensburg, J.B.J. and Van den Berg, J. (2009) 'Perspective on the development of stem borer resistance to Bt maize and refuge compliance at the Vaalharts irrigation scheme in South Africa', *Crop protection* 28(8): 684–689 <https://doi.org/10.1016/j.cropro.2009.04.001>.

Kruger, M., Van Rensburg, J.B.J. and Van den Berg, J. (2012) 'Transgenic Bt maize: farmers' perceptions, refuge compliance and reports of stem borer resistance in South Africa', *Journal of Applied Entomology* 136(1–2): 38–50 <https://doi.org/10.1111/j.1439-0418.2011.01616.x>.

Latour, B. (2004). Why has critique run out of steam? From matters of fact to matters of concern. *Critical Inquiry*, 30(2), 225–248 <https://www.journals.uchicago.edu/doi/epdf/10.1086/421123> .

Latour, B. (2007) 'The recall of modernity: anthropological approaches', *Cultural Studies Review* 13(1): 11–30 <https://doi.org/10.5130/csr.v13i1.2151>.

Latour, B. and Woolgar, S. (1979) *Laboratory Life: The Construction of Scientific Facts*, Princeton University Press, Princeton, NJ.

Law, J. (2015) 'What's wrong with a one-world world?', *Distinktion: Scandinavian Journal of Social Theory* 16(1): 126–139 <https://doi.org /10.1080/1600910X.2015.1020066>.

Mahlase, M.H. (2017) *Exploring the Uptake of Genetically Modified White Maize by Smallholder Farmers: The Case of Hlabisa, South Africa*, doctoral dissertation, University of Cape Town <http://hdl.handle.net/11427/24452>.

Mayet, M. (2007) *Regulation of GMOs in South Africa: Details and Shortcomings. Biosafety, Biopiracy and Biopolitics Series: 2*, African Centre for Biodiversity, Johannesburg <https://www.acbio.org.za/sites/default/files/2015/02/gmo_regulations_in_sa.pdf>.

Merchant, C. (1990) *The Death of Nature: Women, Ecology, and the Scientific Revolution*, Harper & Row, New York.

Merton, R.K. (1973) *The Sociology of Science: Theoretical and Empirical Investigations*, University of Chicago Press, Chicago, IL.

Mignolo, W. (2011) *The Darker Side of Western Modernity: Global Futures, Decolonial Options*, Duke University Press, Durham, NC.

Moore, J.W. (2015) 'Putting nature to work: Anthropocene, Capitalocene, and the challenge of world-ecology', in C. Wee, J. Schönenbach and O. Arndt (eds), *Supramarkt: A Micro-Toolkit for Disobedient Consumers, or How to Frack the Fatal Forces of the Capitalocene*, pp. 69–117, Irene Books, Gothenburg <https://doi.org/10.13140/RG.2.1.3703.0248>.

Morris, E.J. (1995) 'Biosafety regulations in South Africa', *African Crop Science Journal* 3(3): 303–307 <https://doi.org/10.4314/acsj.v3i3.54531>.

Muzhinji, N. and Ntuli, V. (2020) 'Genetically modified organisms and food security in Southern Africa: conundrum and discourse', *GM Crops & Food* 12(1): 25–35 <https://doi.org/10.1080/21645698.2020.1794489>.

Pavone, V., Goven, J. and Guarino, R. (2011) 'From risk assessment to in-context trajectory evaluation: GMOs and their social implications', *Environmental Sciences Europe* 23(1): 1–13 <https://doi.org/10.1186/2190-4715-23-3>.

Preston, C.J. and Wickson, F. (2016) 'Broadening the lens for the governance of emerging technologies: care ethics and agricultural biotechnology', *Technology in Society* 45: 48–57 <https://doi.org/10.1016/j.techsoc.2016.03.001>.

Quijano, A. (2007) 'Coloniality and modernity/rationality', *Cultural Studies* 21(2–3): 168–178 <https://doi.org/10.1080/09502380601164353>.

Reddekop, J. (2014) *Thinking Across Worlds: Indigenous Thought, Relational Ontology, and the Politics of Nature; or, If Only Nietzsche Could Meet a Yachaj*, PhD thesis, University of Western Ontario <https://ir.lib.uwo.ca/cgi/viewcontent.cgi?article=3410&context=etd>.

Republic of South Africa (RSA) (1997) *Genetically Modified Organisms Act 15 of 1997*, Government Gazette 18029 <https://www.gov.za/sites/default/files/gcis_document/201409/act15of1997.pdf>.

RSA (2006) *Genetically Modified Organisms Amendment Act 23 of 2006*, Government Gazette 29803 <https://www.gov.za/sites/default/files/gcis_document/201409/a23-060.pdf>.

Reuters (2002) 'Parliament to review law on GM foods', *Creamer Media's Engineering News*, 22 October.

Rosenow, D. (2019) 'Decolonising the decolonisers? Of ontological encounters in the GMO controversy and beyond', *Global Society* 33(1): 82–99 <https://doi.org/10.1080/13600826.2018.1558181>.

Schnurr, M.A. (2019) *Africa's Gene Revolution: Genetically Modified Crops and the Future of African Agriculture*, McGill-Queen's University Press, Montreal.

Scoones, I. (2008) 'Mobilizing against GM crops in India, South Africa and Brazil', *Journal of Agrarian Change* 8(2–3): 315–344 <https://doi.org/10.1111/j.1471-0366.2008.00172.x>.

Seth, S. (2009) 'Putting knowledge in its place: science, colonialism, and the postcolonial', *Postcolonial Studies* 12(4): 373–388 <https://doi.org/10.1080/13688790903350633>.

Sismondo, S. (2011) *An Introduction to Science and Technology Studies*, John Wiley & Sons, Chichester.

Stirling, A. (2007) 'Risk, precaution and science: towards a more constructive policy debate. Talking point on the precautionary principle', *EMBO Reports* 8(4): 309–315 <http://sro.sussex.ac.uk/1591/1/20070831_e1_stirling.pdf>.

Subramaniam, B. and Willey, A. (2017) 'Introduction to science out of feminist theory part one: feminism's sciences', *Catalyst: Feminism, Theory, Technoscience* 3(1): 1–23 <https://doi.org/10.28968/cftt.v3i1.28784>.

Thomson, J.A. (2008) 'The role of biotechnology for agricultural sustainability in Africa', *Philosophical Transactions of the Royal Society B* 363: 905–913 <https://doi.org/10.1098/rstb.2007.2191>.

Tilley, H. (2011) *Africa as a Living Laboratory: Empire, Development, and the Problem of Scientific Knowledge, 1870–1950*, University of Chicago Press, Chicago.

United Nations Environment Programme (UNEP) (2003) *The Cartagena Protocol on Biosafety: A Record of the Negotiations*, Secretariat of the Convention on Biological Diversity, UNEP, Montreal <https://www.cbd.int/doc/publications/bs-brochure-03-en.pdf>.

Van Rensburg, J.B.J. (2007) 'First reports of field resistance by the stem borer, *Busseola fusca* (Fuller) to Bt-transgenic maize', *South African Journal of Plant and Soil* 24: 147–151 <https://doi.org/10.1080/02571862.2007.10634798>.

Wickson, F., Preston, C., Binimelis, R., Herrero, A., Hartley, S., Wynberg, R. and Wynne, B. (2017) 'Addressing socio-economic and ethical considerations in biotechnology governance: the potential of a new politics of care', *Food Ethics* 1(2): 193–199 <https://doi.org/10.1007/s41055-017-0014-4>.

Wynne, B. (2007) 'Indigenous knowledge and modern science as ways of knowing and living nature: the contexts and limits of biosafety risk assessment', in T. Traavik and L.L. Ching (eds), *Biosafety First: Holistic Approaches to Risk and Uncertainty in Genetic Engineering and Genetically Modified Organisms*, pp. 287–302, Tapir Academic Press, Trondheim.

Box G Agroecology in the curriculum: Biowatch engages with the Owen Sitole College of Agriculture

Vanessa Black

Introduction

Biowatch South Africa is a small NGO formed in 1999 with the promotion of biodiversity and social justice at its heart. Our current focus is on agroecology and food sovereignty, and we therefore target industrial agriculture and the industrialized food system for its devastating impacts on biodiversity, our climate, land and water, livelihoods, health, and nutrition.

Imposed on the country and region through centuries of colonization, this 'modern' agriculture values productivity and trade over smallholder agriculture that is attuned to local ecosystems and centred on local culture and knowledge, and provides diverse materials, nutritious food, health, and resilience: what we and others call agroecology.

This case study focuses on Biowatch's intervention at a tertiary agricultural college, the Owen Sitole College of Agriculture (OSCA), located outside Empangeni, in KwaZulu-Natal, the province where we work. OSCA, one of two colleges run by KwaZulu-Natal's Department of Agriculture and Rural Development, is an hour's drive from Biowatch's farmer support office in Mtubatuba. Many agricultural extension officers in our area are the products of OSCA training, just as many of their successors will be.

Background

Industrialized agriculture is supported and promoted by the government and propagated generation after generation in the tertiary education institutions that produce agricultural extension officers. KwaZulu-Natal's provincial Department of Agriculture and Rural Development sees smallholder agriculture as the main means for the rural poor to attain food and nutrition security, 'which should be promoted and nurtured through *agricultural extension* [our emphasis]' (Witt, 2018). But food and nutrition security, with ecological integrity, is not the main focus of the industrial agriculture system, which is focused on increasing and simplifying the process of production and distribution into the global market. Agricultural extension officers therefore often lack community development skills and the knowledge needed to support complex and diverse smallholder farming systems, having been trained to produce commercial monocultures relying on agrochemicals, which do not necessarily support food and nutrition security.

Civil society organizations working with smallholder farmers have for many years criticized the support provided by the government's extension and advisory services. Feedback from farmers and engagements with some extension officers point to a top-down approach, extension officers not being able or willing to help agroecological producers, and industrial inputs such as GM seed, fertilizers, and pesticides being foisted on farmers. Farmers are pressed to conform and even ridiculed: 'Throw away your *gogo* [grandmother] seeds, they are no good.'

Since 2017, Biowatch has aspired to develop a structured education programme on agroecology, industrial agriculture, and environmental and social justice. The purpose of this is to change the discourse and narrative at agricultural training institutions, with interns from the colleges accompanying Biowatch staff in the field. Through such educational interventions, future agriculture extension officers graduating from these institutions would be better equipped to promote agroecology at a much larger scale. Our plan was to either develop a short course ourselves or work with an institution to develop a curriculum in partnership with Biowatch.

(Continued)

Box G Continued

Fortuitously, in 2018, two lecturers from OSCA visited the Biowatch office in Durban, having been referred by a departmental colleague and after inspecting the Biowatch website. They came to seek information and explore ways in which they could prompt questions about the approach to agriculture being taught at their college, which seemed to be counterproductive to good health and nutrition. Their interest, and OSCA's proximity to our office, gave Biowatch the opportunity to pilot an intervention. Staff agreed that the best intervention would be a 40-hour short course spanning a few weeks that would make it possible to link theory with practical experience.

In February 2018 Biowatch was invited to meet the college principal, who enthusiastically supported the idea of a partnership between Biowatch SA and OSCA to integrate learning on agroecology with the college curriculum. The principal tasked the meeting with drafting a memorandum of understanding that would include Biowatch delivering its pilot 40-hour course to second-year students taking the Food Security and Nutrition module and the establishment of an agroecology practical and demonstration plot at the college.

The intervention

A consultative process followed to reconcile contextual challenges, available resources, and learnings from joint reflection and planning meetings. This led to the following interlinked engagements:

- a one-day participatory workshop with college staff in August 2018;
- a two-day student course in 2018 with the establishment of a practical demonstration site;
- two presentations at OSCA's first ever agriculture symposium in May 2019;
- a two-day student course in 2019;
- employment of an intern from May 2019, extended to the end of April 2020, to implement and manage the agroecology demonstration plot at the college.

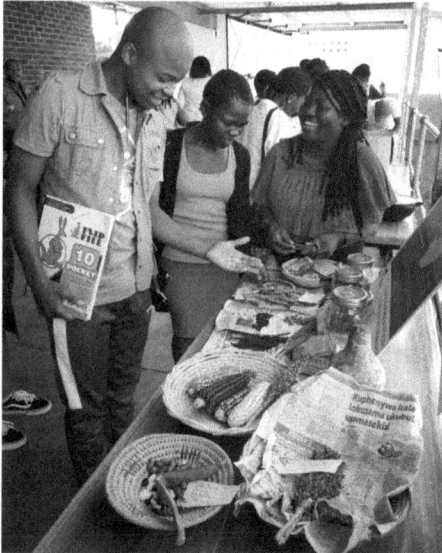

Photo G.1 Students engaging the intern Sthulile Mgwaba at a Biowatch display table during the 1st OSCA Agriculture Symposium in May 2019

(Continued)

Box G Continued

Photo G.2 Students tending their raised fertility beds in the demonstration garden

Photo G.3 Students reflecting on the intern's productive fertility beds in the demonstration garden

(Continued)

Box G Continued

Photo G.4 Students preparing a bed to trial a different preparation method in the student demonstration garden

Photo G.5 Students adding organic material to the bed they are making in the student demonstration garden

Box G Continued

COVID-19: a time for reflection

The global COVID-19 pandemic led to a national hard lockdown in South Africa from 16 March 2020. Lectures had to stop (to be resumed only in 2021) and work came to a halt on the demonstration plot. It was an unprecedented disruption, but it did give Biowatch an opportunity to pause and reflect on the impact of its work, particularly in the intervention and partnership with OSCA. An independent review was commissioned in 2020, drawing on records of the process and interviews with Biowatch and OSCA staff, the intern, and four students. Some key learnings from the process follow.

Challenges and learnings

Contextual challenges

The challenges in both agricultural education and the extension services have been government concerns for some time. A National Education and Training Strategy for Agriculture and Rural Development in South Africa was launched in 2005 to address inequalities in the provision of training across the country and strategically align training to 'support an environmentally and economically sustainable agriculture' (NDA, 2005). In 2015 this was revised to respond to significant changes in the agriculture sector including the drive to modernize, mechanize, and integrate value chains requiring different skills. The strategy was also extended to include training in the fisheries and forestry subsectors (DAFF, 2015). In 2012 the government addressed the shortage of extension services and their level of professionalism with a National Extension Recovery Plan. Extension science was 'professionalized' through its inclusion in 2014 as one of the fields of practice under the South African Council for Natural Scientific Professions, a regulatory body for science professionals. Consultation about the shortcomings of the extension service continued.

In 2017, coinciding with Biowatch's decision to become involved in extension, the Department of Agriculture, Forestry and Fisheries published its National Policy on Extension and Advisory Services (DAFF, 2016). The policy notes that extension services 'lack a developmental and systems approach, where practitioners have a holistic view'. It points to the need for improvement in the areas of efficiency, accountability, relevance, and sustainability and for 'innovative and climate resilient production practices to respond to rising food prices, food and nutrition security, poverty eradication, diversifying market demands, export opportunities and environmental concerns'. Noting the narrow service focus and limitations in the extension education system as a cause, the policy proposes extension training based on a 'multidisciplinary approach' to capacitate extension practitioners with 'relevant and diverse' education and tools to address the 'wider rural livelihood context' of extension support. It also seeks to strengthen the linkages between extension, producers, and researchers with the aim of promoting 'locally viable technologies and indigenous knowledge systems' and 'demand-driven research and extension' (DAFF, 2016).

Even before the focus on extension articulated in the policy, the national Department of Agriculture, Forestry and Fisheries had carried out an evaluation of agricultural education and training (DAFF, 2008), which noted the need to align curricula with the critical policy shift 'from an almost exclusive focus on commercial agriculture to a more rural development and poverty eradication orientation', with the inclusion of 'food security, nutritional issues, land care, sustainability and natural resource management, rural development and water harvesting' as core courses across training institutions (DAFF, 2008).

Despite the congruences between the expressed intent of the policy and Biowatch's engagement with OSCA, the intervention did not proceed smoothly at the college. Part of

(Continued)

Box G Continued

the reason could be conflicting policies and practices: previous policy pronouncements that extension should be a multidisciplinary and participatory developmental process for agricultural and rural development were overshadowed by the focus on organizing producers in commodity groups supplying commercial commodity value chains, and on supporting production through input subsidy programmes favouring the suite of 'Green Revolution' inputs and technologies. This framing continues to shape the form and content of agricultural training and extension in the country. These tensions surfaced in numerous ways.

Scepticism about agroecology's ability to 'feed South Africa' persisted among several non-involved OSCA staff and many students, and the agroecological intervention was consistently pigeonholed as a household food security issue.

Grappling with institutional learning and structure

It was difficult to integrate agroecology content in the college curriculum. Biowatch sees agroecology as a holistic and multidisciplinary approach to transforming the food system to ensure we can continue to live within our planetary means with fair social relations. Ideally it is an approach that cuts across all aspects of agriculture. However, the college curriculum remained structured in discrete course subjects. To be included in the college timetable, the agroecology intervention had to be incorporated with courses in food and nutrition security and soil science managed by interested lecturers who were willing to give up lecture slots. Biowatch was keen to share its holistic understanding of agroecology and engage students on flaws in the food system and current modes of production that were causing hunger, malnutrition, climate change, and ecosystem destruction. However, the students complained that this was not relevant to the specific courses.

Biowatch had conceived a participatory course building theory onto experiential learning and practice. Instead we had to comply with a highly structured timetable and were required to present the theory in the form of lectures. Experiential aspects were confined to work in the demonstration plot during designated practical slots in the timetable, while the college did not support planned nature immersion and field trips to farmers, despite students wanting hands-on experiences. In hindsight it would have been better to deliver the intervention as a discrete short course even if that meant fewer students could attend.

Also problematic was the rule that any produce grown on college land had to be sold at set prices and the monies given to the department, even though the seed was provided by Biowatch and the lecturers. Procurement contracts prevented the produce from being given to the college canteen for the students to eat, so the lecturers became the market. On one hand this exposed sceptical lecturers to the abundance of the demonstration plot, but on the other it limited what was grown to the familiar basics lecturers were keen to buy.

Practice is proof

The academic year is at odds with the growing seasons in northern KwaZulu-Natal: the mid-year exams and breaks tend to fall in the main vegetable growing season, while end-of-year exams and the long summer holiday coincide with field preparation and growing seasons for grains and legumes. Not only did this prevent the students from experiencing the full seasonal growing cycle from soil preparation to harvest, but it also meant that the demonstration plot was not maintained consistently. By the end of the first year of student engagement it became clear that the agroecology plot had to be more consistently managed. Biowatch employed an intern to ensure that the plot was a good showcase for agroecology and to provide hands-on mentoring to students. This was also a

(Continued)

Box G Continued

new process for Biowatch, giving us much to reflect on, but it turned out to be one of the most important outcomes of the intervention at OSCA.

We had conceptualized the course as starting with *why* we need to transition to agroecology, as a response to the scepticism that we were aware of, so that the *how* would be met with less resistance. A key learning is that our focus should be the other way around. The intern's presence enabled the demonstration plot to thrive and produce abundant, chemical-free vegetables and herbs, which became the evidence leading some of the more sceptical lecturers to show more interest in agroecology. Coordinating lecturers noted that sales of the produce also allowed students to integrate lectures on finance and budgeting.[1] In addition, the demonstration plot dispelled the notion held by many students that farming could only be undertaken as a large-scale commercial enterprise and that they could only embark on farming as employees. The agroecology intervention represented choice and encouraged students to re-imagine themselves as farmers, to believe that even with limited resources they could prepare their own land and plant, harvest, finance, and sell produce themselves.

Learning is developmental

Through this process Biowatch was strongly reminded that learning is a developmental process, a journey of transition. Biowatch tends to lead with an ideological approach, advocating for agroecology while pointing to what is wrong with conventional industrial agriculture. In this context especially, that approach tended to alienate some lecturers and students. As a result, although lectures also covered the theory that applied to practices such as methods for water conservation, building healthy soil, pest management, and seed systems, the students had already dismissed the value of the lectures. In future we need to take this into account, encouraging exploration and dialogue, and presenting contextual evidence to elicit self-awareness, rather than actively critiquing the system that is entrenched in the college.

The practical application of agroecology in the garden and field, and a greater awareness in the broader public domain of the devastation brought about by the industrialized food system, whether in the form of climate change or degraded nutrition, will become the routes by which students might arrive, in time, at their own perceptions of what agroecology has to offer them and the farmers that they will work with.

Note

1. During the three months from mid-July to mid-October 2019, the intern recorded sales of 77 bunches of spinach, 47 bunches of herbs, 42 bunches of kale, 30 heads of cabbage, 12 heads of lettuce, 11 heads of broccoli, and a 3 kg bag of eggplants from a 20X30 metre area.

References

Department of Agriculture, Forestry and Fisheries (DAFF) (2008) 'Evaluation of agricultural education and training curricula in South Africa, October 2008', DAFF, Pretoria, Republic of South Africa.

DAFF (2015) 'National Education and Training Strategy for Agriculture, Forestry and Fisheries', DAFF, Pretoria, Republic of South Africa.

DAFF (2016) 'National policy on extension and advisory services', DAFF, Republic of South Africa, Part A <https://www.kzndard.gov.za/images/Documents/PolicyDocuments/National-Policy-on-Extension-and-Advisory-Services---PART-A.pdf>; Part B <https://www.kzndard.gov.za/images/Documents/PolicyDocuments/National-Policy-on-Extension-and-Advisory-Services---PART-B.pdf>; Part C <https://www.kzndard.gov.za/images/Documents/PolicyDocuments/National-Policy-on-Extension-and-Advisory-Services---PART-C.pdf>.

(Continued)

Box G Continued

National Department of Agriculture (NDA) (2005) 'Agricultural education and training strategy for agricultural and rural development in South Africa: Executive Summary', DALA, Pretoria, Republic of South Africa.

Witt, H. (2018) 'Policy impacts: the impact of governmental agricultural and rural development policy on small-holder farmers in KwaZulu-Natal', Biowatch Research Paper, Durban, South Africa: Biowatch South Africa <https://biowatch.org.za/download/research-paper-policy-impacts-the-impact-of-government-agricultural-and-rural-development-policy-on-smallholder-farmers-in-kwazulu-natal/?wpdmdl=511&refresh=6526616cb507b1697014124>.

Box H Developing Southern Africa's first diploma programme in agroecology

Shepherd Mudzingwa and Jaci van Niekerk

Largely based on the industrial agriculture model, Zimbabwe's agriculture extension services have in the past facilitated a robust food production system from field to fork. In recent years, however, the system has failed to sustain the production of safe and nutritious food in adequate quantities to alleviate hunger and foster food security. The system's ineffectiveness results from its dependence on often scarce external agricultural inputs as key production elements. Extension staff in the country, as well as the region, have been equipped with skills designed for the exclusive promotion of conventional (i.e. industrial) agriculture, which is highly reliant on external inputs. This approach offers a one-size-fits-all prescriptive way of managing farm operations which, due to its high cost, is out of reach for most farmers. Moreover, limited capacity and knowledge around ecologically sensitive methods of production have contributed to the system's failure to present working solutions for large numbers of small-scale farmers (Fambidzanai Training Centre, 2019).

The impacts of industrial agriculture, such as loss of biodiversity, degradation of cropping lands, and human health issues, have led to the need for a fundamentally different model of agriculture. This alternative model needs to be based on diversifying farming landscapes, replacing synthetic chemical inputs, optimizing biodiversity, and stimulating interactions between different species. These need to be part of holistic strategies which build long-term soil fertility, and support healthy agroecosystems and secure livelihoods; that is, diversified agroecological systems (Lin et al., 2011). The new path of agricultural development needs to recognize and redefine the role of rural societies using criteria that not only view agriculture as having an economic and food-producing role, but also emphasize environmental, cultural, and social roles (FAO, 2018).

Table H.1 SWOT analysis of the current government extension system in Zimbabwe

Strengths	Opportunities
• Extensive grassroots coverage with district- and/or village-level representation • Amalgamation of the Department of Research and Specialist Services with Zimbabwe's agricultural extension system (AGRITEX) ensures collaboration between technology generators and disseminators	• Improved extension-service delivery and efficiency through capacity-building initiatives • Collaboration opportunities among line ministries, departments, and other system actors
Weaknesses	**Threats**
• Limited efforts, if any, towards the extension of sustainable transformative agriculture • Lagging technical knowledge about new enterprises and concepts	• Prevailing economic situation: it is unlikely that the government is going to increase budgetary allocations towards agriculture extension support services • Unstable macroeconomic and political environment

(Continued)

Box H Continued

• Poor logistical support • Conflicts between line ministries and departments at the expense of rural development programmes and intended beneficiaries	• Limited intellectual capacity to promote and upscale alternative concepts such as agroecology

Agroecology – a dynamic concept that is frequently discussed – has not been fully embraced by those who live on the African continent. Yet it is, without doubt, one of Africa's solutions towards the betterment of rural livelihoods, and many actors in the development field, government, and the private sector have advocated its implementation. Moreover, many Africans, particularly resource-limited, small-scale farmers – who constitute more than 70 per cent of the farming population – have been practising these principles by default, as many are embedded in traditional agricultural practices.

In view of the SWOT (strengths, weaknesses, opportunities, and threats) analysis (Rukuni et al., 2006) of Zimbabwe's extension system shown in Table H.1, Fambidzanai Permaculture Centre, in collaboration with Bindura University of Science Education, set out to develop a capacity-building programme packaged as a Diploma in Agroecology, a qualification recognized both nationally and internationally. The aim of the programme is to develop proficient agriculturalists and development agents whose approach to agriculture resonates with local communities. Trainees acquire the capacity to apply sound scientific principles alongside indigenous knowledge in elevating sustainable agricultural production. They are equipped with the skills and knowledge best suited to the context of small-scale rural farming.

One of the key steps in the structuring of the diploma was curriculum development. The development of an effective curriculum is a multi-step, ongoing, and cyclical process which becomes more meaningful if input is gathered from a range of stakeholders. The curriculum development process started in 2014, when drafts were circulated among actors from various disciplines around the globe for their input. A key comment from stakeholders was that the programme should be able to address resource-poor farmers' challenges and respond to their needs. These observations were incorporated, and the final diploma programme curriculum then went through formal university approval channels before its adoption in 2017 by the Zimbabwe Council for Higher Education.

As the diploma course evolves, and trainees graduate, the numbers of extension personnel with practical and entrepreneurial skills in sustainable agriculture will escalate. This new crop of extensionists has the potential to revolutionize agriculture by lobbying for sustainable agriculture production frameworks and will have the ability to manage agricultural enterprises and agroecological landscapes in a sustainable way.

Agroecology Diploma curriculum and implementation

The programme has practical, hands-on, and experiential learning at its core. During the two-year course, students spend around 70 per cent of their time doing field and practical work and around 30 per cent of their time on lectures and plenary sessions in a classroom environment. The course is designed in such a way that each of the modules (see Table H.2 for a summary of the modules) contains practical components which allow the students to apply the theory they have learned. Students engage in work-related learning while they are studying in a block-release mode, so the work-related learning runs concurrently with the semester-based learning. As part of the learning, students working closely with farming communities have to identify major- challenges faced by the community and together go through a research process to come up with potential solutions. This model provides for co-learning among students, farmers, and teaching

(Continued)

Box H Continued

staff, thus providing unique opportunities for teaching staff to assess the relevance and effectiveness of their teaching.

As an affiliate of Bindura University of Science Education, Fambidzanai Training Centre offers the Diploma in Agroecology under the guidance of the university, particularly in areas of quality assurance and compliance with national standards.

Fambidzanai and Bindura University have developed a working model for strengthening extension training which can be scaled up across the region and beyond. The programme has drawn the attention of extension officers, both private and government-based. To date, a total of 50 people have graduated from the diploma course. Those graduates who are working as government extension officers have garnered a number of accolades, among them 'Best Extension Officer of the District' and runner-up in the 'Best Provincial Extension Worker' category. Upon graduation, two students were promoted to supervisory posts and another now works in the district office.

Table H.2 Agroecology diploma course outline

Part 1	Introduction to agroecology
	Principles of organic agriculture
	Principles of plant and soil science
	Introduction to communication and computers
	Agricultural practice
	Principles of agricultural economics
	Organic crop production
	Organic livestock production
	Climate change management
	Agricultural extension
	Appropriate technologies in agroecology
Part 2	Industrial attachment (internship)
Part 3	Introduction to statistics
	Sustainable seed production
	Organic beekeeping
	Sustainable land use planning
	Plant propagation and management
	Health and safety education
	Sustainable rangeland and pasture management
	Post-production management and value addition
	Organic aquaculture
	Sustainable energy systems
	Watershed and field water management

Box H Continued

More evidence of the programme's success can be found on the ground. The following remarks come from students themselves and from a farmer who had been assisted by a newly trained extensionist.

> *Building this movement will start with me and I promise to use the skills and knowledge acquired during the course of my studies on this diploma in agroecology –* Agroecology Diploma graduate

Photo H.1 The first group of students to receive their Diploma in Agroecology from Fambidzanai, 2019
Credit: Godfrey Tsele

Photo H.2 Agroecology Diploma students observing a practical demonstration
Credit: Shepherd Mudzingwa

(Continued)

Box H Continued

I wish the programme could be expanded to national and mandatory training for extension officers working in rural communities – Agroecology Diploma graduate

The increase in crop and species diversity in our ward is attributed to the efforts by the extension officers, who are at the forefront of promoting diversity and sustainable production systems – Farmer

References

Fambidzanai Training Centre (2019) *Baseline survey report on the role of agricultural extension in promoting agroecology in the face of climate change in different agroecological regions of Zimbabwe*, Harare: Fambidzanai Permaculture Centre.

Food and Agriculture Organization of the United Nations (FAO) (2018) 'Scaling Up Agroecology Initiative: transforming food and agricultural systems in support of the SDGs', Rome: FAO <www.fao.org/3/i9049en/i9049en.pdf>.

Lin, B., Jahi Chappell, M., Vandermeer, J., Smith, G., Quintero, G., Bezner-Kerr, R., McGuire, K.L., Nigh, R., Rocheleau, D., Soluri, J. and Perfecto, I. (2011) 'Effects of industrial agriculture on climate change and the mitigation potential of small-scale agro-ecological farms', *Animal Science Reviews* 6(20): 1–18 <http://dx.doi.org/10.1079/PAVSNNR20116020>.

Rukuni, M., Tawonezvi, P., Munyuki-Hungwe, M. and Matondi, P.B. (eds) (2006) *Zimbabwe's Agricultural Revolution Revisited*, Harare: University of Zimbabwe Publications.

Box I A community of practice: a Southern African journey towards co-creating a transformative social learning space for seed practitioners

Elfrieda Pschorn-Strauss

'What does a seed champion/expert/knowledge holder need in order to be fully equipped in terms of knowledge, skills, and attitude?' is the question around which a small group of people interested in community seed systems brainstormed at a gathering in Domboshawa just outside Harare, Zimbabwe, in September 2015. At the gathering, the idea of forming a community of practice (CoP) to pursue this question was put forward – and this community has since played an important role, at first in advancing community seed systems, and now also in agroecology in Southern Africa.

'Whatever the problem, community is the answer'

This principle, recognizing communities as the most effective locus of change, acknowledges also that solutions for complex problems are to be found in collective processes of learning and doing.[1]

Communities of practice are defined as 'groups of people who share a concern or a passion for something they do and learn how to do it better as they interact regularly' (Wenger-Trayner and Wenger-Trayner, 2015). Three qualities make a community of practice different from other groups such as neighbourhoods and networks:

- a *shared domain* or field of practice that outlines the identity of the community of practice;
- a *distinct community* built on relationships and intentional commitment to advance the field of practice beyond the individual;
- a *shared practice* that sustains community and deepens interaction over time (ibid.).

A CoP can thus become a living curriculum, but because it creates the conditions for trust and confidence to flourish, it can also lead to the social formation of a person, rather than just the acquisition of knowledge.

Margaret Wheatley's theory on the role of communities of practice in catalysing social change and transformation is important for the environmental and agroecology movement in that it presents a framework for understanding emergence, reaching tipping points, and bringing social innovations to scale. 'When separate, local efforts connect with each other as networks, and then strengthen as communities of practice, suddenly and surprisingly a new system emerges at a greater level of scale' (Wheatley and Frieze, 2015: 45). This places communities of practice as having huge potential in advancing agroecology and seed sovereignty, as both are domains that are recognized as knowledge-intensive and context-specific.

Seed and knowledge as a domain, an area of interest

The first thing the SKI CoP did was to conceptualize a curriculum and pedagogical approach that would start filling the knowledge and confidence gap created by years of policies and agricultural extension that devalued community seed systems and farmers' know-how.

A learning-by-doing approach was adopted that included exchange visits, trainings, and practical experiences. But importantly, it also included ample space to dive deep

(Continued)

BOX I A COMMUNITY OF PRACTICE   **269**

Box I Continued

into issues that participants were keen to understand better. Many questions were put up for discussion: 'How can we maintain traditional varieties to avoid genetic erosion in a changing climate? How can farmers be encouraged to grow small grains and crops that do well in their area? How does the introduction of new varieties through participatory varietal selection impact on local varieties? What seeds do we bring to the seed fair? What principles should we adopt about hybrids? How can this process influence policy? How can we chase Monsanto?'

Participants and farmers shared their embodied expertise of the complex and interdependent systems they dealt with every day, and from time to time external experts were invited to invigorate the thinking of the group.

A community of intent

'This is a special group of kindred spirits where I belong and can tap into my gifts'.[2] People were invited to join on the basis of their stance on agriculture and seed, and all were practitioners, the people in partner organizations who interacted with farmers every day. Continuity and a depth of inquiry in the group were nurtured by a core group of participants and facilitators committed to the process. By 2022, around 40 people considered themselves part of the SKI CoP. Although the community was created with intent, great effort was put into nurturing a self-organizing spirit, which showed up in how some would partner after a meeting or exchange to learn from each other and implement joint activities.

'We don't feel that we know something until we know it together'

This quotation from Patton (2017: 57) reflects the principle that consciousness and expertise reside in communities of people, not just individuals. In the words of a participant in the SKI CoP: 'More than anything, the SKI CoP has shown the need to connect with similar minded colleagues that share similar values. [It] has offered opportunities to interrogate my own views and the people who are interacting with the environment and farmer communities. It makes me question everything around us'.[3] This community of practice created a space to think deeper, what Hannah Arendt calls 'to gain experience in how to think' (Patton, 2017: 60).

The very practical and urgent need to learn as much as possible about community seed systems, as well as the transformative learning potential of communities of practice, is what has inspired the shared practices to thrive.

Challenges for the community of practice

Communities of practice can be an enigma for those on the 'outside' or for managers or funders that may feel uncomfortable with the mercurial nature of social learning processes. 'However, the very characteristics that make communities of practice a good fit for stewarding knowledge – autonomy, practitioner-orientation, informality, crossing boundaries – are also characteristics that make them a challenge for traditional hierarchical organizations' (Wenger-Trayner and Wenger-Trayner, 2015). To be successful, communities of practice need support. They need funds to pay for meetings and communications, for good coordination and facilitation, and for implementing their innovative ideas.

To ensure continuation, the SKI CoP will need to stay relevant to the practitioners' everyday realities and learning needs and ensure ongoing support by funders and managers. Determining the impact of a CoP can be tricky, as it is not outcomes-based. Wenger-Trayner and Wenger-Trayner (2015) propose a value-creation cycle as a tool through which the refinement and insight of stories can explain how participation contributes to the transformation of the individual or organization, moving from immediate impact to strategic and transformative impact.

(Continued)

Box I Continued

The value of a community of practice to a young seed activist: an interview with Juliet Nangamba

Juliet Nangamba was a new and young staff member at the Community Technology Development Trust (CTDT) organization in Zambia, when she was invited to attend a SKI CoP meeting in Malawi in 2017. 'I really enjoyed that CoP meeting. I think when I went the first time, I was not very informed about the issues around seed, but I found that the CoP gave me an opportunity to learn, to hear different approaches and get experience from so many people.'

Juliet explained how being part of the CoP transformed not only her but also her organization: 'I came into the CoP as a timid person, not very confident, but I have grown as a person and as a result can also benefit the growth of my organization. That has been the beauty of the CoP for me.' She added that 'it helped me to think more deeply and critically and not just scratch the surface of the issue'.

For her and her organization the CoP has added strategic value in that she is now able to engage on seed issues with farmers and with policymakers at national level and beyond. 'I have been exposed to so much and I am now much more knowledgeable and confident. I can write, and contribute to many issues and spaces. I am also able to participate and lead delegations to international meetings on farmers' rights and agriculture.

Juliet readily explained what it was that had boosted her confidence. 'You can only be confident about a matter if you are informed. CoP meetings are very detailed, they include theoretical discussions as well as experiences in the field where you can reflect on implementation.' Juliet emphasized that the success of the CoP hinged on facilitation that was responsive and ensured equal participation. As a participant in the SKI CoP affirmed, 'The opportunity to interact in an informal way is valued, there is a simplicity in these interactions and a safe space to learn and explore new ideas'.

Expanding minds and expanding focus

For some, the most valuable aspect of the CoP has been their personal growth, opportunities for leadership development, relationship building, and a 'reignited passion and zeal to learn more'. Over time the strategic impact has become more visible.

After a field visit to thriving finger millet farmers in the midst of a drought in Gutu, Zimbabwe, small grains were actively promoted, leading to increased adoption by farmers. The use of biofertilizers for soil and seed health became popular among farmers and they began training each other. Through ongoing interaction with their peers, the vital importance of regenerating soils to give local seed systems a fighting chance was deeply grasped by at least two organizations previously focused largely on the conservation of agrobiodiversity. Holistic livestock management as an approach to regenerate communal grazing areas emerged as a shared interest and is now being put into practice by five organizations that were exposed to this idea through the CoP. A group of seven organizations decided to focus on implementing agroecology on a landscape level, expanding the focus of the CoP, and also expanding the understanding that resilient communities and seed systems are integral to the regeneration of landscapes.

For pioneering efforts in seed and agriculture systems to emerge as systems of power and influence, CoPs must have the space and support to play their crucial role in rapidly spreading new knowledge, thinking, and practices. The true potential of the network is not simply through its members, but through the trustful relationships, the creative conversations, and flourishing knowledgeability among them.

'Only through deep relationships can we bring the transformation we want'.

(Continued)

BOX I A COMMUNITY OF PRACTICE **271**

Box I Continued

Notes

1. Berkana Institute website <https://berkana.org/home/>.
2. Seed and Knowledge Initiative (SKI) Unpublished report, 2021.
3. Seed and Knowledge Initiative (SKI) Unpublished report, 2022.

References

Patton, M.Q. (2017) 'Pedagogical principles of evaluation: interpreting Freire', *Pedagogy of Evaluation: New Directions for Evaluation* 155: 49–77 <https://doi.org/10.1002/ev.20260>.
Wenger-Trayner, E. and Wenger-Trayner, B. (2015) *Introduction to Communities of Practice: a brief overview of the concept and its uses* <https://wenger-trayner.com/introduction-to-communities-of-practice/>
Wheatley, M. and Frieze, D. (2015) 'Lifecycle of emergence: using emergence to take social innovation to scale', *Kosmos* 14(2): 45–47 <https://margaretwheatley.com/wp-content/uploads/2014/12/KosmosJournal-WheatleyFrieze-SS15.pdf>.

PART 4

Transitioning towards agroecology: working together and moving the struggle forward

CHAPTER 13

Cuba's participatory seed system: insights for South Africa

Mvuselelo Ngcoya

Introduction

> I have to say that we are faced with unprecedented challenges – and climate change is at the top of the list. A farmer today has to know more than *granjeros*[1] of yesteryear. Most importantly, you have to know your plants inside out. Yes, inside, as in the genetics of your tomatoes. There is no doubt that the *elbita* tomato is the most resilient to strenuous environmental conditions, to pests, diseases … But that is here, in our region. You need to know what works for your *finca*.[2] But that requires you become a geneticist, it requires you to experiment, try, fail, try again until you find what works for you. Nobody will save you, it's just you, your soil and your seed. Work, work, work! (Cuban farmer)

We were at an agrobiodiversity fair, hosted by a local farmer and the organizers of Cuba's plant breeding programme. The speaker, a farmer from the area, concluded his address with a soaring appeal to self-determination on the farm and the spirit of the revolution. A wildly enthusiastic round of applause followed. This was not his farm, nor was he a scheduled speaker. In the subsequent fog of excitement that surrounded him, I could not ask his name. He was a typical *granjero*, saying what needed to be said. His oratory, delivered in the good old performative tradition of Cuban public speaking, deserved all the applause it drew from our crowd of about 70. The fair was organized by members of the National Institute of Agricultural Sciences (INCA), a research organization I was affiliated with in San José, in the Mayabeque province just outside Cuba's capital city, Havana. It was an untypical cold Monday morning, but the spirits of the farmers, researchers, and municipal workers were high.

I was in Cuba on a five-month study visit (October 2018 to March 2019) to immerse myself in that island's unique experience with sustainable farming methods. Cuba's success with decentralized plant breeding is attributed to a significant shift from conventional industrial agriculture to agroecological methods (Rosset, 2000; Funes Monzote et al., 2012). Part of a larger project, this chapter reflects on the insights I gleaned from Cuba's transition to this plant breeding programme. My account is based on visits to more than 20 farms and two prestigious agricultural research institutions in six of the island's provinces, namely, Pinar del Rio, Artemisa, Habana, Mayabeque, Matanzas, and Sancti Spiritus. In addition to farm stays, visits, and farmer interviews, I also interviewed 13 key informants in Cuban agroecology, including agronomists, scientists, government representatives, extension officers, academics, members of civic organizations, market traders, and the general public. I benefited tremendously from my affiliation with INCA.

My guide was one of the leading thinkers on agroecology in Cuba, Dr Ángel Leyva. We had long, formal discussions as well as conversations over meals or while travelling to different sites. He cleared up many of my superficial impressions and shared his wide-ranging views on the history of Cuban agriculture and debates about agroecology in Cuba. As part of my affiliation with INCA, I also attended its national conference, which was attended by more than 200 participants, most of them from Cuba, but also many from all over the world. Aside from INCA, I spent two days visiting another research organization in the Matanzas province, the Estación Experimental de Pastos y Forrajes (Pastures and Forages Research Station) Indio Hatuey. I also participated in two agrobiodiversity fairs and interviewed key developers of Cuba's participatory breeding programme.

Why Cuba?

I grew up in what we called 'the buttocks of the land' – forsaken areas of apartheid South Africa, where agriculture and life itself were a grim grind. My folks produced scraps of food on unforgiving terrain in KwaZulu-Natal. What I thought was farming back then, was, I now realize, a broken proletariat merely '[scratching] about the land', to use Murray's (1981: 19) apt term, as it never produced enough to support the family. Today, I have joined my parents in their agricultural pursuits. Our community has a large farmers' association and there is a firm commitment to producing for both domestic consumption and the market. Yet there is a sense of foreboding that all is not well.

At one of our farmers' association meetings, members expressed concern that the local nursery was engaging in racial discrimination when selling seedlings. 'How do you know?' I asked. A woman responded: 'You would have to be blind not to notice that the cabbage that white farmers grow only a few kilometres from us looks healthy and ours ragged. We all buy our seedlings from the same source but our cabbage droops like the ears of a sick dog.' A young member reported that he knew someone working in the nursery who confirmed that large-scale commercial farmers ordered their seedlings in large amounts ahead of time and were given the best seedlings, while black farmers who ordered seedlings only in the hundreds got the dregs. For their next order, I asked a white friend to call on their behalf. The order was processed and collected, and when we delivered it, the farmers were stunned: 'This is what we were talking about. It's true. These are vigorous seedlings, not the famished yellowing crap they often sell us.' We repeated this experiment a few times with similar results. I checked on prices as well. If you buy fewer than 250 seedlings per crop (which is the case for a lot of our farmers), the seedling price almost doubles. I asked the farmers why they do not grow their own seedlings. They often shake their heads and say that it never crossed their minds. My visit to Cuba arose out of these concerns. How can small-scale farmers extricate themselves from the stranglehold of seed companies? How has Cuba managed to support its small-scale producers?

There are many reasons why Cuba offers potentially profitable opportunities for comparison with South Africa. First, like South Africa, until recently Cuba had an unworkable land model that denied large portions of society ownership of land and security of tenure. But in the 1990s, the Cuban government embarked on an unprecedented land reform programme that revolutionized small-scale farming and redistributed land to smaller farmers. For example, in 1992, 75 per cent of cultivable land fell under large-scale state farms, but by the end of that decade, over 3,000 smaller cooperatives (called basic units of cooperative production, or UBPCs) had been created, reducing state-owned land to 34 per cent (Wright, 2012).

Again, much like present-day South Africa, prior to 1989, the Cuban agricultural sector was characterized by a large-scale agricultural model that prioritized extensive monocropping, high levels of external industrial inputs, widespread mechanization, and large-scale irrigation (Rosset, 2000; Funes Monzote, 2010). Indeed, the Cubans were recognized as having the most industrialized agricultural sector in all of Latin America (Rosset, 2000). However, following the dramatic geopolitical changes of the early 1990s that reversed Cuba's economic relationships with trading partners in the Eastern European socialist trading bloc, the Cuban economy went through unprecedented food shortages, or what the Cubans dubbed *la temporada de vaca flaca* ('the period of the skinny cow', sometimes simply referred to as 'the Special Period' (Wright, 2012)).

Regarding seeds, prior to the Special Period, the dominant seed system was anchored in a formal model that relied on an expert-led seed management system in which farmers were mere consumers of seeds produced elsewhere – much like South Africa's contemporary corporate-led seed system. The crisis of the Special Period forced Cuba to fundamentally decentralize its seed production and management system. As a result, there is growing evidence that Cuba has reversed the negative effects of the food crisis of the early 1990s and gained a degree of food sovereignty (Rosset, 2000; Funes Monzote et al., 2012). More importantly, the country has revolutionized its agricultural sector and substituted a conventional industrial agricultural system with a low-energy, sustainable agroecological model (Funes Monzote et al., 2012). One of the catalyst programmes that helped transform the country towards agroecology was its famous farmer-to-farmer plant breeding programme, the Programa de Innovación Agrícola Local (Local Agricultural Innovation Programme), or PIAL.

The next part of this chapter deals with participatory plant breeding (PPB) in general and is followed by a section more specifically focused on PIAL itself. I conclude the chapter by examining some lessons South African researchers, policymakers, and farmers can draw from the Cuban experience.

Participatory plant breeding

Plant breeding refers to the application of genetic knowledge in selecting plants with particular desirable characteristics, such as higher yields, ecological suitability, cultural and economic value, and drought and pest tolerance, among many others. While there are different definitions of PPB, Martínez Cruz et al. (2017) identify its main features as involving a strategy of genetic plant improvement that engages key actors (researchers, producers, organizations, and others) in the production chain, who collaborate to develop plant varieties that strengthen local seed systems. However, there is a limitation to this definition that we will revisit below.

Until the 1990s, plant breeding was a specialized scientific field confined to scientific institutions and laboratories in Cuba. Expert plant breeders and other scientists experimented with seeds and technologies in experimental stations and substations throughout the island. They would select the seeds they found suitable for particular environments and distribute them through state institutions in various provinces following national directives and principles. As Humberto Ríos Labrada (2016), one of the founders of Cuba's decentralized plant breeding system[3], recounts, the main benefits of this model were increased monocultural production on a wide scale, high agricultural mechanization, and increased inputs. However, all this was driven by a select few, well-placed individuals in research institutions and the government. During the economic crisis of the 1990s, when Cuba lacked key agricultural inputs such as equipment, fertilizers, chemicals, and energy, this model hit a brick wall.

The old model gave way to an amalgam of old and new: traditional and modern methods of plant breeding and seed conservation. At the genesis of the PPB in Cuba, Ríos Labrada was doing his doctoral research at the Agrarian University of Havana on low-input pumpkin breeding in Cuba. Recounting the early years, he told me:

> We had to fight against an old mindset of conventional scientists and state officials who could not fathom that farmers themselves could be trusted with such highly technical and scientific processes. But the socio-economic circumstances of the time and the urgent needs for food security overwhelmed old-fashioned methods and required innovation. But it took a few of us crazy people to sell the idea not only to the white coats but to the farmers themselves, to instil faith in their own habits and methods but to also question their traditional methods at the same time. (Interview with H. Ríos Labrada, 20 December 2018).

Cuba's decentralized PPB arose from these attempts. It supports seed production for the island's various climatic regions. As a result, scientists and farmers work together to develop cultivars that are adapted to specific conditions and climates rather than a centralized universal system. Instead of prioritizing high yields and uniformity (which characterizes centralized plant breeding systems), the participatory method used in Cuba prioritizes decentralized plant selection on *fincas* and regions where the seeds will be used, favouring genetic diversity and adaptation to specific ecological and social environments (Leitgeb et al., 2011; Lorigados et al., 2013; Martínez Cruz et al., 2017).

The Special Period therefore opened the door for researchers and farmers to collaborate. In his account of the history of PPB, Ríos Labrada (2016) states that they started with a few maize seeds collected in various Cuban communities that had historically maintained their capacity to develop their own native seeds. Together with some commercial seeds, they were planted at one of INCA's trial sites. The four commercial varieties that were selected were important because they had been the dominant seeds until that point and there was more knowledge about their characteristics and behaviour than about those of the traditional seeds. These commercial varieties were donated by professional breeders who were based at public institutions (Ríos Labrada, 2009). Cuban researchers note that public institutions, such as the Institute for Fundamental Research in Tropical Agriculture and the Rice Research Institute, have been a pillar of strength for the participatory seed diffusion system by providing a range of commercial seeds for the process (Ríos Labrada, 2009; Martínez Cruz et al., 2017).

During this first trial, the national crisis prohibited the use of any fertilizers, and the scarcity of equipment made irrigation virtually impossible. As a result of these deficiencies, the researchers lost all hope. However, they were surprised when their array of maize seeds produced plants of differing height, leaf, and yield. Farmers were then invited to select seeds according to their own criteria. This is how the idea of agrobiodiversity fairs emerged, which formed the fulcrum of the new programme, PIAL.

PIAL: Local Agricultural Innovation Programme

The genesis of Cuba's PPB programme can be traced back to a group of radical researchers, who initiated the process in 1999 at INCA. They collaborated with researchers at other institutions in four municipalities. PIAL has a two-pronged main function: to revitalize the agrarian sector through increased active participation of small-scale farmers in the food production and environmental system in Cuba, and to strengthen the resilience of the food system through enhanced biodiversity (Interview with Acosta Roca, 8 November 2018).

The programme has now completed two phases. The first phase (2001–2006) introduced the concept of PPB and showed that increased farmer participation in seed production and maintenance strengthened local seed conservation, improved yields, and ensured the well-being of farmers, significantly elevating their knowledge and social role (Ríos Labrada, 2009, 2016). The second stage (2007–2011) extended the concept of participatory selection and development of technologies to other national institutions and centres of higher learning. By 2012, the network consisted of more than 50,000 farmers working with 12 Cuban institutions of science and technology (Ortiz Pérez, 2013).

Formal informality

An important feature of PIAL is its dual formal–informal nature. As many scholars have pointed out, the label 'informal seed system' may be deceptive, as these informal systems are not entirely closed local systems and may be interwoven with seeds, networks, and other materials from the formal system (Soleri et al., 2013; Bezner Kerr, 2013; Vernooy et al., 2013; African Centre for Biodiversity, 2016; Soleri, 2018). Indeed, as Van Niekerk and Wynberg (2017) show in their research on traditional seed systems in northern KwaZulu-Natal, what are described as informal systems are often quite formal ones, including rituals, rules, and customs that govern seed exchange. However, in Cuba I did not observe any customs and rules in the fairs I participated in. What they call 'informal' in PIAL is an alternative system to the conventional formal approach that traditionally relies on variety production tested and approved by national and provincial scientific boards (Ríos Labrada, 2009). The informal part of PIAL relies on farmers as key players in the breeding, conservation, improvement, and distribution of seeds. In this model, farmers, cooperatives, and other groups informally test and distribute varieties of special interest through a process of chain reaction (Ríos Labrada, 2003a). This process generates a seed diversity nucleus at the local level, which increases exponentially as farmer participation widens. Farmer organize themselves around this nucleus, sharing and promoting new knowledge, and establishing exchange networks and local innovation.

Yet the process has formal features in that it is coordinated at the national level and involves international funders. The formal part is funded by national and international organizations, relies on commercial networks and national and regional research institutions, and has formal structures designed to monitor and evaluate the processes (Leitgeb et al., 2011). This formal–informal nature of PIAL is demonstrated in the variety of seeds available in Cuba. For example, in their study of bean seeds in Cuba, Lorigados et al. (2013) state that within their heterogeneous collection of 174 bean cultivars, 125 were traditional varieties collected in 10 municipalities through agrobiodiversity fairs, conservation centres, and germplasm banks.

As demonstrated in Figure 13.1, PIAL has two key anchors, Agricultural Biodiversity Distribution Centres (CDBAs) and Local Centres of Agricultural Innovation (CLIAs). The former are sites (*fincas* or groups of production stations) that centre on production, experimentation, technologies, and the conservation and distribution of high levels of diverse crops. They focus on

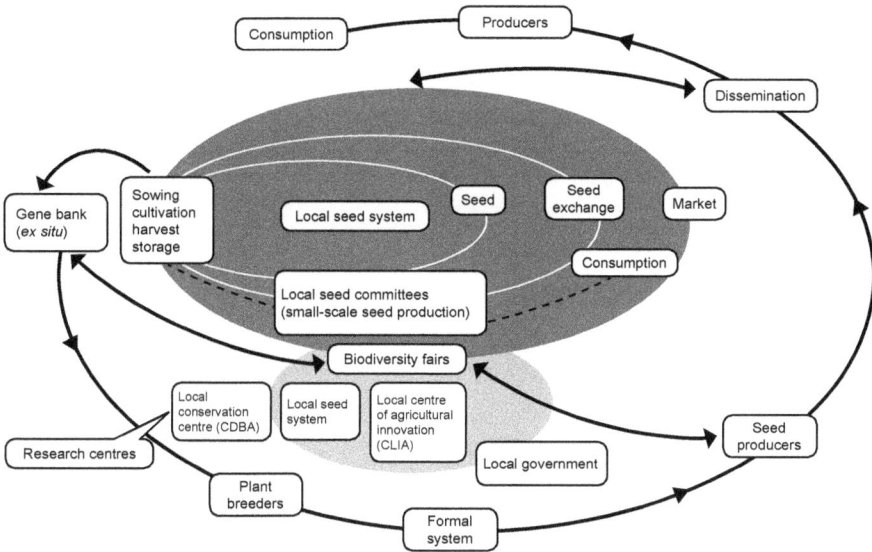

Figure 13.1 Schematic representation of the formal–informal seed management system in Cuba. CDBAs are Agricultural Biodiversity Distribution Centres and CLIAs are Local Centres of Agricultural Innovation.
Source: Adapted from Ortiz Pérez, 2013

minimal cost, sustainable replication of systems, and methods in a particular community. CLIAs, on the other hand, constitute a system of relations among various local, national, and international actors, whose main task is to promote continual changes in the system of production to increase the quantity and quality of economic, environmental, and social benefits of agriculture at the local level (Leitgeb et al., 2011; Ortiz Pérez and Acosta Roca, 2013). The fundamental shift from the conventional model was placing the farmers at the forefront of the seed production and management system, instead of scientific experts and state companies.

Farmer scientists

While scientists, researchers, and municipal officials play important roles, it became clear during my farm visits that relations between scientists and farmers were different from what I was used to in South Africa. I was privileged to travel to farms and agrobiodiversity fairs with researchers from INCA and the Estación Experimental de Pastos y Forrajes Indio Hatuey. On one such visit, the camaraderie and respect between the researcher and the farmer were plain to see. The researcher who was our guide recounted the history of the farm, the types of produce and amounts under production, and the types of innovations with biofertilizers and stone minerals as if he was enthusing about his own project. The farmer had his test tubes (from the institute), ready to collect samples from his various processes, and he took a good hour lecturing about why his cabbages were so impressive. The line between scientist and farmer was blurred. It seemed to me that the old idea of agricultural extension had been turned on its head: it was no longer the research institute extending itself to the farm, but the farmer extending himself to the research institute.

What I witnessed during this visit was the emblematic feature of decentralized PPB. It seemed to affirm Healy and Dawson's (2019: 881, citing Cleveland and Soleri, 2002) assertion that PPB emerged to '"reverse the historical trend of separation between farmers and plant breeders", meet the agronomic needs of marginalized farmers, and bring farmers' knowledge into decisions made about variety improvement'. Indeed, for many of the researchers I spoke with, it was important that farmers maintained control of their own seeds and seed supply. As Ríos Labrada (2003b: 123) puts it, letting farmers take the lead 'gets unforeseen measures and substantially strengthens local seed systems'. Giving farmers stewardship of seeds allows them to continually shift the type and amount of seed to meet dynamic environmental, economic, and cultural conditions of their regions.

This fact was emphasized repeatedly at the agrobiodiversity fair in San José, Mayabeque, that opened this chapter. The host was Señor Ruiz, the soft-spoken and highly respected owner of Finca Amistad, one of the five seed banks in the municipality. Although he grows a host of cultivars, including beans, coffee, fruit, and cassava, on that mid-January morning in 2019 he was demonstrating his diverse tomato seeds, 25 varieties to be exact. Yes, 25 breeds of tomatoes in about 2 hectares of land. The differences were notable, even to my untrained eye: a variety of shapes, sizes, colours, and leaves. It seemed almost too much. I pushed him and the other scientists: 'Surely, this is excessive – a *granjero* doesn't need that much variety to feed his family or to supply the market!' The local seed specialist in the municipality shot back: 'But remember, *compañero* [comrade], he's not doing it just for himself; this is for the whole municipality. He is our local seed bank!' It turns out that Señor Ruiz was an integral component of the national PIAL system.

Four critical phases of participatory plant breeding in Cuba

The PPB methodology in Cuba is organized around four phases: diagnosis, collection of plant genetic resources, agrobiodiversity fairs and the establishment of demonstration plots, and farmer experimentation (Interview with Acosta Roca, 19 December 2018).

Diagnosis

As the initial stage of the process, diagnosis allows farmers to consider the biophysical characteristics of their locality, and their socio-economic and cultural conditions. This is a critical stage to develop relationships, and to identify leaders and create an environment where producers can feel they are active players in the systematization of community knowledge (Interview with Acosta Roca, 6 January 2019). 'How do you identify leaders?' I wondered aloud. The response from Ríos was surprising:

> Oftentimes they are not in the room, or meeting. They are elsewhere, working, doing crazy things. So I always ask those who are present in the room to name some of the crazy farmers they know. Yes, crazy. The doers. Those who do unusual things. I will often ask the participants to name one person they would like to work with. At the end of the meeting, it becomes clear who those people are and that is who forms a local subcommittee. (Interview with H. Ríos Labrada, January 2019)

I asked: 'But it seems like you are driving the process, then, and not the farmers, no?' The response came from Rosa Acosta Roca:

> Well, driving is too strong a description. Our task is to coordinate the process. The farmers are busy with their often punishing work of taking care of their work and they have neither the resources nor patience to coordinate such a process. Someone else has to do it. We guide the process along and make sure the farmer is the centre of our imagination. (Interview with Acosta Roca, 6 January 2019)

Collection of plant genetic resources

Plant genetic resource collection is an integral stage of the process. Here, farmers and researchers improve their knowledge and appreciation of their seeds and of the conservation and distribution of their crop germplasm. While this process valorizes traditional seeds (which they call 'creole seeds' in Cuba), it does not negate the importance of improved seeds. There is an understanding that *in situ* seed production could aggravate some of the disadvantages of traditional seeds. As a result, there is emphasis on improving the quality and quantity of the creole seeds to make them attractive, robust, and commercially viable (Lorigados et al., 2013; Ortiz Pérez, 2013).

Obviously, the role of traditional knowledge holders is critical at this stage. They know the seeds, who holds them, and their cultural relevance, suitability, and economic worth. Still, at the local level, the seed collection strategy emphasizes both cohesive *ex situ* and *in situ* seed conservation strategies. Therefore, farmers are encouraged to identify opportunities for collaboration with institutions that house germplasm and are willing to work with farmers to prioritize, capacitate, and encourage a diverse range of local and other seeds for conservation and multiplication (Lorigados et al., 2013). According to Martínez Cruz et al. (2017), the anonymity of seeds is important during this stage. Each variety planted is identified only by a serial number, and source information is not revealed until after the selection to minimize any bias from the participants.

Agrobiodiversity fairs and demonstration plots

The first agrobiodiversity seed fair was held at INCA in April 1999 as an approach to disseminating maize seeds suitable for low-input agriculture (Ríos Labrada and Wright, 1999). Within a 10-year period, more than 680 seed fairs had been organized for over 40 species, reaching an estimated 600,000 beneficiaries in 45 municipalities and 10 provinces in Cuba (Ortiz Pérez, 2013). Some estimates indicate that about 50 per cent of the plant varieties present on small farms in Cuba correspond to the varieties selected and introduced at the fairs. For example, Lorigados et al. (2013) state that a community of bean farmers in El Tejar-La Jocuma increased the diversity of their beans by adding 34 new varieties after participating in two agrobiodiversity fairs in two provinces. They further report that, prior to participating in the fairs, the community had only six types of grains of four colour tones and two shapes. Agrobiodiversity fairs incorporated new genotypes and increased grain type to 12 new types, 7 new tones, and 4 additional shapes (Lorigados et al., 2013).

The coordinators of the fairs consider various factors when selecting demonstration plots: the location (whether they are public institutions, cooperative farms, or the farmers' own plots), the diversity collected in prior stages, the cultural norms and economic profile of the cultivators, and limiting the necessity of external inputs. The location of the demonstration site is important for other reasons: for example, when farmers are responsible for their sites, the maintenance and logistical costs of the programme are reduced quite remarkably. More importantly, when the producers drive the process from planting to harvest, they can undertake detailed monitoring of the plant performance throughout the cycle. But the site has to be accessible to all participants, and it should represent local conditions as closely as possible in terms of soil type and quality, relief, and other factors. Once the demonstration site is established, the area has to be clearly marked, with a poster or banner at the entrance listing the main features, including the general outline of the area, soil type, cultivars planted, planting date, irrigation, and fertilization (Martínez Cruz et al., 2017: 135).

Agrobiodiversity fairs are essentially an alternative system for plant breeders and producers to share genetic diversity from formal and informal seed systems. Besides injecting genetic diversity into a wider network of producers and increasing community acceptance of seed varieties, agrobiodiversity fairs play an integral role in the conservation of endangered material, and they broaden the spectrum of available seed material (Interview with I. Moreno, 6 January 2019). They are also joyful. I witnessed Señor Ruiz's quiet pride when he was hosting his fair in January 2019. It was a celebratory event, including the media and participants from all over the country and one other international visitor besides me. Señor Ruiz absorbed the appreciation from all of us, and it was clear that he was grateful for the recognition. This was affirmed by another farmer:

> Recognition is the most important. Every time a Cuban visitor or foreigner comes to my farm, I feel recognized. Every time a municipal

manager sends me someone to increase their seed variety, when I sit with him to evaluate what he needs, that excites me. My experiences here in Cuba and elsewhere motivate me. (Interview with A. Alda Cruz by Robaldo Ortiz Pérez of INCA, January 2019)

Farmer experimentation

This is obviously the most important stage of the PIAL. Its main objective is for the farmers to experiment with their chosen varieties in order to assess their adaptability on their farms. This is where their selection criteria meet reality: whether they do indeed get better yields, reduce inputs and associated costs, and respond robustly to local pests and diseases. Here, farmers can also assess the cultural and economic suitability of different varieties. Even relationships are cemented during this period. Researchers and producers work together closely to monitor performance, introduce or modify technologies, affirm or modify existing knowledge, and improve the farmers' abilities in terms of experimentation and diversity management (Ortiz Pérez, 2013; Martínez Cruz et al., 2017: 137).

Lessons for South Africa

There are many laudable features of Cuba's sustainable agricultural model that we would do well to reflect on. These include an effective state and political will, the distribution and management of land, the role of cooperatives, the place of science and innovation, and Cuba's famous farmer-to-farmer model. While Cuba's context is fundamentally different from South Africa's in terms of its size, political and economic systems, and ecological features, there are numerous lessons that are relevant to the South African context.

Radical land reform and family farms

The transformation of the Cuban agricultural sector was facilitated by an aggressive land reform process. Prior to the revolution, 8 per cent of the farmers controlled 70 per cent of farmland (Zeuske, 2000: 29). As I indicated earlier, the picture had changed drastically by the 1990s, when – thanks to multiple agrarian reform programmes – state-owned farms accounted for approximately 80 per cent of Cuban arable land while the remaining 20 per cent was evenly divided between private farms and production cooperatives (Sáez, 1998: 49). Over half of this state land was redistributed to worker cooperatives and family farms. These reforms enhanced family farms and increased production. For example, from 1997 to 2003, vegetable production in Havana increased from about 20 tonnes to about 250 tonnes; and fruit and vegetable production countrywide was 250 per cent higher than 1990 (Koont, 2009; Zepeda, 2003; Wright 2012).

Today, as long as the land remains productive and they meet quotas, farmers have free usufruct rights, and can use the land in perpetuity and even bequeath it (Zebeda, 2003: 2). When I asked Esteban, a member of a cooperative in Alamar, Havana, how difficult it would be to get more land, he responded: 'Easy. We would simply apply for usufruct rights and two months or so later, we would have land. The state would only reappropriate it if we play around. Work and you keep the land' (Interview with E. Gonzales, 12 January 2019).

This contrasts sharply with the South African landscape. At the dawn of democracy in 1994, white farmers owned some 77 million hectares (or 86 per cent of farmland); today that number is about 61 million hectares (Kirsten and Sihlobo, 2022). At this pace, it will take centuries before South Africa's dispossessed population get justice. As the Cuban example shows, equitable agrarian reform is impossible without a radical and thoughtful land redistribution programme, including urban land reform.

The agrarian revolution is urban

The transformation of South Africa's seed production and distribution system will require a radical change of the agricultural system as a whole. Given the tight leash that large-scale agricultural and food enterprises keep on the South African food system, it is clear that the Cuban model would have to be redrawn quite significantly. In 1994, about 55 per cent of the South African population resided in urban areas; this has since increased to about 70 per cent in 2020 (World Bank, n.d. b), yet the agricultural system is firmly rural (Bureau for Food and Agricultural Policy, 2018). In 2019, the urban share of Cuba's population was 77 per cent – 8.7 million of the island's 11 million people (World Bank, n.d. a). Cuba's radical plant breeding programme was therefore ensconced in a national programme that recognized the country's urban demographics. As such, the focus of its national agricultural programme is urban. The government created the National Group for Urban Agriculture (Grupo Nacional de Agricultura Urbana, GNAU) in 1997 to ensure the growth and to strengthen the development of urban agriculture (Leitgeb et al., 2011: 358). Although many of the farms I visited were in rural towns and areas, the extent of urban agriculture in Cuba was impressive. According to the FAO (2013), there was a mere 257 hectares of *organopónicos* (urban agriculture farms) in the country in 1995, but, within a five-year period, this increased to 45,000 hectares of urban agriculture. So extensive is urban agriculture now that there are 28 seed planting units (or stations) in the city of Havana alone, in addition to 10 municipal horticultural seed farms, which supply urban farmers (FAO, 2013). Since the mid-1990s, urban agriculture has contributed to the development of 56 species of vegetables and fresh condiments, and in order to encourage agrobiodiversity, *organopónicos* and intensive orchards are required to plant or produce a minimum of 10 different species annually (Herrera Sorzano, 2009).

This urban feature of Cuban agriculture requires heavy municipal involvement. Each municipality has a specialist focused on coordinating all agricultural activities, including seed banking and plant breeding. According to Leitgeb et al. (2011: 358), urban agriculture specialists typically reside close to the urban farms and have close relationships with the farmers. While there are doubts about the efficacy and desirability of urban agriculture to solve South Africa's food security dilemmas (Battersby and Haysom, 2016), there are enough studies that indicate that urban agriculture in South Africa can indeed offer viable solutions to the complex challenges of feeding the country's burgeoning urban population (Olivier and Heinecken, 2017; D'Alessandro et al., 2018; Chihambakwe et al., 2019). Looking at Cuba's dynamic and effective model, South African farmers, activists, researchers, and policymakers would be foolish to consider a farmer-centred plant breeding programme that does not have an urban strategy.

Linking consumers and producers

Thanks to the central command that characterizes the Cuban economic system, the state there has tighter controls over production and consumption than in South Africa. There are definitional limitations that arise from the Cuban economic model. To recap, Martínez Cruz et al. (2017: 133) define PPB as 'a strategy of genetic improvement of plants where the different actors in the production chain (researchers, producers, organizations and others) work together in the process of developing varieties for strengthening local seed systems'. Note the focus on the production chain and the producers. Because of the state's firm grip on the links between production and consumption, the Cuban model of plant breeding does not give much attention to players beyond the farm gate. In other words, while the model involves farmers, researchers, the government, and other organizations, it gives less attention to consumers and the general public. At the three agrobiodiversity fairs I attended in 2019, I did not meet a single representative of consumer organizations or of the food processing industry. In the South African context, there are a few big players in the dominant food system that would make it difficult for agroecologically minded researchers and farmers to make a dent. However, there are opportunities that can be explored in alternative seed networks and markets.

Interactions would go beyond cooperation between farmers and scientists, and include a host of other players. Representatives of the transport sector, chefs, government departments that procure millions of tonnes of food per year, informal traders, and consumers would be involved in determining and evaluating varieties for breeding purposes. This is what experiments elsewhere in the world have shown to be critical. For example, the promotion of indigenous varieties in Bolivia assigned an important role to the organization representing chefs and restaurants (Delgado and Delgado, 2014). Similarly, Pereira et al. (2019) recount how young chefs in the community of Zimatlán de Álvarez, Mexico, have become the most important promoters of traditional foods and plants in their community, using gastronomy festivals, students, traditional knowledge holders, and local cooks. They also cite the example of a movement of chefs that led to an increase in the use of dune spinach, a type of fynbos plant, by peri-urban farmers and chefs in Cape Town. These are a few instances to show that there are multiple sites and actors in the transformation of a seed and food system.

Legal rights

Again, because of the dominant role played by the state in Cuba, questions of legal rights did not seem to concern the plant breeders I interviewed. In South Africa, the corporate dominance over our food and seed systems makes ownership of the knowledge and material of plant breeding a concern. Who has the right to this knowledge? Who can benefit from it, when and how? These are concerns that plant breeders in South Africa will have to tackle. Some communities may not be entirely comfortable with open-sourcing their seeds because of their cultural significance. At a minimum, community-based plant breeders will have to be involved as active partners in the value chain of seed experimentation, being co-creators and economic beneficiaries of genetic diversity. Policy change and support are required to encourage the emergence of small-scale seed companies, community-based seed organizations, and seed start-ups for a new breeding programme (Lammerts van Bueren et al., 2018).

Forgotten plants, forgotten people

In Cuba, with its impressive array of traditional seeds and foods, I did not witness the sidelining of indigenous seeds and cuisine. This stands in contrast to South Africa's food system, where corporate power over the food system militates against traditional foodways (Shonhai, 2016; Mbhenyane, 2017). In the community in KwaZulu-Natal where I am based (and in numerous rural areas throughout the country), there is an abundance of edible plants that are considered indigenous and therefore unpopular. Yet, as numerous studies have shown, these plants are often more nutritious and more ecologically friendly than the popular commercial varieties (Mbatha and Masuku, 2018; Mabhaudhi et al., 2019). Knowledge about the cultivation of many indigenous crops and their nutritional and cultural values is in decline. In light of this challenge, farmers, NGOs, and research organizations will have to identify individuals who hold knowledge of indigenous crop varieties, so that they can become anchors of activities and programmes promoting the breeding of these varieties. The conservation, improvement, multiplication, and distribution of these indigenous seeds will need to become critical components of agricultural innovation in South African plant breeding programmes.

Financial rewards

When valorizing community seed banking, proponents correctly praise the absence of the profit motive in seed exchanges (Kumarakulasingam and

Ngcoya, 2016). However, we also have to recognize that it is exceedingly difficult for many small-scale farmers to take on the role of community seed banker without financial support or incentive. Some of the leaders of PIAL expressed this concern. In Cuba's economic system, entrepreneurship is not highly rewarded financially. This is a challenge the coordinators of the island's plant breeding programme are seeking to overcome. We will have to develop novel models to motivate South African farmers to engage in small and large networks and value chains of seed production and distribution. But they will have to be financially rewarded in a just and equitable manner to encourage them to do so.

Conclusion

Cuba's farmer-centred and farmer-driven sustainable agriculture model has fundamentally transformed the country's food system. This chapter has demonstrated that the challenges of the Special Period compelled the government, researchers, and farmers to abandon an old agricultural model that relied on expensive external inputs. This required a significant land reform programme that resulted in the redistribution of over half of state-owned land to worker cooperatives and family farms. This is a key lesson for South Africa. An aggressive and equitable land reform programme is a quintessential component of the transformation of our agrarian system. Another important element is the devolution of agricultural decisions to local municipalities. South Africa would do well to learn from the valorization of these localised farmer-centred approaches. In Cuba, the conversion to a low-input, farmer-focused, sustainable farming system was stimulated by a plant breeding programme that was uniquely decentralized and participatory. Working hand in hand with farmers and government entities, agricultural research institutions developed a formal–informal national plant breeding programme that placed greater control over seed production, management, and distribution in the hands of farmers themselves. The impressive results of Cuba's programme include greater crop diversity, improved food security, and better nutritional and health indicators in the population. Another area where Cubans have excelled is through taking urban agriculture seriously. While urban agriculture alone will not solve our food insecurity problems, the Cuban example illustrates that given the fast pace of urbanization in South Africa, the country's plant breeding programmes need to have a well-formulated urban strategy. However, this chapter has also emphasised that we need to take steps that go beyond Cuba's centrally planned economy. These include strong entrepreneurship and financial incentives for small-scale seed producers and tighter links between producers and consumers. While Cuba's experience is unique, the country's PPB programme suggests that with political will, organization, and faith in the ability of the farmers themselves, it is possible to design and implement a seed system that enhances farmers' ability to organize themselves and to develop systems that benefit them, the environment, and society at large.

Notes

1. A *granjero* is anyone who owns a farm or works on a farm.
2. Spanish dictionaries generally define *finca* as 'estate', 'ranch', or 'large rural property'; in Cuba, however, a *finca* is generally a small farm.
3. Although he now lives abroad, Humberto Ríos Labrada is still engaged in the PPB programme and visits Cuba frequently as he owns a small farm outside Havana.

References

African Centre for Biodiversity (2016) 'Integration of small-scale farmers into formal seed production in South Africa: a scoping report' <https://acbio.org.za/seed-sovereignty/integration-small-scale-farmers-formal-seed-production-south-africa/>.

Battersby, J. and Haysom, G. (2016) 'Urban agriculture: the answer to Africa's food crisis?', *Quest* 12(2): 8–9.

Bezner Kerr, R. (2013) 'Seed struggles and food sovereignty in northern Malawi', *The Journal of Peasant Studies* 40(5): 867–897 <https://doi.org/10.1080/03066150.2013.848428>.

Bureau for Food and Agricultural Policy (2018) *BFAP Baseline Agricultural Outlook 2018–2027*, BFAP, Pretoria <www.sagis.org.za/BFAPBaseline-2018.pdf>.

Chihambakwe, M., Mafongoya, P. and Jiri, O. (2019) 'Urban and peri-urban agriculture as a pathway to food security: a review mapping the use of food sovereignty', *Challenges* 10(6) <https://doi.org/10.3390/challe10010006>.

D'Alessandro, C., Hanson, K.T. and Kararach, G. (2018) 'Peri-urban agriculture in Southern Africa: miracle or mirage?', *African Geographical Review* 37(1): 49–68 <https://doi.org/10.1080/19376812.2016.1229629>.

Delgado, F.B. and Delgado, M.A. (2014) *El Vivir y Comer Bien en los Andes Bolivianos: Aportes de los Sistemas Agroalimentarios y las Estrategias de Vida de las Naciones Indígenas Originario Campesinas a las Políticas de Seguridad y Soberanía Alimentarias*, La Paz, Bolivia: AGRUCO.

Food and Agriculture Organization of the United Nations (FAO) (2013) 'Agricultura urbana y peri-urbana en América Latina y el Caribe: compendio de estudios de casos' <http://www.fao.org/ag/agp/greenercities/pdf/Compendium.pdf>.

Funes Monzote, F.R. (2010) 'Cuba: a national-level experiment in conversion', in S.R. Gliessman and M. Rosemeyer (eds), *The Conversion to Sustainable Agriculture: Principles, Processes, and Practices*, pp. 205–235, Boca Raton, FL: CRC Press.

Funes Monzote, F.R., Bello, R., Alvarez, A., Hernández, A., Lantinga, E.A. and van Keulen, H. (2012) 'Identifying agroecological mixed farming strategies for local conditions in San Antonio de Los Baños, Cuba', *International Journal of Agricultural Sustainability* 10(3): 208–229 <http://dx.doi.org/10.1080/14735903.2012.692955>.

Healy, G.K. and Dawson, J.C. (2019) 'Participatory plant breeding and social change in the Midwestern United States: perspectives from the Seed to Kitchen Collaborative', *Agriculture and Human Values* 36: 879–889 <https://doi.org/10.1007/s10460-019-09973-8>.

Herrera Sorzano, A. (2009) 'Impacto de la agricultura urbana en Cuba', *Novedades en Población* 5(9): 1–14 <https://revistas.uh.cu/novpob/article/view/3510>.

Kirsten, J. and Sihlobo, W. (2002) 'Land reform in South Africa: 5 myths about farming debunked', *The Conversation*, 26 November <https://theconversation.com/land-reform-in-south-africa-5-myths-about-farming-debunked-195045>.

Koont, S. (2009) 'The urban agriculture of Havana', *Monthly Review* 60(1): 63–72.

Kumarakulasingam, N. and Ngcoya, M. (2016) 'Plant provocations: botanical indigeneity and (de)colonial imaginations', *Contexto Internacional* 38(3): 843–863 <https://doi.org/10.1590/s0102-8529.2016380300006>.

Lammerts van Bueren, E.T., Struik, P.C., van Eekeren, N. and Nuijten, E. (2018) 'Towards resilience through systems-based plant breeding: a review', *Agronomy for Sustainable Development* 38(42): 1–21 <https://doi.org/10.1007/s13593-018-0522-6>.

Leitgeb, F., Funes-Monzote, F.R., Kummer, S. and Vogl, C.R. (2011) 'Contribution of farmers' experiments and innovations to Cuba's agricultural innovation system', *Renewable Agriculture and Food Systems* 26(4): 354–367 <http://dx.doi.org/10.1017/S1742170511000251>.

Lorigados, S.M., Ortiz Pérez, R., Ponce Brito, M., Rodríguez Miranda, O., Acosta Roca, R., Ríos Labrada, H., Chaveco, O. and Quintero, E. (2013) 'Conservación y diseminación de recursos fitogenéticos de frijol común a nivel local', in R. Ortiz Pérez (ed), *La Biodiversidad Agrícola en Cuba: En Manos del Campesinado Cubano*, pp. 93–108, San José, Cuba: Instituto Nacional de Ciencias Agrícolas (INCA).

Mabhaudhi, T., Chimonyo, V.G.P., Hlahla, S., Massawe, F., Mayes, S., Nhamo, L. and Modi, A.T. (2019) 'Prospects of orphan crops in climate change', *Planta* 250(3): 695–708 <https://doi.org/10.1007/s00425-019-03129-y>.

Martínez Cruz, M., Ríos Labrada, H., Ortiz Pérez, R., Lorigados, S.M., Acosta Roca, R., Moreno Moreno, I., Ponce Brito, M., De la Fé Montenegro, C.F. and Martin, L. (2017) 'Metodología del fitomejoramiento participativo (FP) en Cuba', *Cultivos Tropicales* 38(4): 132–138.

Mbatha, M.W. and Masuku, M.M. (2018) 'Small-scale agriculture as a panacea in enhancing South African rural economies', *Journal of Economics and Behavioural Studies* 10(6): 33–41 <http://dx.doi.org/10.22610/jebs.v10i6(J).2591>.

Mbhenyane, X.G. (2017) 'Indigenous foods and their contribution to nutrient requirements', *South African Journal of Clinical Nutrition*, 30(4): 5–7.

Murray, C. (1981) *Families Divided: The Impact of Migrant Labour in Lesotho*, New York: Cambridge University Press.

Olivier, D.W. and Heinecken, L. (2017) 'Beyond food security: women's experiences of urban agriculture in Cape Town', *Agriculture and Human Values* 34: 743–755 <https://doi.org/10.1007/s10460-017-9773-0>.

Ortiz Pérez, R. (2013) 'Sistema formal e informal de semillas: nuevos horizontes', in R. Ortiz Pérez (ed), *La Biodiversidad Agrícola: En Manos del Campesinado Cubano*, pp. 122–131, San José, Cuba: Instituto Nacional de Ciencias Agrícolas (INCA).

Ortiz Pérez, R. and Acosta Roca, R. (2013) 'Los centros de diseminación de la biodiversidad agrícola en el contexto del programa de innovación agropecuaria local', in R. Ortiz Pérez (ed), *La Biodiversidad Agrícola en Cuba: En Manos del Campesinado Cubano*, pp. 40–56, San José, Cuba: Instituto Nacional de Ciencias Agrícolas (INCA).

Pereira, L.M., Calderón-Contreras, R., Norström, A.V., Espinosa, D., Willis, J., Guerrero Lara, L., Khan, Z., Rusch, L., Correa Palacios, E. and Pérez Amaya, O. (2019) 'Chefs as change-makers from the kitchen: indigenous knowledge and traditional food as sustainability innovations', *Global Sustainability* 2(16): 1–9 <https://doi.org/ 10.1017/S2059479819000139>.

Ríos Labrada, H. (2003a) 'Logros en la implementación del fitomejoramiento participativo en Cuba', *Cultivos Tropicales* 24(4): 17–24.

Ríos Labrada, H. (2003b) 'Nuevas luces del fitomejoramiento participativo en Cuba', *Cultivos Tropicales* 24(4): 123–134.

Ríos Labrada, H. (2009) 'Participatory seed diffusion: experiences from the field', in S. Ceccarelli, E.P. Guimarães and E. Weltzien (eds), *Plant Breeding and Farmer Participation*, pp. 589–612, Rome: FAO.

Ríos Labrada, H. (2016) 'Fitomejoramiento participativo e innovación local', in F. Funes Aguilar and L.L. Vazquez Moreno (eds), *Avances de la Agroecología en Cuba*, pp. 183–198, Matanzas, Cuba: Estación Experimental de Pastos y Forrajes Indio Hatuey.

Ríos Labrada, H. and Wright, J. (1999) 'Early attempts at stimulating seed flows in Cuba', *ILEA Newsletter* No. 15: 38–39.

Rosset, P. (2000) 'Cuba: successful case study of sustainable agriculture', in F. Magdoff, J.B. Foster and F.H. Buttel (eds) *Hungry for Profit: The Agribusiness Threat to Farmers, Food and the Environment*, p. 203–213, New York: Monthly Review Press.

Sáez, H. (1998) 'Resource degradation, agricultural policy and conservation in Cuba', *Cuban Studies* 27: 40-67.

Shonhai, V. (2016) *Analysing South African Indigenous Knowledge Policy and its Alignment to Government's Attempts to Promote Indigenous Vegetables*, PhD dissertation, University of KwaZulu-Natal <https://researchspace.ukzn.ac.za/bitstream/handle/10413/14927/Shonhai_Venencia_F_2016.pdf?sequence=1&isAllowed=y>.

Soleri, D. (2018) 'Civic seeds: new institutions for seed systems and communities – a 2016 survey of California seed libraries', *Agriculture and Human Values* 35: 331–347 <https://doi.org/10.1007/s10460-017-9826-4>.

Soleri, D., Worthington, M., Aragón-Cuevas, F., Smith, S.E. and Gepts, P. (2013) 'Farmers' varietal identification in a reference sample of local *Phaseolus* species in the Sierra Juárez, Oaxaca, Mexico', *Economic Botany* 67(4): 283–298 <https://dx.doi.org/10.1007/s12231-013-9248-1>.

Van Niekerk, J. and Wynberg, R. (2017) 'Traditional seed and exchange systems cement social relations and provide a safety net: a case study from KwaZulu-Natal, South Africa', *Agroecology and Sustainable Food Systems* 41(9–10): 1099–1123 <https://doi.org/10.1080/21683565.2017.1359738>.

Vernooy, R., Sthapit, B., Tjikana, T., Dibiloane, A., Maluleke, N. and Mukoma, T. (2013) *Embracing Diversity: Inputs for a Strategy to Support Community Seedbanks in South Africa's Smallholder Farming Areas: Report of Field Visits to Limpopo and Eastern Cape*, Rome, Italy: Bioversity International, and Pretoria, South Africa: Department of Agriculture, Forestry and Fisheries <https://alliancebioversityciat.org/publications-data/embracing-diversity-inputs-strategy-support-community-seedbanks-south-africas>.

World Bank (no date, a) 'Urban population (% of total population) – Cuba' <https://data.worldbank.org/indicator/SP.URB.TOTL.IN.ZS?end=2019&loc ations=CU&start=1960&view=chart>.

World Bank (no date, b) 'Urban population (% of total population) – South Africa' <https://data.worldbank.org/indicator/SP.URB.TOTL.IN.ZS?locations=ZA>.

Wright, J. (2012) 'The little-studied success story of post-crisis food security in Cuba: does lack of international interest signify lack of political will?', *International Journal of Cuban Studies* 4(2): 130–153.

Zepeda, L. (2003) 'Cuban agriculture: A green and red revolution', *Choices Magazine* 18(4): 1–4.

Zeuske, M. (2000) 'Notas retrospectivas sobre la sociedad agraria cubana en el siglo XIX y XX', in H. Burchardt (ed) *La Última Reforma Agraria del Siglo: La Agricultura Cubana Entre el Cambio y el Estancamiento,* p. 28–42, Caracas, Venezuela: Editorial Nueva Sociedad.

CHAPTER 14
We are what we eat: nurture nature

Kristof J. Nordin

A disconnect

'Agriculture', as defined by the *American Heritage Dictionary*, is: 'The science, art, and business of cultivating soil, producing crops, and raising livestock.' Modern agriculture, however, has focused too much attention on the 'science' and 'business' components, while losing sight of the 'art'. This change began as early as the 19th century. Writing in 1854, Henry David Thoreau remarked: 'Ancient poetry and mythology suggest, at least, that husbandry was once a sacred art; but it is pursued with irreverent haste and heedlessness by us, our object being to have large farms and large crops merely' (Thoreau, [1854] 1995: 107). Since the dawn of agriculture, the main purpose of growing crops and raising animals has been to provide the nutrition required for sustaining life. The phrase 'You are what you eat' is much more than a simple aphorism; it is a universal truth akin to saying that humans are mortal. This was recognized as far back as 400 BC, when the Greek physician Hippocrates was credited with saying: 'Let food be thy medicine, and let medicine be thy food.' At the Second International Conference on Nutrition hosted by the Food and Agriculture Organization of the United Nations (FAO, 2014), there was a call for 'nutrition-sensitive agriculture', a food-based approach to agricultural development placing nutritionally rich foods, dietary diversity, and food fortification at the heart of overcoming malnutrition and micronutrient deficiencies. Calls for 'nutrition-sensitive' agriculture, along with the need to medicinally 'fortify' our food, seem to be indicative of the world's escalating disconnect between agriculture and health.

An estimated 300,000 plant species occur on Earth, with at least 12,000 (4 per cent) thought to be edible by humans. However, only 150 to 200 are utilized as food (United Nations, 2019). As agriculture continues to shift from the production of diversified foods towards the large-scale monocropping of commodity crops for fuel, livestock feed, and processed products, access to highly nutritious food options is ever more compromised. Currently, just three crops – rice, wheat, and maize – supply more than half of the world's plant-derived calories, and only 12 crop and 5 animal species provide 75 per cent of the world's food (Bioversity International, 2017).

In Malawi, traditional agricultural systems once gave farmers access to nutritional diversity and seasonal harvests throughout the year. From 1938 to 1943, a wide-ranging survey of Malawi was conducted by a team which included a medical officer, an agriculturalist, a food specialist, an anthropologist, and a botanist. The data and conclusions of this survey, known as *The Nyasaland Survey Papers*, were sidelined due to World War II, but eventually published in 1992. The information is as pertinent today as it was over 80 years ago, and gives unique insights into some of the historical changes that have taken place. Particularly significant was the team's description of their difficulties in trying to quantify data due to the vast amount of agricultural diversity observed. They noted that the usual practice of relating agricultural observations and measurements such as seed rates, yields, and labour expenditure per unit area to a particular crop was not feasible because there was so much mixed planting in local communities. For purposes of recording, therefore, the team had to relate their observations and measurements to the gardens as a whole, rather than to single crops (Berry and Petty, 1992). The study also revealed that indigenous knowledge was being lost even then, with many people in the survey villages not remembering many of the wild plants formerly used for food and now abandoned. In terms of nutrition, however, the survey team notably remarked that the introduction of new vegetables or improved varieties was not necessary due to the fact that the existing strains and varieties needed little improvement (ibid.).

Food, the source of nutrition

At its most basic, a 'nutrient' is a substance which is used by an organism to live, grow, heal, and reproduce. At its most complex, 'nutrition' is an ever-evolving science. New discoveries are constantly being made about components of food that have important effects on human health. There are currently recommended dietary allowances set for 45 different nutrients believed to be essential for the healthy growth and development of the human body. These include 16 different macronutrients (and their subcomponents), including water, carbohydrates, fibre, fat (plus two fatty acids), and protein (plus nine amino acids), plus micronutrients, which include 14 different vitamins and 15 different minerals (Otten et al., 2006).

The prefix of the term 'malnutrition' comes from the Latin *malus*, meaning 'bad'. So 'mal-nutrition' is, in essence, 'bad nutrition'. Malnutrition is caused by imbalances in the form of both overnutrition and undernutrition. Overnutrition generally occurs when the amount of energy-giving macronutrients (proteins, carbohydrates, and fats) exceeds that required for normal growth and development. Global obesity has nearly tripled since 1975, and the majority of the world's population now live in areas where deaths related to obesity outnumber those related to being underweight (WHO, 2021). The health risks associated with obesity include several major non-communicable diseases, such as type 2 diabetes, coronary heart disease, stroke, asthma, and several cancers (Nyberg et al., 2018). As societies have changed their diets towards the consumption of highly processed foods containing more sugars, fats, and sodium, it is now estimated that almost half of the world's population will be overweight or obese by 2030. If current trends continue, the health care costs attributed to obesity – already ranging in the billions of dollars – are projected to double each decade (Rocha, 2017). Malawi seems to be following a similar trajectory, with the number of women in the country who are overweight or obese doubling from 10 per cent in 1992 to 21 per cent by 2015–16 (Ntenda and Kazembwe, 2019).

Undernutrition occurs when there is a deficiency in the consumption of overall nutrients required to meet an individual's needs to maintain good health, resulting in the body becoming *underweight* (low weight for age), *wasted* (low weight for height), or *stunted* (reduced growth and development) (WHO, 2021). Globally, 149 million children under the age of 5 are nutritionally stunted and, within this group, 45 per cent of deaths are linked to undernutrition (ibid.). This can intensify the severity of certain diseases, limit cognitive development, and increase the chance of death during specific stages of the life cycle. Undernutrition in children is especially devastating as they become more susceptible to recurring illness. Data for Malawi has shown that underweight children under 5 years of age had an increased risk of anaemia, diarrhoea, respiratory infection, and fever/malaria (WFP and AU, 2015).

In Malawi, the problems associated with malnutrition are sobering. A publication on the economic impacts of undernutrition in Malawi reported that an estimated 60 per cent of adults in Malawi suffered from stunting as children, which represents some 4.5 million people of working age who are not able to achieve their potential as a consequence of child undernutrition. As a result of these negative effects on health, education, and productivity, an estimated 147 bn Malawian kwacha (US$597 m) was lost in 2012. These losses were equivalent to 10.3 per cent of Malawi's gross domestic product, with the highest costs associated with the loss in potential productivity as a result of undernutrition-related mortalities (WFP and AU, 2015).

Apart from nutrients, there are numerous components of food which, although not thought to be essential to life, may have extremely beneficial effects for the human body. Phytochemicals, for example, are the components of plants which help to account for attributes of colour, taste, and smell. They also help plants protect themselves from pathogens and predators. Thousands of phytochemicals have been discovered to date, and researchers are only beginning to understand their importance in human health. This includes stimulating the immune system, blocking substances we eat, drink, and breathe from becoming carcinogens, reducing the kind of inflammation that makes cancer growth more likely, preventing DNA damage and helping with DNA repair, reducing the kind of oxidative damage to cells that can spark cancer, slowing the growth rate of cancer cells, and helping to regulate hormones (Collins, 2015). Considering the amount of knowledge which continues to be discovered regarding the medicinal and healing properties of food, it would appear that Hippocrates was very far ahead of his time.

Food security is not maize security

The expression 'food and nutrition security' is often used by governments and development agencies to provide an all-encompassing label for activities aimed at alleviating hunger and malnutrition. The use of this terminology, however, is slightly contentious as 'food security' and 'nutrition security' have historically carried different meanings. In 1995, the International Food Policy Research Institute defined 'nutrition security' as 'adequate nutritional status in terms of protein, energy, vitamins, and minerals for all household members at all times' (CFS, 2012). The 1996 World Food Summit defined 'food security' as existing 'when all people at all times have physical and economic access to sufficient, safe and nutritious food to meet their dietary needs and food preferences for an active and healthy life' (ibid.). This definition continues to be used today with two additional pillars of food security relating to agency and sustainability recommended in 2020 by the High Level Panel of Experts on Food Security and Nutrition of the Committee on World Food Security (HLPE, 2020: 7–11).

'Food and nutrition security' is now commonly referred to by those who wish to highlight the linkages between food and nutrition, with the ultimate goal of raising levels of nutrition (ibid.). The distinction between these terms is important, as many countries have been shifting food production away from nutritional diversity towards an over-reliance on a handful of staple foods, often resulting in a situation where food is available, but nutrition remains compromised.

Malawians are the third-highest per capita consumers of maize in the world through direct human consumption of white maize, and although the government of Malawi has noted that achieving national food security has been a major objective of agricultural policies adopted since independence in 1964, it has also stated that national food security is mainly defined in terms of access to maize, the main staple food (Conrad, 2014a).

Maize (*Zea mays*) is historically a Central American crop, first recorded under cultivation in Malawi just over 200 years ago, but well into the 20th century crops like bulrush millet, finger millet, and sorghum were still the predominant staple grains of East and Central Africa (GTZ, 1991). As maize dependency has intensified through the promotion of Green Revolution-style monocropping, it has also given rise to the problem of chronic 'hungry seasons'. When Malawian farmers plant their maize seeds during the onset of the rains (generally around December), they have to wait four to five months until those seeds mature into food. During this time, the maize reserves from the previous year often run short, creating a time of seasonal hunger, with households having limited access to food during the months prior to the annual harvest. In a typical year, up to 57 per cent of rural Malawian households and 36 per cent of urban households experience hunger during these pre-harvest months (Anderson et al., 2018). After harvest, the crop residue is gathered up and burned, and the fields lie barren until the next planting season. Malawi – a tropical country that could be capitalizing on the production of food for 12 months of the year through the integration of perennial and seasonal crops – is using just one month to harvest primarily one type of food. Even when there are bumper yields of maize, they still represent only one type of food from one food group, providing an insufficient diversity of nutrients. Reducing the world's agricultural and dietary needs down to only a handful of high-carbohydrate, low-nutrient staple foods is causing a wide range of problems. Genetic simplicity, with a narrow focus on high-energy cereal staples, promotes poor dietary diversity, thus contributing to undernutrition, overnutrition, and non-communicable diseases (Nordin et al., 2013).

In 2001, Malawi's government launched a six-food-group model for nutrition (Figure 14.1). Up to that point, the country had been using a three-food-group model, comprising 'body energy', 'body protection', and 'body building'. The new model was launched to promote an increase in nutritional diversity and split 'body energy' into 'staple foods' and 'fats'; 'body protection' into 'fruits' and 'vegetables'; and 'body building' into 'legumes and nuts' and 'animal

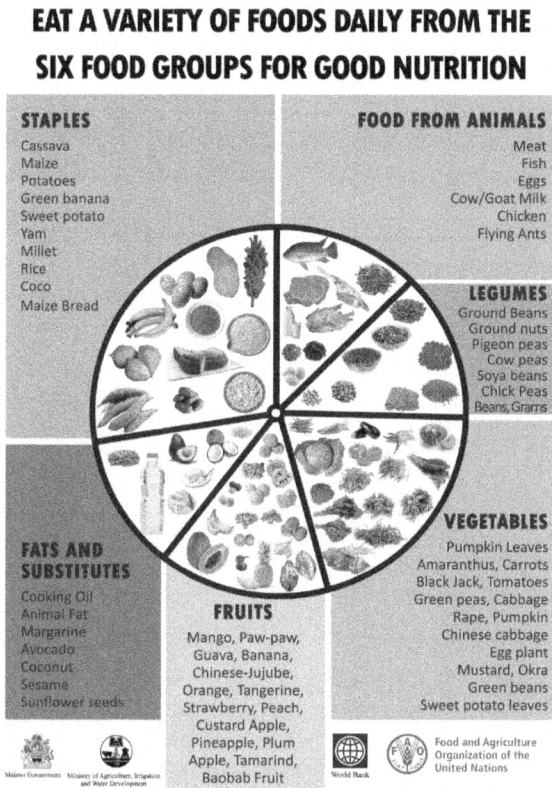

EAT A VARIETY OF FOODS DAILY FROM THE SIX FOOD GROUPS FOR GOOD NUTRITION

STAPLES
Cassava
Maize
Potatoes
Green banana
Sweet potato
Yam
Millet
Rice
Coco
Maize Bread

FOOD FROM ANIMALS
Meat
Fish
Eggs
Cow/Goat Milk
Chicken
Flying Ants

LEGUMES
Ground Beans
Ground nuts
Pigeon peas
Cow peas
Soya beans
Chick Peas
Beans, Grams

FATS AND SUBSTITUTES
Cooking Oil
Animal Fat
Margarine
Avocado
Coconut
Sesame
Sunflower seeds

FRUITS
Mango, Paw-paw,
Guava, Banana,
Chinese-Jujube,
Orange, Tangerine,
Strawberry, Peach,
Custard Apple,
Pineapple, Plum
Apple, Tamarind,
Baobab Fruit

VEGETABLES
Pumpkin Leaves
Amaranthus, Carrots
Black Jack, Tomatoes
Green peas, Cabbage
Rape, Pumpkin
Chinese cabbage
Egg plant
Mustard, Okra
Green beans
Sweet potato leaves

Malawi Government   Ministry of Agriculture, Irrigation and Water Development   World Bank   Food and Agriculture Organization of the United Nations

Figure 14.1 Malawi six-food-group nutritional model
Source: Government of Malawi, 2001

foods'. This was a step in the right direction, released by Malawi's Ministry of Agriculture, but the majority of the nation's agricultural policies continue to focus solely on the monocropping of maize. As the lack of agricultural diversity has contributed to a lack of nutritional diversity, many countries now resort to treating nutrition medicinally. In 2016, global aid from donors and multilateral agencies to provide for 'basic nutrition' programmes (which include micronutrient interventions such as providing vitamin A, iodine, and iron) was $856 m. This is a fraction of the $7 bn per year estimated to be needed to meet global targets of reducing stunting, anaemia, and wasting (Development Initiatives, 2018). Moreover, some of these programmes promote unhealthy food choices by fortifying products like cooking oil and sugar. For example, in 2012 the Malawian government received a $5 m donation from Irish Aid, UNICEF, and USAID to fortify sugar with vitamin A (*Nyasa Times*, 2012). A high dietary intake of sugars has been associated with high blood cholesterol, high blood pressure, and type 2 diabetes (Rocha, 2017).

Nature's nutrition

Current industrialized and commercialized methods of food production have come to be known as 'conventional' agriculture (Photo 14.1).

When we take a look at the definition of 'conventional', however, we find that it simply means 'conforming or adhering to accepted standards'. This should prompt us to take a closer look at what these 'accepted standards'

Photo 14.1 Monocropped maize in bare soil in Malawi
Credit: K. Nordin/NEF

truly are, and to consider whether we really want to 'conform or adhere' to them. Perhaps a better term would be 'compensation' agriculture. When ecosystems are disrupted, imbalances typically emerge. When natural habitat is cleared and burned to make way for the monocropping of a limited handful of crops, natural predators are often removed. This creates an imbalance in pest populations which requires farmers to *compensate* for the loss through the use of chemical pesticides. For example, the fall armyworm, first reported on the African continent in 2016, has now been confirmed in over 30 African countries. The rapid spread of this one insect species, which has the potential to cause maize yield losses of 21 to 53 per cent, has been associated with a loss of natural predators (Prasanna et al., 2018). Rather than promoting ecologically balanced forms of integrated pest management strategies to deal with the fall armyworm, conventional approaches have primarily focused on the application of synthetic pesticides, further contributing to the loss of natural predators (ibid.).

Similarly, when organic matter is cleared and burned, farmers are essentially clearing and burning the soil's food. To compensate for this loss of nutrients, farmers resort to buying expensive synthetic fertilizers. In 2005, the Malawi government launched the Farm Input Subsidy Programme (FISP), aiming to increase local farmers' access to agricultural inputs. Primarily, this was money spent to subsidize the high costs of synthetic fertilizers and commercially hybridized maize seeds. From 2005 to 2009, the supply of synthetic fertilizer in Malawi drastically increased: from 14,237 tonnes to 216,553 tonnes (Mutegi et al., 2015). This increase came at a cost: in the first five years of its implementation, from the 2005–06 growing season to 2009–10, the Malawi government invested $571.3 m in fertilizer subsidies, with an additional $74.6 m invested by donor organizations, totalling $645.9 m dollars. These enormous recurring financial investments, geared primarily towards the intensification of maize production, represented over 50 per cent of the entire nation's annual agricultural budget (Dorward and Chirwa, 2010). FISP is not a sustainable coping strategy and does not address the underlying problems of soil infertility and inadequate rainfall that make these inputs necessary on an ongoing basis. In addition, they are primarily used to increase maize yields, failing to address the issues of food quality and malnutrition associated with predominately maize-based diets (Conrad, 2014a). Despite a national maize surplus since 2007–08 after FISP implementation in 2005–06, there was an increase in rural poverty and food insecurity in 2011–12 (ibid.).

The truth of the saying 'We are what we eat' is not limited to the nutrition that we receive from food, but also pertains to the nutrition that our food receives from the soil. At the time that FISP was launched in Malawi in 2005–06, the synthetic fertilizer which was being subsidized by the government was labelled as 23:21:0: +4S, which meant that it contained 23 per cent nitrogen, 21 per cent phosphorus, 0 per cent potassium, and 4 per cent sulfur. In 2018, however, the government of Malawi changed the formulation of the synthetic fertilizer to reduce the amount of phosphorus, but increase the amount of potassium and sulfur, with a new addition of zinc (23:10:5+6S+1.0Zn). The reason for this change is troubling. Soil analysis conducted throughout Malawi showed that soils were lacking in potassium due to repetitious use of the 23:21:0: +4S fertilizer (Sangala, 2018), and human nutritional deficiencies of zinc in Malawi ranged from 60 to 66 per cent in all groups studied (men, women, and children) (Government of Malawi, 2017). It was proposed that applying zinc to the soil through zinc-enriched fertilizer would increase the concentration of consumable zinc in maize, thereby reducing the loss of disability-adjusted life years by up to 10 per cent (Joy et al., 2015). What seems to be missing, however, is recognition of the correlation between a predominately maize-based agriculture and diet and the lack of access to nutritional diversity within Malawi. There are many high-zinc crops that could be promoted and grown in Malawi, including legumes, wholegrains, nuts, and seeds (along with fruits and vegetables that help to enhance the body's absorption of zinc) (Saunders et al., 2013). The same can be said for the promotion of foods that could help to minimize other nutritional deficiencies in Malawi, such as vitamin A, iron, and selenium.

The expensive subsidization of synthetic fertilizers containing only four or five nutrients is more of a short-term, sticking plaster approach rather than a long-term sustainable solution. Plants require up to 20 different essential nutrients, which include *macronutrients*: carbon, hydrogen, oxygen, nitrogen, phosphorus, potassium, calcium, magnesium, and sulfur; and *micronutrients*: iron, manganese, boron, molybdenum, copper, zinc, chlorine, nickel, cobalt, sodium, and silicon (LibreTexts, n.d.). This means that the government of Malawi is spending millions of dollars to subsidize only a fraction of the nutrients that are necessary for crops to thrive. It also does not take into account soil structure, water retention, and the fact that the immense biodiversity of microorganisms found in healthy soil also need good nutrition. The more organic diversity which is returned to the soil, the more likely it is that the soil will receive the nutrients it needs to help us sustain life on this planet. Likewise, the more diversity that is consumed in terms of dietary choices, the more likely it is that humans will receive the nutrients they need to sustain healthy and active lives.

Nurture nature

The Oxford dictionary defines 'nature' as the 'basic or inherent features, character, or qualities of something'. One of the underlining characteristics of the current environmental crisis is humanity's disconnect from nature: the inherent character and qualities of our own nature as well as those of our environment. As Pope Francis wrote in his encyclical on climate change: 'We are faced not with two separate crises, one environmental and the other social, but rather with one complex crisis which is both social and environmental' (Francis, 2015).

Nutrition systems are out of balance

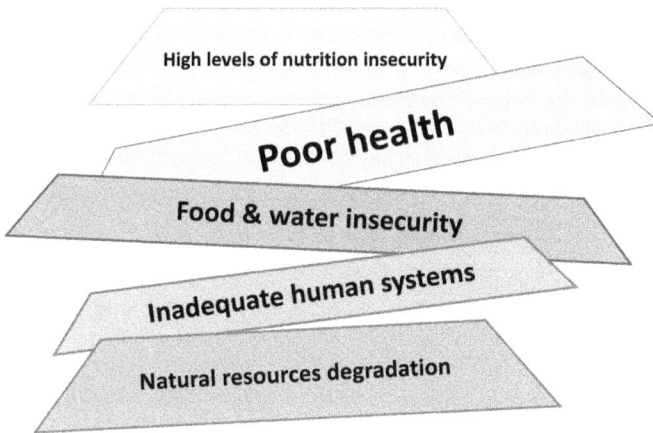

High levels of nutrition insecurity

Poor health

Food & water insecurity

Inadequate human systems

Natural resources degradation

Figure 14.2 Systems out of balance

It is increasingly recognized that food systems which do not facilitate healthy diets are an underlying cause of malnutrition (FAO, 2016). When food systems and health are viewed as part of an interconnected framework, it has profound implications for the way that knowledge is developed and deployed in our societies. Concepts such as 'sustainable diets' and 'planetary health' help to promote holistic scientific discussions and pave the way for integrated policy approaches (Rocha, 2017).

Environmental degradation, social injustice, and economic inequalities are all contributing factors to issues of food insecurity, lower standards of living, and health risks (D'Odorico et al., 2019). When ecosystems are degraded, many aspects of society are destabilized, including human nutrition (Figure 14.2).

Conversely, when natural resources are managed in an ecologically sustainable manner, when there is equity within social and economic systems, and when agriculture is designed to adhere to the definition of food security by providing diversified access to nutrition to all, then balance is restored (Nordin and Nordin, 2017) (Figure 14.3).

Viewed from a holistic point of view, the saying 'We are what we eat' expresses a concept that is greater than the sum of its parts, encompassing both 'what we eat' and 'what we are'. When we are able to come to a well-informed understanding of the laws of nature, we quickly recognize that humanity is part of – not above, nor removed from – these very same laws. The American philosopher Ralph Waldo Emerson reminded us of this when he wrote: 'The violations of the laws of nature by our predecessors and our contemporaries, are punished in us also' (Emerson [1841] 1979: 147).

Many of the current agricultural practices in Malawi have detrimental environmental implications. Poor soil stewardship practices, such as the inefficient construction of annual planting ridges (only 12 per cent of

Agriculture is the <u>SOURCE</u> of nutrition

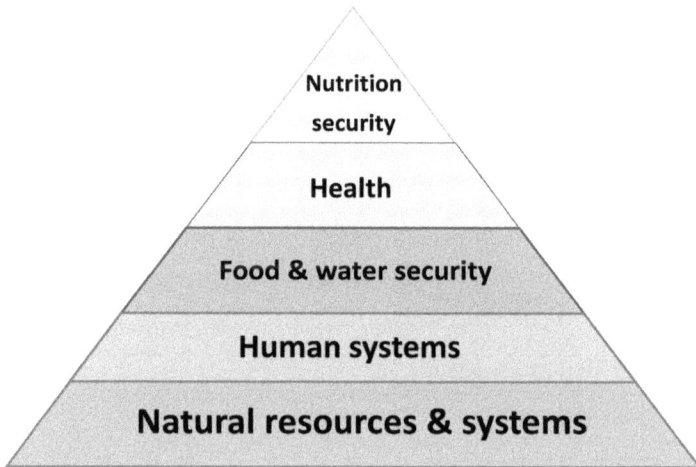

Figure 14.3 Systems in balance

cultivated land in Malawi has ridges on contour), the cutting down and clearing of forests, and the burning of crop residues, are contributing to alarming rates of soil erosion and nutrient loss (World Bank, 2019). According to a soil nutrient loss assessment conducted in Malawi, the nation lost an estimated 26 tonnes of topsoil per hectare in 2010. By 2017, this had risen to 30 tonnes per hectare per year. Soil erosion has been shown to be responsible for annual losses of over 2,000 metric tonnes of fertilizer throughout the country (Omuto and Vargas, 2018). These fertilizers often drain off fields into wetlands, rivers, and lakes, causing bodies of water to become overly enriched with nutrients such as nitrogen and phosphorus. This effect is known *eutrophication*, and can lead to a depletion in the water's oxygen levels, toxic algal blooms, a loss of biodiversity, and even health problems in humans and animals (Bassem, 2020). Human exposure to cyanobacteria in eutrophic water, through swimming, bathing, or drinking, can cause nausea and vomiting, skin irritation, diarrhoea, fever, throat irritation, headache, mouth blisters, muscle and joint aches, eye irritation, and allergic reactions, as well as more serious and chronic effects such as cancer of the liver and colon (Mchau et al., 2019).

Eutrophication caused by phosphorus run-off has become a particularly serious problem in many countries, with negative impacts extending to health, food security, tourism, ecosystems, and economies (Ngatia and Taylor, 2018). In 1997–98, sedimentary core samples were taken from the bottom of Lake Malawi, the world's ninth-largest freshwater lake, to study the historical impacts of nutrient levels in the lake. These samples demonstrated that in 1940, about the time when fertilizer-dependent agriculture was introduced, there was a quantifiable change in the core samples (Otu et al., 2011).

Population growth, deforestation, and intensive agriculture were noted as accelerating soil erosion, causing rivers to transport greater sediment and nutrient loads into Lake Malawi. The researchers looked specifically at diatoms, algae of a type that have silica shells and respond actively to phosphorus. They found that Lake Malawi had experienced increased nutrients at its south end since 1940. By 1980, water quality had changed such that the diatom types dominant in the 1940s were replaced by different diatoms that prefer higher phosphorus relative to silica. These changes in the diatom taxa are similar to those observed in Lake Victoria in East Africa, which has suffered a dramatic loss of fish diversity due to eutrophication (ibid.). Nutrient inputs from soil erosion have already affected the base of the food web in southern Lake Malawi, and this study highlights the need to mitigate the effects of land use in order to protect water quality and biodiversity.

An ethical approach

Permaculture is a holistic agroecological design philosophy first established in Australia in the 1970s. It focuses on the maximization of resources to create ecologically resilient systems. Although it emphasizes sustainable and nutritious food production, permaculture is more aptly described as a 'cornucopia' of best practices encompassing agroecology, organics, ecology, ethical economics, community organizing, conservation of energy, green architecture, zero-waste, regenerative land-stewardship, indigenous knowledge, and much more (Mollison, 1988). It is for this reason that the term 'permaculture' is generally described as a fusion between the two words 'permanent' and 'culture' (Conrad, 2014a).

One of the things that make permaculture unique when compared to many other agricultural methodologies is that it is based upon a set of ethics (Figure 14.4). Ethics help us make decisions about our behaviour and activities based on what is 'right' and 'wrong'. In permaculture, the ethics are simple: *Earth Care, People Care,* and *Fair Share* (or a return of surplus) (Aiken, 2017). First and foremost, we need to ask ourselves whether our actions are caring for the Earth and all of its living and non-living systems; second, we need to assess whether or not our actions serve to care for ourselves, our families, our communities, and the greater needs of humanity; and, third, we need to ensure that our actions are promoting an equitable sharing of resources so that we don't end up with waste or social and economic inequalities. This third ethic was exemplified by Mahatma Gandhi when he wrote: 'The world has enough for everybody's need, but not enough for everybody's greed.'

Issues related to nutrition and health problems are often grouped into three main categories: environmental (earth care), socio-cultural (people care), and political-economic (fair share). Given that nutritional strategies are determined by government policies, the 'politics' of food and nutrition security cannot be overstated. Well-strategized, long-term, and cross-sectoral political strategies can have a positive effect on global and domestic

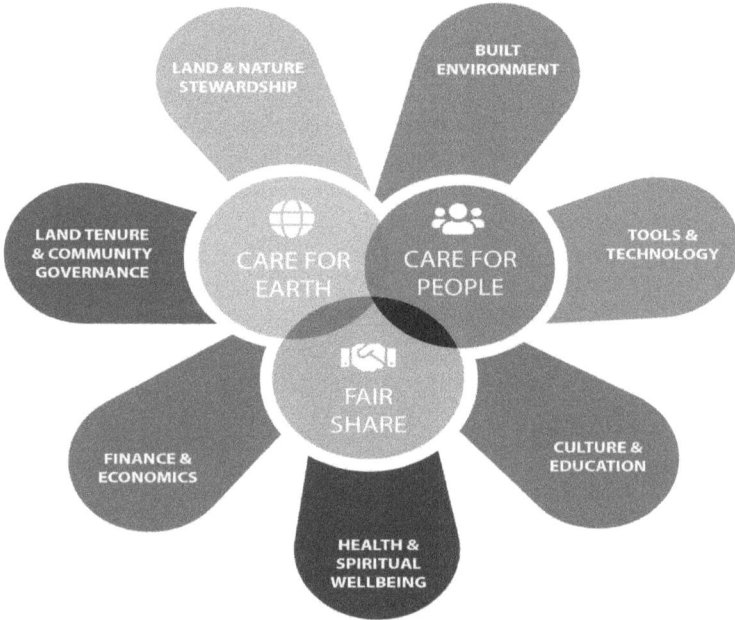

Figure 14.4 Permaculture ethics
Source: Holmgren, 2002; graphic by Darren Roberts (Creative Commons/ShareAlike)

nutritional problems. In its position paper on nutrition security in developing nations, the Academy of Dietetics and Nutrition calls for nutrition interventions and practices to be 'governed by national policy and be integrated into sector policies and programs such as health, agriculture, food security, education, gender equality, environment, habitat, water, sanitation, and energy' (Nordin et al., 2013).

Sustainable approaches to food security are often ignored, overshadowed, and criticized due to the political nature of many agricultural and nutritional initiatives. For example, research on permaculture conducted in Malawi demonstrates that practitioners grow on average three times more crops and more crop varieties per food group than conventional farmers, spend less on inputs, eat a wider diversity of food groups, have increased food security, and benefit from permaculture because they use practices that address household constraints and expand their adaptive capacity (Conrad, 2014b). This study also addressed political obstacles presenting a barrier to wide-scale implementation of such strategies. With regard to nutritional fortification and medicinal approaches, Conrad notes that a sick body can be treated with medicine in isolation from other factors and 'implicates no one', whereas a 'hungry body exists as a potent critique of the society in which it exists' and that 'scientists, governments, and agribusinesses have often devalued and eroded indigenous farming knowledge, like that used in permaculture with the imposition of monocropping and Green Revolution technologies' (Conrad, 2014a).

Due to the fact that agriculture is an essential human activity, agricultural practitioners and ethicists need to work together to nourish and strengthen the aspects of agriculture that are beneficial and change those that are not (Zimdahl, 2018). In Malawi a wide range of initiatives are using the ethical principles of permaculture to help address issues of earth care, people care, and fair share. For example:

- Malawi's capital city of Lilongwe is home to the country's largest permaculture training centre, the Kusamala Institute of Agriculture and Ecology. This centre has become one of the nation's leaders in training and consulting on how to use permaculture design to maximize land productivity, reduce the need for expensive inputs, and increase agricultural diversity to improve nutrition, food security, and livelihoods (Kusamala, n.d.).
- The Schools and Colleges Permaculture (SCOPE, n.d.) programme uses the Integrated Land-Use Design process as a tool to assist schools to redesign their grounds in an ecologically sound manner. SCOPE works with schools in the areas of natural resources management, environmental education, sustainable agriculture, climate change adaptation and mitigation, school health and nutrition, and the functional landscaping of school grounds, and on issues of social injustice, stigma, and discrimination.
- Never Ending Food is a community-based initiative in Malawi which teaches, demonstrates, and advocates for the use of permaculture principles, primarily through the utilization of locally available resources, indigenous foods, and low-input technologies (NEF, n.d.). At the national level, Never Ending Food has been influential in introducing permaculture into government-level programmes through various development partners. These programmes include the Ministry of Education's School Health and Nutrition Programme, which has piloted permaculture implementation in 8 districts in 40 primary schools, 10 teacher development centres, and 1 teacher training college (AFSA, 2015).

In 2018, the government of Malawi approved the use of the *Sustainable Nutrition Manual: Food, Water, Agriculture and Environment* (Nordin, 2016; see Figure 14.5). This is an extension tool which was first developed for the World Food Programme in 2005 and was piloted throughout the country. The response was impressive and the manual was used extensively by extension workers in health, agriculture, and education in sites such as clinics, nutritional rehabilitation units, early childhood centres, schools, and colleges, and within various food security, nutrition, and HIV-support programmes. Since that time, the manual has been revised to include three sections. The first section, *Healthy Humans*, takes an in-depth look at the importance of diversified nutrition and the fundamental need to link good nutrition directly to Malawi's six-food-group model through the creation of more nutritious systems of agricultural production. Approximately

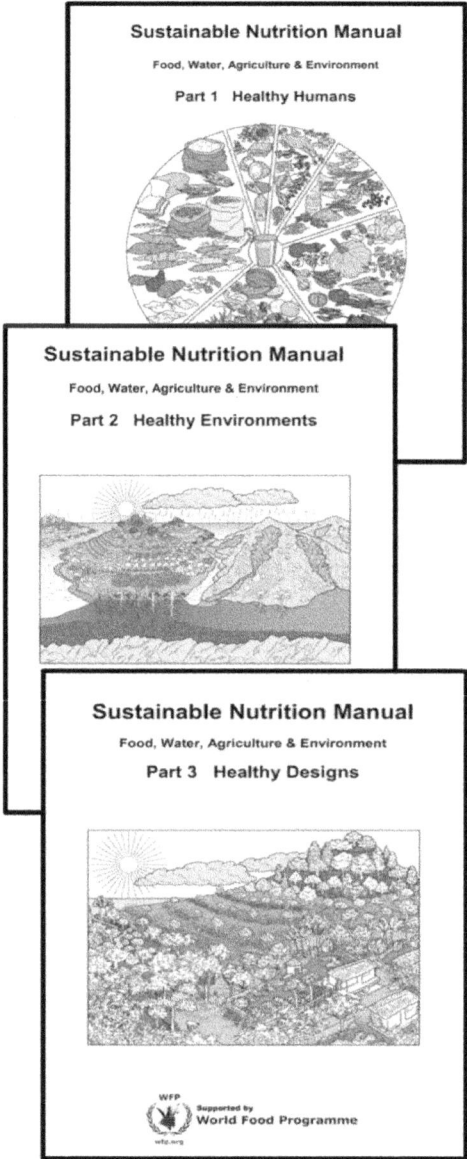

Figure 14.5 *Sustainable Nutrition Manual*, as used throughout Malawi
Source: Never Ending Food, n.d.

600 indigenous or naturalized foods in Malawi are highlighted to be used to meet the nutritional requirements from all of Malawi's six food groups through all the months of the year. The second section, *Healthy Environments*, explores the agroecological components of healthy ecosystems, including soil, water, plants, animals, and human activities. The third section,

on *Healthy Designs*, shows people how to use principles of agroecology and permaculture in their own lives, work, and community outreach to improve food security and create more resilient and sustainable living systems.

Conclusion

The global implementation of sustainable and agroecological solutions is sometimes referred to as a 'silent revolution', due to the fact that it is quietly but steadily growing in both size and popularity. In the media, these praiseworthy undertakings often get drowned out by headlines about agribusiness mergers and calls to scale up industrialized agriculture in order to 'feed the world'. In economics, governments continue to invest in large-scale subsidies to agribusiness approaches, while organic farmers are required to pay out of their own pockets to certify the fact that they are farming in an eco-friendly manner. Far too often the high-input approaches to food production are promoted as 'progress and development' while sustainable solutions are ignored or shunned. But we are finding, as this revolution grows, that individuals, families, and entire communities are joining together to reap the benefits of working towards the creation of an ecologically sustainable future.

The introduction to a 1977 handbook entitled *Teaching Conservation in Developing Nations* remarks: 'People will not preserve and protect a natural environment which they do not understand or respect. When people learn about the relationship of all forms of life to each other and to the earth, they begin to have a responsible attitude toward natural resources and their wise use' (Brace et al., 1977). Throughout the world, there are many individuals, projects, and programmes aimed at helping people 'understand and respect' the natural environment. Unfortunately, far too many agricultural programmes remain overly focused on the 'how' rather than the 'why'. If agricultural and nutritional diversification is a goal, then people need to understand *why* diversity is so important – to our health, to the health of our communities, and to the health of the environmental ecosystems which sustain life. When the 'why' is grasped, the 'how' becomes relatively easy, especially in a country like Malawi, where almost everybody is a farmer at some level.

Malawi is not a poor country: it is rich in locally available resources essential for solving the vast majority of the nation's problems, yet many of these resources are being ignored. People often point to population pressure being one of the world's more critical problems. More people means more pressure on natural resources, more mouths to feed, and more limitations on the development of equitable societies. But, as Margaret Mead famously stated, 'Never doubt that a small group of thoughtful, committed citizens can change the world; indeed, it's the only thing that ever has.' If this is true, then imagine the power of a large group united behind the common goal of creating a sustainable and resilient future. Malawi currently has a population nearing 19 million people. If leaders from every level (household, community, and political) can get this group of 'thoughtful, committed citizens' acting sustainably, we would have nearly 38 million hands working to implement solutions: planting trees, harvesting water, mulching the soil, diversifying crops, protecting wildlife, saving seeds, designing green buildings, harnessing renewable energy, sharing indigenous knowledge, and marketing sustainable products. If we can get people committed to restoring ecosystems, we could easily increase access to perennial, seasonal, diversified, and highly nutritious sources of food in public parks, at churches, health centres, businesses, along roadsides, at homes, and on farms. Consider the fact that if everybody in Malawi agreed to plant even one tree, we'd have nearly 19 million more trees *today* – along with the immeasurable potential for achieving healthy, nutritious, and sustainable solutions long into the future.

References

Aiken, G.T. (2017) 'Permaculture and the social design of nature', *Geografiska Annaler: Series B, Human Geography* 99: 172–191 <https://doi.org/10.1080/0 4353684.2017.1315906>.

Alliance for Food Sovereignty in Africa (AFSA) (2015) *Never Ending Food in Malawi*, AFSA, Kampala <https://afsafrica.org/wp-content/uploads/ 2015/11/Never-Ending-Food-in-Malawi.pdf>.

Anderson, C.L., Reynolds, T., Merfeld, J.D. and Biscaye, P. (2018) 'Relating seasonal hunger and prevention and coping strategies: a panel analysis of Malawian farm households', *Journal of Development Studies* 54(10): 1737–1755 <https://doi.org/10.1080/00220388.2017.1371296>.

Bassem, S.M. (2020) 'Water pollution and aquatic biodiversity', *Biodiversity International Journal* 4(1): 10–16 <https://medcraveonline.com/BIJ/BIJ-04-00159.pdf>.

Berry, V. and Petty, C. (eds) (1992) *The Nyasaland Survey Papers 1938–1943: Agriculture, Food and Health*, Academy Books, London.

Bioversity International (2017) *Mainstreaming Agrobiodiversity in Sustainable Food Systems: Scientific Foundations for an Agrobiodiversity Index*, Bioversity International, Rome <https://hdl.handle.net/10568/89049>.

Brace, J., White, R.R. and Bass, S.C. (1977) *Teaching Conservation in Developing Nations*, Peace Corps Information Collection & Exchange, Washington, DC <https://books.google.co.za/books?id=mgrIYQDEzcMC&printsec=fro ntcover>.

Collins, K. (2015) 'Difference between antioxidants and phytochemicals?', American Institute for Cancer Research, 16 November <https://www.aicr. org/resources/blog/healthtalk-whats-the-difference-between-an-antiox-idant-and-a-phytochemical/>.

Committee on World Food Security (CFS) (2012) *Coming to Terms with Terminology: Food Security, Nutrition Security, Food Security and Nutrition, Food and Nutrition Security*, CFS <http://www.fao.org/3/MD776E/MD776E.pdf>.

Conrad, A. (2014a) *We are Farmers: Agriculture, Food Security, and Adaptive Capacity among Permaculture and Conventional Farmers in Central Malawi*, PhD thesis, American University, Washington, DC <https://doi.org/10.17606/kk5w-5v87>.

Conrad, A. (2014b) *Conventional vs Permaculture Farming in Malawi: What is the Difference?* [website] <http://www.abigailconrad.com/research-findings-infographic>.

Development Initiatives (2018) 'The fight against malnutrition: commitments and financing', in *2018 Global Nutrition Report*, Development Initiatives, Bristol <https://globalnutritionreport.org/reports/global-nutrition-report-2018/the-fight-against-malnutrition-commitments-and-financing/>.

D'Odorico, P., Carr, J.A., Davis, K.F., Dell'Angelo, J. and Seekell, D.A. (2019) 'Food inequality, injustice, and rights', *BioScience* 69(3): 180–190 <https://doi.org/10.1093/biosci/biz002>.

Dorward, A. and Chirwa, E. (2010) 'The Malawi agricultural input subsidy programme: 2005/06 to 2008/09', *International Journal of Agricultural Sustainability* 9(1): 232–247 <http://dx.doi.org/10.3763/ijas.2010.0567>.

Emerson, R.W. ([1841] 1979) 'Essay VIII: Heroism', in J. Slater (ed.), *The Collected Works of Ralph Waldo Emerson, Volume II*, pp. 145–156, Harvard University Press, Cambridge, MA.

Food and Agriculture Organization of the United Nations (FAO) (2014) 'Nutrition-sensitive agriculture', fact sheet, *Second International Conference on Nutrition, 19–21 November*, Rome <https://www.fao.org/3/as601e/as601e.pdf>.

FAO (2016) *Influencing Food Environments for Healthy Diets*, FAO, Rome <http://www.fao.org/3/a-i6484e.pdf>.

Francis (2015) *Encyclical letter Laudato si' of the Holy Father Francis on Care for Our Common Home*, The Holy See, Vatican City <https://www.vatican.va/content/francesco/en/encyclicals/documents/papa-francesco_20150524_enciclica-laudato-si.html>.

Government of Malawi (2001) 'Eat a variety of foods daily from the six food groups for good nutrition', poster, Ministry of Agriculture, Irrigation and Water Development, Lilongwe, Malawi.

Government of Malawi (2017) *Malawi Micronutrient Survey: Key Indicators Report 2015–16*, National Statistical Office, Zomba; Community Health Sciences Unit, Lilongwe; Centers for Disease Control and Prevention, Emory University, Atlanta, GA <https://dhsprogram.com/pubs/pdf/FR319/FR319m.pdf>.

GTZ (1991) *The Context of Small-Scale Integrated Agriculture-Aquaculture Systems in Africa: A Case Study of Malaŵi*, International Center for Living Aquatic Resources Management, Manila; Deutsche Gesellschaft für Technische Zusammenarbeit, Frankfurt am Main <https://digitalarchive.worldfish-center.org/bitstream/handle/20.500.12348/3113/Pub%20SR76%2018.pdf?sequence=1&isAllowed=y>.

High Level Panel of Experts on Food Security and Nutrition (HLPE) (2020) *Food Security and Nutrition: Building a Global Narrative towards 2030*, HLPE, Committee on World Food Security, Rome <https://www.fao.org/3/ca9731en/ca9731en.pdf>.

Holmgren, D. (2002) *Permaculture: Principles and Pathways Beyond Sustainability*, Permanent Publications, East Meon, UK.

Joy, E.J.M., Stein, A.J., Young, S.D., Ander, E.L., Watts, M.J. and Broadley, M.R. (2015) 'Zinc-enriched fertilisers as a potential public health intervention in Africa', *Plant Soil* 389(1): 1–24 <https://doi.org/10.1007/s11104-015-2430-8>.

Kusamala (no date) 'Working with nature for a better future' [website], Kusamala Institute of Agriculture and Ecology <https://kusamala.org/>.

LibreTexts (no date) '31.1C: essential nutrients for plants', LibreTexts, Biology, University of California, Davis, CA <https://bio.libretexts.org/@go/page/13781?pdf>.

Mchau, G.J., Makule, E., Machunda, R., Gong, Y.Y. and Kimanya, M. (2019) 'Harmful algal bloom and associated health risks among users of Lake Victoria freshwater: Ukerewe Island, Tanzania', *Journal of Water and Health* 17(5): 826–836 <https://doi.org/10.2166/wh.2019.083>.

Mollison, B. (1988) *Permaculture: A Designer's Manual*, Tagari Publications, Tyalgum, NSW.

Mutegi, J., Kabambe, V., Zingore, S., Harawa, R. and Wairegi, L. (2015) *The Status of Fertilizer Recommendation in Malawi: Gaps, Challenges and Opportunities*, Soil Health Consortium of Malawi <http://ssa.ipni.net/ipniweb/region/africa.nsf/0/F7501955BAE1F4F085257F080026F963/FILE/Malawi%20soil%20fertility%20recommendations.pdf>.

Never Ending Food (NEF) (no date) 'Permaculture: designing a sustainable culture', Never Ending Food <https://www.neverendingfood.org/b-what-is-permaculture/>.

Ngatia, L. and Taylor, R. (2018) 'Phosphorus eutrophication and mitigation strategies', in T. Zhang (ed.), *Phosphorus: Recovery and Recycling*, pp. 45–61, IntechOpen, London <https://doi.org/10.5772/intechopen.79173>.

Nordin, K. and Nordin, S. (2017) 'Food, the source of nutrition', *World Nutrition Journal* 8(1): 87–94 <https://doi.org/10.26596/wn.20178187-94>.

Nordin, S. (2016) *Sustainable Nutrition Manual: Food, Water, Agriculture and Environment*, World Food Programme, Malawi <https://www.neverendingfood.org/sustainable-nutrition-manual/>.

Nordin, S.M., Boyle, M. and Kemmer, T.M. (2013) 'Position of the Academy of Nutrition and Dietetics: nutrition security in developing nations: sustainable food, water, and health', *Journal of the Academy of Nutrition and Dietetics* 113(4): 581–595 <https://doi.org/10.1016/j.jand.2013.01.025>.

Ntenda, P.A.M. and Kazembwe, J.F. (2019) 'A multilevel analysis of overweight and obesity among non-pregnant women of reproductive age in Malawi: evidence from the 2015–16 Malawi Demographic and Health Survey', *International Health* 11(6): 496–506 <https://doi.org/10.1093/inthealth/ihy093>.

Nyasa Times (2012) 'Malawi spends $5m for vitamin A sugar fortification programme', *Nyasa Times*, 9 October <https://www.nyasatimes.com/malawi-spends-5m-for-vitamin-a-sugar-fortification-programme/>.

Nyberg, S.T., Batty, G.D., Pentti, J., Virtanen, M., Alfredsson, L., Fransson, E.I., Goldberg, M., Heikkilä, K., Jokela, M., Knutsson, A., Koskenvuo, M., Lallukka, T., Leineweber, C., Lindbohm, J.V., Madsen, I.E.H., Magnusson Hanson, L.L., Nordin, M., Oksanen, T., Pietiläinen, O., Rahkonen, O., Rugulies, R., Shipley, M.J., Stenholm, S., Suominen, S., Theorell, T., Vahtera, J., Westerholm, P.J.M., Westerlund, H., Zins, M., Hamer, M., Singh-Manoux, A., Bell, J.A., Ferrie, J.E., and Kivimäki, M. (2018) 'Obesity and loss of disease-free years owing to major non-communicable diseases: a multicohort study', *Lancet Public Health* 3(10): e490–e497 <https://doi.org/10.1016/s2468-2667(18)30139-7>.

Omuto, C.T. and Vargas, R. (2018) *Soil Nutrient Loss Assessment in Malawi*, technical report, Food and Agriculture Organization of the United Nations,

UNDP-UNEP Poverty-Environment Initiative, Ministry of Agriculture, Irrigation and Water Development, Malawi <http://www.fao.org/3/CA2666EN/ca2666en.pdf>.

Otten, J.J., Hellwig, J.P. and Meyers, L.D. (2006) *Dietary Reference Intakes: The Essential Guide to Nutrient Requirements*, The National Academies Press, Washington, DC <https://doi.org/10.17226/11537>.

Otu, M.K., Ramlal, P., Wilkinson, P., Hall, R.I. and Hecky, R.E. (2011) 'Paleolimnological evidence of the effects of recent cultural eutrophication during the last 200 years in Lake Malawi, East Africa', *Journal of Great Lakes Research* 37(Supplement 1): 61–74 <https://doi.org/10.1016/j.jglr.2010.09.009>.

Prasanna, B.M., Huesing, J.E., Eddy, R. and Peschke, V.M. (eds) (2018) *Fall Armyworm in Africa: A Guide for Integrated Pest Management*, Feed the Future, Washington, DC <https://reliefweb.int/sites/reliefweb.int/files/resources/FallArmyworm_IPM_Guide_forAfrica.pdf>.

Rocha, C. (2017) *Unravelling the Food-Health Nexus: Addressing Practices, Political Economy, and Power Relations to Build Healthier Food Systems*, Global Alliance for the Future of Food and IPES-Food <http://www.ipes-food.org/_img/upload/files/Health_FullReport(1).pdf>.

Sangala, T. (2018) 'Government phasing out 23:21:+4s fertilizer', *The Times* (Malawi), 7 May <https://times.mw/government-phasing-out-23214s-fertilizer/>.

Saunders, A.V., Craig, W.J. and Baines, S.K. (2013) 'Zinc and vegetarian diets', *Medical Journal of Australia* 199(4): 17–21 <https://doi.org/10.5694/mja11.11493>.

SCOPE (no date) 'Welcome to SCOPE Malawi' [website], SCOPE Malawi, Lilongwe <https://scopemalawi.com/>.

Thoreau, H.D. ([1854] 1995) *Walden, or Life in the Woods*, Dover Publications, Mineola, NY.

United Nations (2019) 'UN Biodiversity Convention partners with Slow Food International in celebrating the International Day for Biological Diversity', United Nations Decade on Biodiversity <https://www.cbd.int/doc/press/2019/pr-2019-05-22-idb-en.pdf>.

World Bank (2019) *Malawi Country Environmental Analysis*, World Bank, Washington, DC <http://documents1.worldbank.org/curated/en/508561550587004266/pdf/AUS0000489-WP-P162772-PUBLIC-18-2-2019-13-4-24-MalawiCEAReportWeb.pdf>.

World Food Programme (WFP) and African Union (AU) (2015) *The Cost of Hunger in Malawi: Social and Economic Impacts of Child Undernutrition in Malawi, Implications on National Development and Vision 2020*, World Food Programme, African Union <https://documents.wfp.org/stellent/groups/public/documents/newsroom/wfp274603.pdf>.

World Health Organization (WHO) (2021) 'Malnutrition' [fact sheet], 9 June <https://www.who.int/news-room/fact-sheets/detail/malnutrition>.

Zimdahl, R. (2018) 'Agriculture's moral dilemmas and the need for agroecology', *Agronomy* 8(7): 116 <http://dx.doi.org/10.3390/agronomy8070116>.

CHAPTER 15
A movement for life: African food sovereignty

Haidee Swanby

Introduction

This chapter takes its cue from African farmer leaders and civil society, who insist that food producers and their acts of resistance, constituted through their daily practice, should be acknowledged as the primary bastion of food sovereignty in Africa. It distinguishes between 'food sovereignty' and 'Food Sovereignty' in Africa: 'food sovereignty' as an unbroken thread of local knowledge and practices that have nourished the continent and co-created an astounding diversity of agricultural resources; and 'Food Sovereignty' as a political movement struggling for the rights of producers to shape and control their food systems, as well as for 'cognitive justice' (Visvanathan, 2006; Belay, 2012) to end the racist bias against Indigenous ontologies. The chapter begins by unravelling the ontologies[1] that inform and underpin industrial food systems and African food systems. The chapter then describes the emergence of the political Food Sovereignty movement, globally and in Africa, looking at key events and players. It goes on to consider the evolution of global development practices and the ways in which this shapes the Food Sovereignty movement, and then hints at an aspect of the movement which is still nascent – a feminist take.

Decolonial and critical feminist studies have led us to deeply interrogate our own positionality and power, moving away from the so-called objectivity and neutrality of reductionist ontologies (Haraway, 1988; Visvanathan, 2006). My personal experience, embedded in the African and global food movement over the past two decades, has been extraordinarily rich. I have been privileged and humbled to learn from men and women who think out of the box, informed by a deep humanity; I have been taught by farmers and scientists about ecology and how humans partner with an astounding variety of human and non-human actors in the production of food; I have had the honour of staying in the homes of farming families in many African countries while tasting the fruits of their labour; I have learnt about the multitude of ways that people operate within the politics of their communities, and about the rituals and taboos that maintain pristine islands of biodiversity in landscapes destroyed by extractivism. My friends have taken me to the source of the Nile, to sacred forests and lakes, to see the little red foxes of the Ethiopian

Bale mountains; they have shared from their home seed banks and treated me to home-cooked cuisine. From my coordinating roles in networks and organizations, I know what people need when they travel from their homes to participate in activist environments and in policy spaces. Food Sovereignty assemblages are expansive and complex. I have been exposed to many new environments, cultures, knowledges, and worldviews that push me to perpetually examine my own history and beliefs, and my social positioning as a white South African woman on a continent where locations based on gender, race, nationality, class, and sexuality are incredibly complex. I am deeply grateful to be a part of this movement and of all the visceral, structural, political, and discursive assemblages that are constantly emerging.

Food production in a living, sacred world

> Walking a little bit in front of me as if to assert his authority, he was saying, 'Do you see this stream? It is sacred. We used to sacrifice a lamb to make sure that it continues to flow.' A few steps later, 'do you see that little lake? It is sacred and we have a ritual every other year to honour it and to thank our ancestors for keeping our land safe.' He glances and points to a tall tree, 'Do you see that tree? It is sacred and we keep it in our prayers when we pray to our God.' When we reach a small hill, he points at a part of a forest which looked thicker. 'Do you see that forest? It is sacred and nobody can even cut a grass from it, except putting their bee hives.' Every stream, every pond and every patch of forest are sacred to him and the community. (Belay, 2020)

The term 'Food Sovereignty' is often associated with the political movement championed by peasant farmers around the globe under the auspices of the vibrant La Via Campesina movement. In Africa, however, any account of food sovereignty must begin with those who are producing food and continuing daily with their own practices, despite multiple pressures to modernize and to conform to Western values (activist personal interview, Harare, 29 May 2020; farmer leader personal interview, 21 July 2020). By some estimates small producers generate up to 80 per cent of Africa's food, relying primarily on their own agricultural resources, such as farmers' seed varieties and traditional livestock (FAO, 2021; FAO et al., 2021). While societies of the Global North have largely ceded their food sovereignty to corporations, the lived and networked experiences of small producers still provide the blueprint for food sovereignty in Africa.

The FAO's (2021) online family farming knowledge portal describes family farmers as:

> smallholder to medium-scale farmers … [including] peasants, indigenous peoples, traditional communities, fisher folks, mountain farmers, pastoralists and many other groups representing every region and biome of the world. They run diversified agricultural systems and preserve traditional

food products, contributing both to a balanced diet and the safeguarding of the world's agro-biodiversity. Family farmers are embedded in territorial networks and local cultures, and spend their incomes mostly within local and regional markets, generating many agricultural and non-agricultural jobs.

It is no accident that 80 per cent of the world's biodiversity is sustained within Indigenous territories (Sobrevila, 2008) by peoples who deeply value collective rights and community. The stewarding of biodiversity and agrobio-diversity is a function of the profound connection that Indigenous peoples have to their territories and the intergenerational knowledge that has been passed down to 'observe, adapt and incorporate traditional knowledge to ever-changing ecosystems, and harmoniously reside within the biological diversity of Mother Earth' (FAO et al., 2021: ix). Local food sovereignty emerges from this interaction and embeddedness in territory, and astute observation of the natural world. Food sovereignty concerns may therefore go beyond the bounds of what might usually be considered within the realm of food systems. Many traditional cultures base their views of nature on spiritual worldviews, whereas industrialized cultures tend to base their beliefs on science and the teachings of formal education (Milton, 1999). For example, sacred natural sites, as they are commonly known, are places of governance as well as spiritual practices. These places are linked with the management and governance of the natural systems around them, including agricultural systems, livelihoods, and community cohesion and governance (Belay, 2012). This experience of being embedded in a living universe, which is conceived of as a moral space where actions have moral consequences (Smith et al., 2018), stands in stark contrast to modern food production paradigms, which are based on efficiency and technology. The former is forward-looking precisely because it considers future generations in its approach to the land and life on the land. It embraces our moral obligation to each other and the living world, coupled with practical knowledge and science on generating healthy soil, food, water, and climate change adaptation and mitigation. The African Food Sovereignty movement embraces Indigenous knowledge and ways of being, while advocating for this right to be autonomous and to farm for the future. While such ideological approaches are often critiqued as backward-looking, romanticizing the past, or anti-science and anti-progress, they are in fact indicative of socio-ecological transformation, of the dismantling of exploitative and destructive hierarchies and a 'reaching towards a socio-ecological self and new society which strives to embody us as deeply interconnected and dependent' (Andrews, 2020: 4).

Food production in a dead world and the prejudices of the Green Revolution

Scrutinizing the logic of the Green Revolution helps us to understand why Indigenous ontologies have been falsely labelled as primitive, resulting in a drive to supplant related agricultural practices with modern technology.

In his 1970 Nobel Peace Prize Lecture, the father of the Green Revolution, Norman Borlaug, spoke with passion and fervour of the power of science and technology to bring peace and an end to hunger, most especially for those in the 'forgotten world … [who] live in poverty with hunger as a constant companion and fear of famine as a continual menace'(Borlaug, 1970) (see Chapters 4 and 8 for further analyses of the Green Revolution).

Borlaug was a product of his time, place, and social position. Decolonial and feminist scholars, as well as Indigenous activists and scholars, have since deepened and developed discourses to interrogate the so-called neutrality of the scientific method and have begun to unpick the damaging social impacts of this domineering approach (Haraway, 1988; Merchant, 2006; Rosenow, 2018). These discourses allow us to expose why social and environmental harm has accompanied the undeniably remarkable scientific achievements of the Green Revolution. Borlaug exposes his inherent prejudice and ontological bias in his Nobel lecture, laying out an evolution of agricultural practice over the ages, in which he identifies the starting point of 'civilization' as emerging from settled agricultural practice. He also characterizes nomadic peoples and hunter-gatherers as relics of the Neolithic Age, 'wandering' people who were probably unable to develop 'village civilizations' due to frequent food shortages. This is a familiar refrain that persists today: a 'linear notion of human development as progressing from savage to civilized' (Durante et al., 2021: 25). The colonial practice of reorganizing decimated cultures and societies in reservations, villages, settlements, or towns was framed as a process of civilization, allowing for '"honest" labor, education, evangelical services supported by agricultural surplus, and the acquisition of proper manners' (Wolford, 2021: 1627). The organizing principle of colonial plantations 'propelled colonial exploration, sustained an elite, perpetuated a core-periphery dualism within and between countries, organized a highly racialized labor force worldwide, and shaped both the cultures we consume and the cultural norms we inhabit and perform' (Wolford, 2021: 1623). This fragmenting and dispossessing process had a profound impact on African agriculture and the social systems in which it is embedded. The dynamics of the plantation still define the social, ecological, and political characteristics of new commodity frontiers across the globe, sustained by 'naked power' and 'government preferences' (Wolford, 2021: 1623).

The impact of centuries of colonization of the Global South, and the extraction of wealth and labour that resulted in much of the luxury and abundance of the Global North, seems to fall outside the frame of analysis of Green Revolution proponents. They advance a narrative that the agricultural practices and knowledges of the Global South are simply inferior, and that modernity and well-being are inherently entwined with technological progress. If technical inputs and credit were the physical and structural package from which the Green Revolution evolved, the notion of the superiority of Western thinking, and the need for Western patronage, was the psychological package. This kind of psychological violence, or psychological oppression, has been a defining and enduring feature of colonization (Fanon, 1962; Biko, 1987) and remains potent even in the postcolonial era. It has accompanied the project of Western culture, conceived in the days of the Enlightenment, to master nature through science, technology, and capitalism, and effectively exploit the Earth and its natural resources at the expense of nature, women, minorities, and Indigenous peoples (Merchant, 2006).

Today it is accepted that industrialized agriculture contributes significantly to greenhouse gas emissions, is the most significant polluter of inland and coastal waters (Mateo-Sagasta et al., 2017), destroys soil, kills wildlife, and costs the environment the equivalent of about US$3 trillion every year (UNEP, 2020). Taxpayers pick up the bill of externalized costs: for example, funds required for water purification and health bills stemming from poor nutrition and diet-related non-communicable diseases (UNEP, 2020).

However, Borlaug's powerful dream has deeply transformed the global approach to agriculture, and this is buttressed by the international discourse on food security and international trade regimes under the World Trade Organization (WTO). While more than 70 per cent of global nutrition is still produced by peasant agriculture (FAO, 2021), industrial agriculture has commandeered policy and institutional space through powerful alliances of financial capital, agribusiness, the state, and mass media, violently displacing family farmers and rural peoples and deepening structural violence against rural women (Shiva, 1991; Andrews et al., 2019).

The global rise of a Food Sovereignty movement

'Food sovereignty' critiques 'food security', the mainstream approach to global hunger. Food security is a developmental neoliberal approach that seeks to address hunger through increased production, thus bringing smallholders into the capitalist circuit and providing welfare measures where food is lacking (Holt Giménez and Shattuck, 2011). Food security emphasizes efficiency and the primary roles of the state and market (Patel, 2009). Measures of success are quantitative, such as increased yield and profits, which are spurred on by technological progress. In contrast, Food Sovereignty is concerned with power and control in the food system, while calling for structural and redistributive reforms around land, water, and agricultural resources such as seed and markets (Holt Giménez and Shattuck, 2011). Global discourse has marginally responded to the framing of the Food Sovereignty movement, broadening the definition of food security to include 'agency' and 'sustainability' (HLPE, 2020).

Although the roots of Food Sovereignty are varied, it was the Latin American peasant movement La Via Campesina that popularized the concept when countries of the Global North placed agriculture on the trade agenda through the WTO (Patel, 2009; Edelman et al., 2014.

This development in the WTO, beginning from the mid-1990s, radically shifted the agricultural landscape, pitting farmers across the globe against one another in a free trade market where only the fittest and biggest would survive (De Schutter, 2015).

'Food Sovereignty' can be defined as the right of people to democratically control or determine the shape of their food system, and to produce sufficient and healthy food in culturally appropriate and ecologically sustainable ways in or near their territory (Edelman et al., 2014). The six pillars that commonly define Food Sovereignty are: food for people (as opposed to commerce); placing producers at the centre; localization of food systems; localization of decision-making; building skill and knowledge (in contrast to industrial agriculture, which tends to de-skill and enclose knowledge commons); and working with nature (agroecology) (Nyéléni, 2007). Since its inception, Food Sovereignty as a concept has evolved in diverse ways and in diverse contexts all over the globe. It is certainly not a strict cohesive discipline, but rather a 'dynamic process' that has served to 'galvanize broad-based and diverse movements around the need for radical changes in agro-food systems' (Edelman et al., 2014: 911).

Food Sovereignty is now part of the international discourse on hunger, nutrition, and agriculture, and has found its way, too, into regional and national policy frameworks. For example, the Economic Community of West African States (ECOWAS) has adopted Food Sovereignty as an important goal for the region and has incorporated measures to achieve it in an array of policy instruments, including the ECOWAS Agricultural Policy (De Loma-Ossorio et al., 2014). In 2013, the FAO formalized a relationship with La Via Campesina (FAO, 2013) under the Civil Society Mechanism of the UN's Committee on World Food Security. The International Year of Family Farming in 2014 created a political opportunity to push for the recognition of agroecology by the FAO, something that had been vehemently blocked by some member states (Canada, Australia, Argentina, and especially the USA) who felt agroecology did not promote their own national interests (Anderson and Maughan, 2021). Another notable triumph for the recognition of Food Sovereignty as a concept is the UN Declaration on the Rights of Peasants and Other People Working in Rural Areas (UNDROP), championed by La Via Campesina and adopted in 2019 (Hubert, 2019).

The rise of Food Sovereignty in Africa

> Hear the footsteps from the receding market squares
> Are you too far gone to hear?
> Hear the rumblings of resistance to naked market forces
> That roasted habitats and habitations
> Lands, seas and skies grabbed yet dreams cannot be corralled 'cause
> Daughters of the soil are ever alert, awake, hoisting the sky
> And its watery dusts

Knowledge demonized by demons of market environmentalism and
brazen extractivism
As the hunter's bag becomes a weapon of mass destruction
Bulging pockets hack horns and tusks and an array of idiotic aphrodi-
siacs for limp brains
Slithering across the Savannah, stomping on our ancestral hearths
Shall we look, exiled, silent, sullen, sunk and annihilated as our trees
metamorphose into carbon sinks?

From 'Return to Being' by Nnimmo Bassey (2021)

History of the movement and key actors

The African Food Sovereignty movement emerges from multiple actors and
networks of actors engaging in daily production, food provision, capacity
building, grant making, and movement building. In describing the growth of
social movements, Dianne Rocheleau (2015) invokes metaphors of rhizomes
or fungal networks periodically producing mushrooms above the ground that
are evidence of the active and connected life below the soil. She helps us
see the 'networked phenomenon of grassroots groups simultaneously rooting,
localizing, linking and globalizing' (Rocheleau, 2015: 73). This dynamic and
nested network of relationships which converge, fruit, and subside at different
levels and at different times beautifully captures African food movements,
evoking the 'relational logic and complexities of the living worlds we inhabit'
(Rocheleau, 2015: 71). These 'rooted networks' (Rocheleau, 2015: 73) are
based in African fields and informal settlements, in kitchens and local seed
fairs, extending their reach to affiliate with ever-larger networks in solidarity
across issues but with shared perspectives and values. An example of major
convergences of rooted networks is the vibrant social network platforms
that grew from NGO parallel meetings at the 1992 Rio Summit and later
World Social Forum actors, linking Indigenous people's networks with peace,
human rights, alternative economics, and sustainable agriculture networks
(Rocheleau, 2015).

The fertile ground for the African Food Sovereignty movement was built
during the negotiations of the Convention on Biological Diversity, stemming
from the 1992 Rio Earth Summit. At that time, opposing forces were
emerging globally. The profound problem of economic growth in a world of
finite resources was acknowledged, as was the role of Indigenous lifestyles in
addressing environmental collapse, including the loss of agricultural genetic
resources (AFSA, 2017). At a similar time, the WTO was established, bringing
agriculture onto the international trade agenda as well as obligating members
to domesticate intellectual property rights regimes for agricultural genetic
resources. The ability to patent life became the engine for novel technologies
such as genetically engineered crops. This was a time when African leaders
banded together against powerful forces to fight for the protection of African
agricultural systems, related knowledge and resources, and the lifestyles in

which these are encoded. In particular, the negotiations on the International Treaty on Plant Genetic Resources for Food and Agriculture (the 'Seed Treaty') and the Cartagena Protocol on Biosafety served to politicize biodiversity and agriculture, building capacity and planting the seeds of social movements resisting the push for genetically modified organisms, trade agreements, and land grabbing in Africa. African civil society solidarity and support were essential in these 'David and Goliath' battles, where elected leaders negotiated in hostile and inequitable international policy spaces (Tewolde, 2007). Two model laws were adopted by the then Organization of African Unity: the African Model Law for the Protection of the Rights of the Local Communities, Farmers and Breeders and for the Regulation of Access to Biological Resources (2001) and the African Model Law on Safety in Biotechnology (2003). Figure 15.1 gives an overview of key events that galvanized the African Food Sovereignty Movement, followed by a brief description of some of these events.

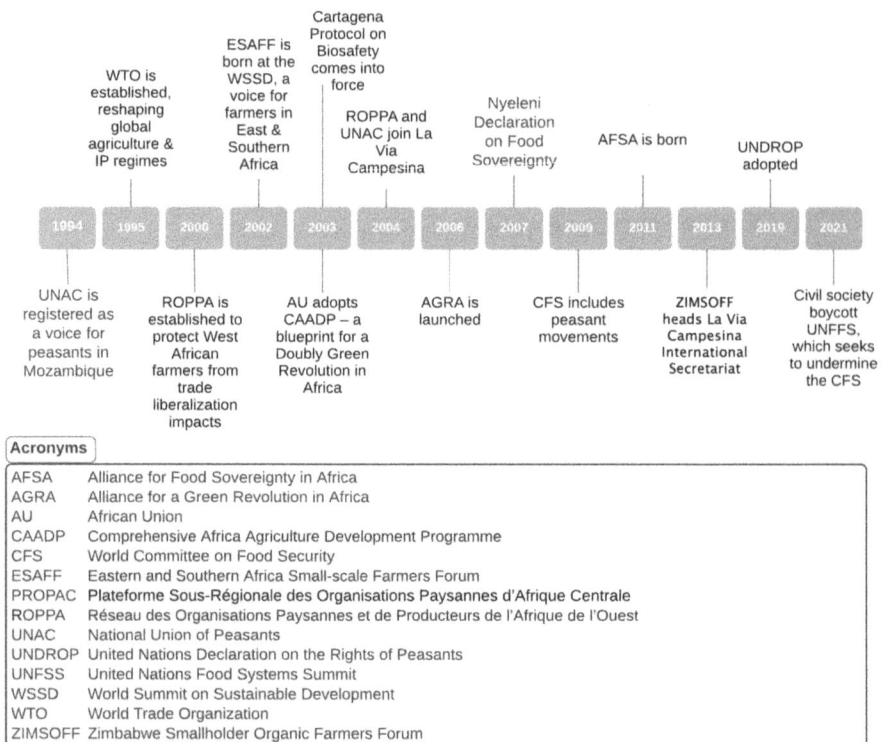

Acronyms

AFSA	Alliance for Food Sovereignty in Africa
AGRA	Alliance for a Green Revolution in Africa
AU	African Union
CAADP	Comprehensive Africa Agriculture Development Programme
CFS	World Committee on Food Security
ESAFF	Eastern and Southern Africa Small-scale Farmers Forum
PROPAC	Plateforme Sous-Régionale des Organisations Paysannes d'Afrique Centrale
ROPPA	Réseau des Organisations Paysannes et de Producteurs de l'Afrique de l'Ouest
UNAC	National Union of Peasants
UNDROP	United Nations Declaration on the Rights of Peasants
UNFSS	United Nations Food Systems Summit
WSSD	World Summit on Sustainable Development
WTO	World Trade Organization
ZIMSOFF	Zimbabwe Smallholder Organic Farmers Forum

Figure 15.1 Food sovereignty: timeline of key events

Some scholars put the emergence of the Food Sovereignty movement on the African continent at 2004, when the West African network of producers (Réseau des Organisations Paysannes et de Producteurs de l'Afrique de l'Ouest, ROPPA) and Mozambique's peasant union (União Nacional de

Camponeses, UNAC) joined the Latin America-based La Via Campesina Movement (Shilomboleni, 2017). UNAC grew out of Mozambique's civil war, in which millions were killed or displaced. The peasant union was registered in 1997 to ensure that the country's food producers had a voice in Mozambique's development (UNAC, 2021). ROPPA was registered in 2000, having emerged as a response to trade liberalization and the destruction of local markets in the face of cheap, subsidized imports (Shilomboleni, 2017). At this time, there was renewed interest from the Global North in agricultural investment in Africa. In 2003 the African Union adopted the Comprehensive Africa Agriculture Development Programme (CAADP), an explicit strategy to attract foreign direct investment by creating investor-friendly policies and partnering with the private sector to bring in its expertise through development programmes. The role of African states was conceptualized as taking on the responsibility to create appropriate infrastructure and institutional frameworks so that private companies could easily develop and deploy their products (Greenberg, 2013). The Alliance for a Green Revolution in Africa, a project spearheaded by the Rockefeller and Gates foundations, was launched in 2006, modelled on the Asian Green Revolution of the 1970s (Daño, 2007).

In 2007, a group including Friends of the Earth International, La Via Campesina, the World March of Women, and ROPPA brought 500 delegates from five continents with an interest in agricultural and food issues together in Sélingué, Mali. Participants at the gathering noted the central role of family producers and women in feeding the world and the value of their knowledge and practices. However, it was clear that 'neo-liberalism and global capitalism' threatened their 'capacity to produce healthy, good and abundant food' (Nyéléni, 2007). This gathering clarified the economic, social, ecological, and political aims of Food Sovereignty with the resultant Nyéléni Declaration. At a follow-up meeting held in Sélingué in February, 2015, agroecology was validated as a 'key form of resistance to an economic system that puts profit before life' (International Forum for Agroecology, 2015:4). The forum also raised alarm over the co-option of agroecology by the proponents of Green Revolution approaches, through the promotion of false solutions such as 'climate smart agriculture' and 'sustainable intensification' (International Forum for Agroecology, 2015).

Numerous organizations, networks, and platforms at the national, regional, and continental levels are now actively engaged in Food Sovereignty advocacy in Africa. Their campaigns are shaped by their local or regional contexts and histories (Gyapong, 2017). Regional producer networks such as ROPPA are powerful players in the movement. ROPPA is made up of the 13 national producer organizations in West Africa, from Benin, Burkina Faso, Côte d'Ivoire, Gambia, Guinea-Bissau, Guinea-Conakry, Mali, Niger, Senegal, Togo, Ghana, Sierra Leone, and Liberia. In Central Africa, the producer umbrella organization PROPAC (Plateforme Régionale des Organisations Paysannes d'Afrique Centrale) is the main network for Food Sovereignty in that region. It is made up of farmers' organizations from 10 members of the

Economic Community of Central African States: Angola, Burundi, Cameroon, the Central African Republic, Chad, the Congo, the Democratic Republic of the Congo, Equatorial Guinea, Gabon, and Sao Tome and Principe. The third producer powerhouse for Food Sovereignty on the continent is the Eastern and Southern Africa Small-scale Farmers Forum (ESAFF). ESAFF brings together small-scale crop growers, livestock keepers, and fisherfolk from Tanzania, Kenya, Uganda, Zambia, South Africa, Malawi, Rwanda, Zimbabwe, Lesotho, Burundi, Madagascar, Seychelles, and Mozambique. One of ESAFF's members, the Zimbabwe Smallholder Organic Farmers Forum (ZIMSOFF), took on the hosting of La Via Campesina's International Operational Secretariat in 2013 (ESAFF, 2013). According to the La Via Campesina website, there are a total of 15 African producer organization members (mostly at national level) who represent farmers, landless people, and peasants and support sustainable agriculture and rural development. They represent all the African sub-regions with the exception of North Africa (Gyapong, 2017).

NGO networks and local and international NGOs have also played a vital supporting role in the emergence of a Food Sovereignty movement in Africa. The partnership between the African Biodiversity Network, the Coalition for the Protection of African Genetic Heritage, and Friends of the Earth led to the idea to establish a continental network of networks towards a unified African voice on Food Sovereignty. The consolidation of 21 African networks into the Alliance for Food Sovereignty in Africa (AFSA) at the 2011 Conference of the Parties to the UN Framework Convention on Climate Change in Durban, South Africa, marked the birth of a pan-African Food Sovereignty movement (Shilomboleni, 2017). The main purpose of AFSA is to influence policies and to promote African solutions for food sovereignty and agroecology. Core members are regional food producer organizations (e.g. farmers, fishers, pastoralists), regional Indigenous peoples' organizations, regional consumer movements, and regional NGO networks. Associate members include specialist NGOs and national networks, among them organizations based outside Africa that are supportive of AFSA's vision. Key international partners have helped African players keep a finger on the pulse of international policy and stakeholder agendas that have a direct impact on Africa's Food Sovereignty agenda. These important allies have watched and analysed technological trends and related international laws, followed the money and hidden agendas, opened spaces in international negotiations, given technical support to African leaders in negotiation spaces, and stand in solidarity with African campaigns. These key players have included the ETC Group (the Action Group on Erosion, Technology and Concentration, formerly the Rural Advancement Foundation International, RAFI), the Community Alliance for Global Justice, Global Justice Now, GRAIN International, the Association for Plant Breeding for the Benefit of Society, the Third World Network and many others.

Space for the movement to grow: feminism in Food Sovereignty

The radical inclusivity that is engendered by critical feminist approaches to Food Sovereignty remains nascent within Africa's movement, which, as we have seen, is anchored in diverse African cultures across the continent. The Rural Women's Assembly (RWA), a coalition of rural women from 12 Southern African countries, takes Food Sovereignty into the very heart of rural women's private, intimate, and community lives, working on the basis that the 'personal is political' (Hanisch, 2006; Andrews, 2019). They call attention to the many layers of oppression that rural women endure, including inhumane labour demands in fields, households, and waged work; lack of opportunity and access to resources stemming from patriarchy, religion, and culturally imposed norms; the plight of widows; gender-based violence; child marriages; the loss of the natural resources on which women depend for survival and nurturing the family; issues of waste, pollution, and energy; and much more (RWA, 2020). The RWA, in line with ecofeminist discourse, equates the violence inflicted by the neoliberal order on nature with violence against their own bodies (Andrews et al., 2019; Andrews, 2019) and asserts that without sovereign women, there can be no food sovereignty.

A feminist approach critically engages with the continent's food systems and works to promote 'African feminist traditions of critiquing power, re-imagining and re-building a world that is livable and shareable by and for all – especially for those who are deemed disposable and marginalized' (Merino, 2017). According to RWA, 'The methodology that underpins our approach, is that we have to use every opportunity to unpack difficult issues and deal with our prejudices. Homosexuality wasn't treated as a side-issue, or as a waste of time, or as irrelevant to the "main issue" of land struggles' (Andrews, 2019: 55). Other marginalized peoples could include, for example, migrant workers and refugees, who are often subject to xenophobia and multiple survival challenges, tribal or religious minorities, and landless peoples. The global Food Sovereignty movement places the struggles of women at the centre of its campaign, and the African chapters enthusiastically embrace this by proactively electing women leaders and valorizing women's knowledge (Mpofu, 2020). At their fifth congress in Maputo in 2008, La Via Campesina declared their campaign against violence against women. While the struggles of women are being centralized in African Food Sovereignty campaigns, there is little evidence that the movement is taking up the broader, more radical feminist agenda to support other marginalized peoples, or that there is any desire to do so. This is hardly surprising, given the incredible diversity of cultures, beliefs, and ways of being across the continent and the challenge of simply galvanizing common positions on agriculture and food systems. However, radical feminists on the continent have much that is new to offer on building healthy and equitable food systems, which is apparently not yet part of the African Food Sovereignty discourse.

The power of donors and donor-directed development also continues to be a site of conflict within the Food Sovereignty movement. Critiques of 'outcomes-based' funding and the ubiquitous and notorious 'log-frames' are emerging from complexity and decolonial theories. These critiques acknowledge the violence of this reductionist computer-type logic, which is at odds with reality and can undermine lived experience, local knowledge, and problem solving (Lowe and Plimmer, 2019), as well as the dance between human and non-human actors through which co-being emerges. Some donors are taking their cue from complexity theory and theories of learning to develop new funding mechanisms based on relationships of trust and devolving decision-making to local levels, as opposed to requiring checks on outcomes and indicator boxes (Lowe and Plimmer, 2019). This is fairly new ground for Western institutions grounded in mechanistic thinking rather than complex relationships and emergent properties, while many Indigenous cultures are adept at working with complex relationships among themselves and with nature, and, as a result, reciprocity and protocols for good relations are fundamental values and practices (Rosiek et al., 2020).

It is interesting to see this critical and reflective work coming out of donor communities in the North, and the use of a complexity lens is showing an encouraging (albeit nascent) willingness to move from hierarchies to relationships, and from universalities to place-based truths. However, organizations are microcosms of the society from which they emerge (Batliwala, 2010). Several frank questions could therefore help deepen cultural reflection toward more emancipatory practices: How do Northern donors deal with the fact that their wealth is undeniably built on imperialism, dispossession, and racism? And, if that question were to contextualize their work, would the donor relationship be transformed to one of solidarity, redress, and emancipation? Can lessons from Indigenous ontology strengthen their complexity work?

Conclusion

When we talk about the Food Sovereignty movement we're thinking about a movement that is fighting. The notion is that a movement is articulating issues and fighting the dominant system. But a food sovereignty movement has always existed in Africa. When the settlers came there was a food movement in Africa and our forbearers were active participants in this movement; they knew how to participate without putting a name to it. At that time communities were vibrant and producing sufficient food, and exchanging food. Before colonization, seed would go from Zimbabwe to Zambia without thinking of borders. Food sovereignty in Africa is rooted in practice, is deeply connected to nature and embedded in the more-than-human world. Non-human agency is part of the movement; it shapes and catalyses and participates in our movement. Nature is the one that speaks to us [humanity] in a profound way. The trees and forests inspire us to a very different conversation. We just practise, we do, we be who we are, and that is the characteristic of our movement, that is our resistance. (Anonymous interview with a leading African activist, 2020)

The rhizomatic nature of the African Food Sovereignty movement at grassroots level is dynamic and emergent, revelling in relationships with family and community and the living territory in which we are embedded. Grassroots food producers are supported by many allies, including producer alliances, NGOs, donors, technical expertise, and social movements across continents. The Food Sovereignty movement has been described in terms of 'big tent politics' (Patel, 2009), in which disparate groups can recognize themselves in this sprawling egalitarian project. It could be both the blessing and the curse of the movement, in that the 'big tent' embraces complexity and diversity on the one hand, but, on the other hand, the project of food sovereignty is opened so wide that it 'becomes everything and nothing'. Can the complexity and diversity of the 'big tent' we see at the grassroots level emerge as a coherent movement at key moments? Can the movement stand in unison to protect the fecund spaces for alternatives? Can the political project of the movement maintain the complex assemblages and aspirations of African food sovereignty in mainstream policy arenas, where the parameters of negotiation continue to be set by white capitalist patriarchy?

The movement has grown from strength to strength since the heady days of the 1990s, when elected African leaders fought for African dignity and the protection of agricultural resources in international arenas. African voices are now an integral part of the powerful La Via Campesina movements, and African farmer leaders are acknowledged in regional, continental, and international policy spaces, resulting in, for example, the adoption of UNDROP at the UN or the constant reframing of the notion of food security to include issues of power. The power of African Food Sovereignty can be seen in many extraordinary moments of convergence of rooted networks, moments when 'assemblages and rhizomes running silent and deep like the movable malleable stuff of roots and shoots … push through the stuff of the world to send up a bright shining fruit of a mushroom that will reproduce, scatter spores and fall back into the ground from which it came' (Rocheleau 2015: 75).

Note

1 Ontology relates to our experience of being in the world. Chapter 12 explores how 'modern scientific rationalism' is based on an ontology of the world as a dead machine over which thinking humans rule, while Indigenous ontologies arise from human participation in a living, intelligent universe.

References

Alliance for Food Sovereignty in Africa (AFSA) (2017) *Resisting corporate takeover of African seed systems and building farmer managed seed systems for food sovereignty in Africa*, Kampala, Uganda <https://afsafrica.org/wp-content/uploads/2018/09/seed-policy-eng-online-single-pages.pdf>.

Anderson, C. and Maughan, C. (2021) '"The innovation imperative": the struggle over agroecology in the international food policy arena', *Frontiers in Sustainable Food Systems* 5 <https://doi.org/10.3389/fsufs.2021.619185>.

Andrews, D. (2020) 'Reflecting on socio-ecological transformation research: critical questions to consider', *Critical Food Studies* <http://www.critical-foodstudies.co.za/wp-content/uploads/2020/02/D-Andrews-Graz-chapter_checked_2020023_11h20.pdf>.

Andrews, D., Smith, K. and Alejandra Morena, M. (2019) 'Enraged: women and nature', in M. Alejandra Morena (ed.), *Right to Food and Nutrition Watch, Issue 11: Women's Power in Food Struggles*, pp. 6–15, Global Network for the Right to Food and Nutrition <https://www.righttofoodandnutrition.org/files/rtfn-watch11-2019_eng.pdf>.

Andrews, M. (2019) 'A case study of the Southern African Rural Women's Assembly: "We can bend the stick"', *Agenda* 33(1): 48–58 <https://doi.org/10.1080/10130950.2019.1598275>.

Bassey, N. (2021) '"Return to Being": a poem by Nnimmo Bassey', International Union for Conservation of Nature, Gland, Switzerland <https://www.iucn.org/news/commission-environmental-economic-and-social-policy/202109/return-being-a-poem-nnimmo-bassey>.

Batliwala, S. (2010) *Feminist Leadership for Social Transformation: Clearing the Conceptual Cloud*, Creating Resources for Empowerment in Action, Bangalore <https://creaworld.org/wp-content/uploads/2020/11/feminist-leadership-clearing-conceptual-cloud-srilatha-batliwala.pdf>.

Belay, M. (2012) *Participatory Mapping, Learning and Change in the context of Biocultural Diversity and Resilience*, PhD thesis, Rhodes University, Grahamstown, South Africa <http://www.iapad.org/wp-content/uploads/2015/07/belay-PhD-TR12-1164.pdf>.

Belay, M. (2020) 'Rethink–biocultural diversity', African Biodiversity Network <https://africanbiodiversity.org/rethink-biocultural-diversity/>.

Biko, S. (1987) *I Write What I Like*, Heinemann, Oxford.

Borlaug, N. (1970) 'The Green Revolution, peace, and humanity', Nobel Lecture, 11 December <https://www.nobelprize.org/prizes/peace/1970/borlaug/lecture/>.

Daño, E.C. (2007) *Unmasking the New Green Revolution in Africa: Motives, Players and Dynamics*, Third World Network, Penang, Church Development Service (EED), Bonn, and African Centre for Biosafety, Richmond, South Africa.

De Loma-Ossorio, E., Lahoz, C. and Portillo, L.F. (2014) *Assessment on the Right to Food in the ECOWAS region*, Food and Agriculture Organization, Rome <https://www.fao.org/3/i4183e/i4183e.pdf>.

De Schutter, O. (2015) 'Food democracy South and North: from food sovereignty to transition initiatives', *openDemocracy*, 17 March <https://www.opendemocracy.net/en/food-democracy-south-and-north-from-food-sovereignty-to-transition-initiatives/>.

Durante, F., Kröger, M. and LaFleur, W. (2021) 'Extraction and extractivisms: definitions and concepts', in J. Shapiro and J.-A. McNeish (eds), *Our Extractive Age: Expressions of Violence and Resistance*, pp. 19–30, Routledge, New York.

Edelman, M., Weis, T., Baviskar, A., Borras Jr, S.M., Holt-Giménez, E., Kandiyoti, D. and Wolford, W. (2014) 'Introduction: critical perspectives on food sovereignty', *The Journal of Peasant Studies* 41(6): 911–931 <http://dx.doi.org/10.1080/03066150.2014.963568>.

ESAFF (2013) 'Zimbabwe selected to host international movement of peasant organization', 11 March, Eastern and Southern Africa Small-scale Farmers Forum <https://esaff.org/index-php/zimbabwe-selected-to-host-international-movement-of-peasant-organization/>.

Fanon, F. (1962) *The Wretched of the Earth*, Grove Press, New York.

Food and Agriculture Organization of the United Nations (FAO) (2013) 'FAO will cooperate with La Via Campesina, the largest movement of small food producers in the world', 5 October <http://www.fao.org/partnerships/civil-society/news/news-article/en/c/201834/>.

FAO (2021) 'Family Farming Knowledge Platform' <http://www.fao.org/family-farming/background/en/>.

FAO, Alliance of Bioversity International and CIAT (2021) *Indigenous Peoples' Food Systems: Insights on Sustainability and Resilience from the Front Line of Climate Change*, FAO, Rome <http://www.fao.org/3/cb5131en/cb5131en.pdf>.

Greenberg, S. (2013) *Capitalist Expansion and Agri-food Systems in Southern Africa: A Study on the Relationship between the Southern African Confederation of Agricultural Unions (SACAU) and Small-scale Farmer Associations*, People's Dialogue <https://www.researchgate.net/publication/274379147_Capitalist_expansion_and_agri-food_systems_in_Southern_Africa_A_study_on_the_relationship_between_the_Southern_African_Confederation_of_Agricultural_Unions_SACAU_and_small-scale_farmer_associations>.

Gyapong, A.Y. (2017) 'Food Sovereignty in Africa: the role of producer organisations', presented at *Icas-Etxalde Colloquium: The Future of Food and Agriculture. Vitoria-Gasteiz, Basque Country, April 2017* <https://www.researchgate.net/publication/327920631_Food_Sovereignty_in_Africa_The_Role_of_Producer_Organisations>.

Hanisch, C. (2006) *The Personal is Political* <http://www.carolhanisch.org/CHwritings/PIP.html>.

Haraway, D. (1988) 'Situated knowledges: the science question in feminism and the privilege of partial perspective', *Feminist Studies* 14(3): 575–599 <https://doi.org/10.2307/3178066>.

High Level Panel of Experts on Food Security and Nutrition (HLPE) (2020) *Food Security and Nutrition: Building a Global Narrative towards 2030*, HLPE, Committee on World Food Security, Rome <https://www.fao.org/3/ca9731en/ca9731en.pdf>.

Holt Giménez, E., and Shattuck, A. (2011) 'Food crises, food regimes and food movements: rumblings of reform or tides of transformation?' *Journal of Peasant Studies* 38(1): 109–144 <https://doi.org/10.1080/03066150.2010.538578>.

Hubert, C. (2019) *The United Nations Declaration on the Rights of Peasants: A Tool in the Struggle for our Common Future*, CETIM, Geneva <https://viacampesina. org/en/wp-content/uploads/sites/2/2020/07/The-UN-Declaration-on-the-Rights-of-Peasants.pdf>.

International Forum for Agroecology (2015) *International Forum for Agroecology, Nyéléni Center, Selingue, Mali, 24–27 February 2015* <https://www.ukfg.org. uk/2015/international-forum-agroecology-nyeleni2015/>.

Lowe, T. and Plimmer, D. (2019) *Exploring the New World: Practical Insights for Funding, Commissioning and Managing in Complexity*, Northumbria University and Collaborate CIC <https://collaboratecic.com/wp-content/ uploads/2023/09/1.-Exploring-the-New-World-Report-MAIN-FINAL.pdf>.

Mateo-Sagasta, J., Marjani Zadeh, S., Turral, H. and Burke, J. (2017) *Water Pollution from Agriculture: A Global Review: Executive Summary*, FAO, Rome, and International Water Management Institute, Colombo <http://www. fao.org/3/i7754e/i7754e.pdf>.

Merchant, C. (2006) 'The scientific revolution and the death of nature', *Isis* 97(3): 513–533 <https://doi.org/10.1086/508090>.

Merino, J. (2017) 'Women speak: Ruth Nyambura insists on a feminist political ecology', *Ms.*, 15 November <https://msmagazine.com/2017/11/15/women-speak-ruth-nyambura-feminist-political-ecology/>.

Milton, K. (1999) 'Nature is already sacred', *Environmental Values* 8(4): 437–449 <http://dx.doi.org/10.3197/096327199129341905>.

Mpofu, E. (2020) 'Presentation prepared for AGRA Watch Webinar and Report Launch: "The Struggle over Agroecology: Mapping and Mobilizing against the Gates Foundation's Influence in African Agriculture"', 8 August, Community Alliance for Global Justice <https://cagj.org/2020/08/11159/>.

Nyéléni (2007) *Synthesis Report*, 2 April, Nyéléni 2007 Forum for Food Sovereignty, Sélingué, Mali <https://www.nyeleni.org/IMG/pdf/31Mar200 7NyeleniSynthesisReport-en.pdf>.

Patel, R. (2009) 'Food sovereignty', *The Journal of Peasant Studies* 36(3): 663–706 <https://doi.org/10.1080/03066150903143079>.

Rocheleau, D. (2015) 'Roots, rhizomes, networks and territories: reimagining pattern and power in political ecologies', in R.L. Bryant (ed.), *The International Handbook of Political Ecology*, pp. 70–88, Edward Elgar, Cheltenham <http:// dx.doi.org/10.4337/9780857936172.00013>.

Rosenow, D. (2018) *Un-making Environmental Activism: Beyond Modern/Colonial Binaries in the GMO Controversy*, Routledge, New York.

Rosiek, J., Pratt, S.L. and Snyder, J. (2020) 'The new materialisms and indigenous theories of non-human agency: making the case for respectful anti-colonial engagement', *Qualitative Inquiry* 26(3–4): 331–346 <https:// doi.org/10.1177/1077800419830135>.

RWA (2020) *The Southern Africa Rural Women's Assembly 10th Anniversary: Towards a Transformative Feminist Agenda for the Country Side*, held at Wits University Sports Centre, 26–28 November 2010.

Shilomboleni, H. (2017) 'A sustainability assessment framework for the African green revolution and food sovereignty models in southern Africa', *Cogent Food & Agriculture* 3: 1 <https://doi.org/10.1080/23311932.2017.1328150>.

Shiva, V. (1991) *The Violence of the Green Revolution; Third World Agriculture, Ecology and Politics*, Zed Books, London.

Smith, L.T., Tuck, E. and Yang, K.W. (2018) *Indigenous and Decolonizing Studies in Education*, Routledge, New York <https://doi.org/10.4324/9780429505010>.

Sobrevila, C. (2008) *The Role of Indigenous Peoples in Biodiversity Conservation: The Natural but Often Forgotten Partners*, World Bank, Washington, DC <https://documents1.worldbank.org/curated/en/995271468177530126/pdf/443000WP0BOX321onservation01PUBLIC1.pdf>.

Tewolde Berhan Gebre Egziabher (2007) 'The Cartagena Protocol on Biosafety: history, content and implementation from a developing country perspective', in T. Traavik and L.C. Lim (eds.), *Biosafety First: Holistic Approaches to Risk and Uncertainty in Genetic Engineering and Genetically Modified Organisms*, pp. 389–405, Tapir Academic Press, Trondheim <http://genok.no/wp-content/uploads/2013/04/Chapter-25.pdf>.

UNAC (2021) 'O Que é a UNAC?' [website] <https://www.unac.org.mz/quem-somos/>.

UNEP (2020) '10 things you should know about industrial farming', United Nations Environment Programme, 20 July <https://www.unep.org/news-and-stories/story/10-things-you-should-know-about-industrial-farming>.

Visvanathan, S. (2006) 'Alternative science', *Theory, Culture and Society* 23: 164–169 <https://doi.org/10.1177%2F0263276406002300226>.

Wolford, W. (2021) 'The Plantationocene: a lusotropical contribution to the theory', *Annals of the American Association of Geographers* 111(6): 1622–1639 <https://doi.org/10.1080/24694452.2020.1850231>.

Box J Connecting generations through celebrations

Million Belay

(Based on a talk given in December 2018 at TEDxEuston, an independently organized TED event.)

Culture, they say, is like a river. The river has a source, and if the source is kept alive, it keeps on flowing. The source for our culture is nature: the rivers, the lakes, the mountains, the land, the wildlife, and the sea. It is our respectful connections with nature that have kept the source functioning. The source is our traditional medicine, which a large part of our people still use. The source is the diversity in our seeds and food, a diversity that we need in this time of uncertainty. The source is the knowledge of our mothers and fathers and our ancestors about nature and life. The source is our language, because the knowledge of our fathers and mothers is coded in our language. The source is how our ancestors managed their relationship with each other and their environment. We need our cultural value, the source, to keep us living and thriving into the future.

But the river is no longer flowing, because its source is in great danger of drying up. I have had the chance to talk to so many communities in Africa, and outside Africa, and they confirm that it is drying up. This is an African and a global disaster.

The river is not flowing because of the gap between elders and youth. Urbanization, globalization, formal education, religion, and bad models of development are some of the causes for this widening gap. The problem is that the new generation does not seem to be interested in the source.

The good news is that it is possible to reverse this draining of knowledge and the cutting of our connection to the source. It is possible to connect the young with the knowledge-holders and nature and let the river of information flow. We can do that through celebrations.

It was not till I started teaching at a remote rural school in Ethiopia that I began to appreciate the source, our culture. I was collecting plant specimens for our school and the National Herbarium. Since I had to know what I was collecting, I used to ask the elders, who gave me a detailed description of the plants and their uses. The extent of their knowledge astounded me. I wondered if my students had this information.

In 2000, I read a book called *Cultural and Spiritual Values of Biodiversity*, and one of the stories in it is about basket-weaving and biodiversity in the Pacific Northwest (Posey, 1999: 86). It talks about how basketry is a profound cultural activity for women, connecting the past, present, and future. It talks about how basketry is so significant to the social and spiritual life of women. The women take the basket as a symbol of their womb. Basket-making is also a social activity. Women put their valuables in baskets, not in containers made of tin. The book triggered a question in me: is it possible to use cultural artefacts to connect children and youth with the source, with their culture and nature? I decided to start a cultural biodiversity programme for schools in Ethiopia. I had experience in working with schools because of the successful environmental club that I had started in one of the rural towns in Ethiopia.

With this in mind, with my colleagues at the Institute for Sustainable Development in Ethiopia and supported by the Gaia Foundation in the United Kingdom, we held the first workshop with teachers from 17 schools coming from all over the country. The workshop, held in a small town called Holeta, was about how to learn from knowledge holders in communities and how to document that learning. After we had discussed the value of culture and biodiversity, the source, the participants were sent to households in the town to record what they saw, so they could come back and share it with us.

(Continued)

Box J Continued

I was astonished at the amount of information that they brought from their study visits in the community. Some visited traditional cloth-makers, some met woodworkers, some interacted with women in households and others visited a traditional herbalist. We were all surprised at the richness of knowledge in that small town.

One of the teachers from Addis Ababa replicated the exercise with his students, and they came back with a lot of cultural information. We could see that culture thrives even in urban areas, and I said to myself, 'The source is there! If only we could make sure that it flows from knowledge-holders to children and youth.'

After a year or so, we organized the first cultural biodiversity celebration in Ethiopia. The 17 participating schools came from all over the country, and from different cultures, with their clothing, music, art, seeds, artefacts, and plays. Some even brought entire dwellings representing their cultures. It was opened by the president of the country and visited by close to 30,000 people. Each day there were cultural displays and schools took turns to entertain and educate the audience. They also visited each other's stalls to learn about other cultures.

The African Biodiversity Network took the programme on as one of its thematic areas, and it became part of my responsibility to spread it in other countries. So celebrations were held in Ghana, Togo, Benin, South Africa, Tanzania, Kenya, and Ethiopia. Togo still holds this celebration every year, as does Ethiopia.

In 2004, we started an Indigenous NGO called the Movement for Ecological Learning and Community Action, or MELCA-Ethiopia. Inspired by a similar programme in South Africa called Imbewu, we started a programme called SEGNI, or seed. This is to connect people with their culture, nature, and themselves. Groups of students are taken by local elders into a forest, where they play games, walk, have solitary time, sit in a circle talking about their lives, and, in the evening, hear stories from the elders. When they go back to their schools, they build traditional houses for cultural centres, do mapping, and organize yearly celebrations.

Over the years, I have experienced a number of deeply touching and invigorating events. Let me report on two of them.

The first demonstrates the value of our seeds to the resilience of farming and livelihoods. It is the story of an agricultural community called Telecho. While we were doing mapping with this community, we were told that out of the 19 barley varieties that they had once had, they were left with only 5. This was due to the introduction of 'improved varieties'. We started a seed restoration and soil and water conservation programme with the community. We took their children to experience the SEGNI programme. After some time, we organized a huge celebration involving the community and the schools. I visited the exhibition stalls along with government dignitaries, and at one of them, I saw 19 varieties of barley arranged in order, with their names in the local language.

I asked the students, 'What is this?' The students replied, 'These are the 19 barley varieties that we have here in Telecho.' 'I heard you lost them,' I said. 'Yes, but we got them back,' they answered. My next question: 'Where did you get them from?' Their response: 'We hunted for them in our locality for this celebration and found them in some households, mostly of older women.' You cannot imagine how elated I was.

The other story is about celebration and its value for peace-building. MELCA-Ethiopia works in Majang Zone, Gambella Region, which is located in the south-west of Ethiopia. It is a beautiful and forested area. The zone has inhabitants who came from other parts of the country as the soil and climate are good for agriculture, especially coffee. It also has its original inhabitants. MELCA-Ethiopia was working to have the area declared a biosphere reserve and had started the SEGNI programme, among other things. We had been mapping the zone, and we finished our work and left the area on 8 September 2015.

(Continued)

Box J Continued

I had heard a lot, while mapping, about a conflict that was brewing. On 11 September, while I was celebrating the Ethiopian new year with my family and friends, I got a call from one of our staff. He was making the call while hiding under his bed. He told me that there was conflict and that people from both communities were being killed and thousands displaced. I was devastated, because I had enjoyed the hospitality of these people when I was there. The following year, we organized a cultural biodiversity celebration at a small town called Tepi and invited students to come from Majang and Sheka zones. While I was busy with my colleagues coordinating the day, I heard a song and was told that the Majang students had come. I turned around and saw them entering the celebration arena dancing and singing. Children of those who had been killing each other only a few months before were singing and dancing together. I felt like crying. That was so powerful!

Celebrations are critical for a number of reasons.

- First, youth and children are motivated to participate in these activities when they prepare for an event. Celebratory events have an amazing galvanizing power.
- Second, they study the names and value of the seeds, medicinal plants, and artefacts that they collect. This is where the learning comes in. They have to know these things so they can explain them to those who come to visit them.
- Third, a celebration enthuses them to participate in the arts. They draw and paint, they write and rehearse plays, they compose songs, poems, and ballads. It is really amazing to hear the poems and see the plays.
- Preparations for celebration teach them to work in teams. They get to know how to solve problems in groups.
- Celebrations bring the community together. People come to visit and learn to connect with their culture.
- Celebrations also bring in decision-makers, and this helps in advocacy for integrating people's knowledge into the school system.

Students dive back into their cultural pool, the source, and come back nourished and enriched, and express that through art and artefacts.

I hope I have convinced you that we can revive the source, our culture and nature, and let the river flow through celebration.

Reference

Posey, D.A. (1999) *Cultural and Spiritual Values of Biodiversity*, Intermediate Technology Publications, London <https://wedocs.unep.org/bitstream/handle/20.500.11822/9190/Cultural_Spiritual_thebible.pdf>.

Box K Transitioning to agroecology: farming for the 21st century

John Wilson

Maize colonizes western Kenya

We bumped along the road on Ferdinand's motorbike. Ferdinand runs a small, community-based NGO promoting agroecology in Vihiga, western Kenya. I was his passenger and had to hold on tightly. It was hilly country, and green. Vihiga is the most densely populated county in Kenya, because it is high-potential farming land.

Very occasionally we passed remnants of the forest that once covered these hills and much of western Kenya. These remnants were small pockets, usually next to a streamline. The big trees were magnificent and the cover thick and dense. I thought to myself: 'That's how nature evolved in this part of western Kenya, which has such a wonderful climate, especially for a visiting Zimbabwean!'

Mostly we saw maize field after maize field with lots of bare soil between the plants. Scattered trees dotted the landscape. It seemed to me like the kind of environment in which trees can't help but grow! *Grevillea*, a fast-growing timber tree, was quite common, as it is across most of the wetter parts of Kenya. There were also some fruit trees.

Maize has colonized Vihiga, just as it has my home country, Zimbabwe. In this case it had turned dense subtropical forest into maize fields. 'Was that a sensible thing to do?' I wondered. Just as I was disappearing into that thought, Ferdinand slowed down. We had come around a bend and he pulled off the road. We climbed off the motorbike and Ferdinand led the way down a path leading off the road.

A food forest

I was immediately struck by what I saw. It was very different from the maize-dominated view we had been passing on the road. This was a forest, but it was also different from the remnants of indigenous forest that we had passed. I looked around, identifying some of the trees. It dawned on me as we approached a modest house in the midst of it that this was a food forest.

Before I could think about this further, a lithe and healthy-looking man emerged from the house with a huge smile. Ferdinand led the introductions. The beaming man was Julius Astiva, a name I haven't forgotten since that day, though generally I have a very bad memory for names.

Photo K.1 Julius Astiva, a farmer full of vitality and health, who doesn't look his years. Is it the diverse, healthy food that his family is eating from their farm?
Credit: John Wilson

(Continued)

Box K Continued

What followed was a memorable couple of hours during which we toured Julius's farm and he told us the story of how it had come about. Julius has a two-acre farm, which slopes fairly steeply from the road down to a stream at the bottom. He inherited the farm many years ago. After school, Julius attended an agricultural college and then became a teacher himself at the college. He used to visit his farm sporadically when he had days off from work. It was then a typical Vihiga maize farm, with a few trees dotted here and there.

Photo K.2 The neighbouring farm, typical of farms throughout Vihiga county in western Kenya. The very high-potential land had been turned to monocrop maize with odd trees here and there, including single stands of eucalyptus
Credit: John Wilson

About 20 years previously Julius had a calling to become a full-time farmer. He thought he understood how he should farm, and it was not along the lines of what he was teaching at the college. To the amazement of his family and friends he gave up the teaching job and became a full-time farmer on two acres. 'Who would do such a thing?' they wondered. Is he crazy?

Julius is a little crazy, I think! Crazy in a brilliant and pioneering way. In my experience innovative farmers have to be a little crazy. Julius had a vision of what he wanted his farm to be, and he worked very hard to bring this about. He realized from the beginning the importance of water. Even though good rains occur in that part of Kenya, Julius understood that every drop of rain is precious, and that water in the ground is a resource and water running over the ground is damaging.

He terraced his sloping land so that he lost no rain as run-off. The ditches forming each terrace harvested all run-off. From the beginning he also ensured as much ground cover as possible to increase water infiltration. Run-off water is the curse of farming across Africa, made worse with the practices of ploughing, monocultures, bare soil, and freely wandering livestock, which all create an ever downward spiral towards desertification.

Julius knew he had to make his farm profitable quickly if he was to survive as a farmer. At the bottom of the slope he dug two fishponds, each about 10 m x 5 m. Helped by his

(Continued)

Box K Continued

water-harvesting up-slope, water seeped into his ponds and they were always full. There he grew fish, which were his main source of income at the beginning. They are still a source of income, but now only one of many.

With the terraces and fishponds in place Julius began to create his vision for the land, a forest of food that would also bring him a decent living and allow him to pay for his children's education. He planted many kinds of trees, all with a view to income as well as food. Some, like the tamarillo, produce fruit to eat and sell within six months. Others, like avocado, take longer, and timber trees even longer.

Photo K.3 The diversity of Julius's food forest farm, including some eucalyptus, mixed in with other plants
Credit: John Wilson

Julius thinks in the short, medium, and long term at the same time. He does not see his farm simply as a piece of land on which to grow a crop and from which to make money, but rather in a holistic way as a piece of nature that he must look after. If he does it will look after him. Just as nature does, he ensures that the ground is covered by both mulch and canopy.

Julius is not following the instructions or ideas of agronomists or extension workers. He is his own agronomist. He is constantly thinking about what he is doing and will do, trying things out all the time. Some trials work, some do not. He further develops those that work. He learns from those that do not.

Above all, farming with nature

Twenty years on, as a visitor you wander around a farm that is packed with productive plants. Many are long-term, perennial species, but he also grows annual crops, including maize. Maize has its place but it doesn't dominate as it does on his neighbours' farms.

(Continued)

Box K Continued

Photo K.4 Julius also grows maize, but as just one crop in the variety cultivated on his food forest farm
Credit: John Wilson

Unfortunately he has not been able to influence those in his immediate vicinity, but through Ferdinand and the NGO he set up in 2009 called Bio Gardening Innovations (BIOGI), Julius is sharing his experience, skills, and knowledge with other farmers in Vihiga.

Photo K.5 At the lower end of Julius's two-acre farm are his fish ponds, benefiting from all the water harvested higher up – initially using ditches, but now enhanced by the constant ground cover and healthy soils that infiltrate water quickly and easily thanks to their healthy structure, instead of letting it run off in Western Kenya's regular heavy storms. Fish were his main early source of income and are now one of many sources
Credit: John Wilson

Box K Continued

Photo K.6 Another farmer in the area who has learnt from Julius. Her first step was digging the ditches to harvest water, and then she planted her food forest
Credit: John Wilson

Julius's farm is an excellent example of a transition to agroecology. It has given him a decent income while he has regenerated the soil and greatly increased the agricultural biodiversity. He is a very intelligent small-scale farmer who has the independence to do his own thing, and is not worried about what the mainstream does or thinks. He knows that the right way of farming is to farm with nature. Everything he does grows out of that understanding.

The 'industrial' approach to farming

Compare this with what has become the conventional and Green Revolution approach to farming. In this you plough the soil. You bring in synthetic fertilizers that feed the plant and not the soil. These fertilizers acidify the soil and have a negative effect on the microorganisms that we now know are critical to a healthy soil, in more ways than we can begin to understand (Bulluck et al., 2002; Hathaway, 2016). Then, if there is a pest or disease, you spray dangerous poisons which usually kill everything, upsetting any balance that was there (Gill and Garg, 2014). More recently there has been a significant move towards minimum tillage in industrial farming, but this is combined with the regular use of herbicides whose damaging effect on health – that of soil and of people – is becoming apparent (Giller et al., 2015).

   Nature functions around having a diversity of species in dynamic balance (Kremen et al., 2012). Things fluctuate all the time, as in any living system, but always move towards achieving some kind of dynamic balance. Pesticides kill the pests they are targeting but throw the balance out badly. What often happens is that the pests rebound and thrive because it is more difficult for their predators to bounce back from the pesticides (Gill and Garg, 2014; Toher, 2018).

(Continued)

Box K Continued

Mimicking nature

At the heart of the transition to agroecology, then, is working to emulate the processes of nature. By looking at how nature works, we see the evidence we need for the kind of farming system we should transition to. Nature's processes have been functioning for millions of years. What more evidence do we need than that? Looked at in economic terms, this means not destroying your capital as you produce. What business destroys its capital in the process of production, year after year? That's what industrial agricultural practices have done and continue to do, and there is plenty of evidence for that (see, for example, Frison and IPES-Food, 2016; TEEB, 2018). Agroecological practices regenerate the soil and biodiversity.

The direction is clear: transforming our practices to be in line with nature's processes. This should guide our every step in the transition to agroecology. We can learn from pioneers like Julius even though they are few. This transition to agroecology will need the support of a very wide range of people, from farmers, of course, to scientists and governments, and others who can support the minimal but very necessary infrastructure development related to agroecology. Such development could include:

- earthworks to harvest water via ditches, ponds, and small dams;
- other water-harvesting structures such as tanks;
- fencing to protect the replanting of mixed woodland at key points in the landscape, and also for nutrition gardens everywhere;
- movable kraal material so as to be able to rotate night kraals for livestock through cropping fields in the dry season;
- small production centres for organic inputs run by local businesses;
- farm produce processing units at the local level;
- equipment for small-scale irrigation, including to keep crops going in the rainy seasons, because of the increasing occurrence of mid-rainy-season dry spells.

This transition will need the backing of consumers and progressive private sector companies with a strong ethical base. It will need to create learning opportunities everywhere since we are at the beginning of a very long learning journey.

Nearly all landscapes across Africa are in decline (Gnacadja and Wiese, 2016). Agroecology means turning this around. It does not mean answers overnight. It means changing direction and beginning the long journey of transition from degradation to regeneration. Agroecology means transition:

- transition from cash monocropping, where farmers are simply cheap labour for a corporate value chain that profits from their cheap labour, to farmers growing crops for sale in a biodiverse farm landscape;
- transition from subsistence practices that enable families to eke out a living in an ingenious way, but mean they are on a survival treadmill, to an increasingly abundant life;
- transition from savannah landscapes that are gradually enclosing the commons to open commons that allow pastoralists and farmers to move their livestock, as wildlife moved for millions of years, in large enough herds and in ways that resemble many of the traditional practices of pastoralist communities in the past;
- transition from hilly landscapes that shed most of the water from heavy rainstorms, which are increasingly likely with the changing climate, to planned landscapes where communities are very deeply committed to the health of the land and water is increasingly harvested into the ground, no matter how heavy the storm;
- transition from a situation where science develops 'magic bullet' answers in isolation from farmers for those farmers to apply, to one where scientists from a range of

(Continued)

Box K Continued

disciplines feed their specialist knowledge into farmer-led research, farmer learning groups, and farmer networks, and where farmers conduct scientist-supported trials, with scientists also helping with the documentation – all of which also implies transition to a situation where different knowledge systems are valued equally;

- transition from today's market value chains, where there is little connection between farmers and consumers, to one in which consumers increasingly know where their food is coming from and how it has been produced;
- transition to a situation where the voices of farmers, particularly smallholder farmers, are heard as a matter of course, and not only when they take to the streets;
- transition towards a situation where countries increasingly recognize that their future depends on how they manage water, and where they develop policies that enable widespread water management based on the water management principles of nature, developed over millions of years.

All these kinds of transitions need a lot of support and joint learning at every stage. The science and local knowledge that must combine to begin this journey do exist. The challenge is to convince enough people that pursuing the old industrial agriculture and land-use paradigm really is a dead end, and that an agroecological direction is the one to move in.

References

Bulluck, L.R., Brosius, M., Evanylo, G.K. and Ristaino, J.B. (2002) 'Organic and synthetic fertility amendments influence soil microbial, physical and chemical properties on organic and conventional farms', *Applied Soil Ecology* 19(2): 147–160 <http://doi.org/10.1016/S0929-1393(01)00187-1>.

Frison, E.A. and IPES-Food (2016) *From uniformity to diversity: a paradigm shift from industrial agriculture to diversified agroecological systems*, Louvain-la-Neuve, Belgium: IPES.

Gill, H.K. and Garg, H. (2014) 'Pesticides: environmental impacts and management strategies', in M.L. Larramendy and S. Soloneski (eds), *Pesticides: Toxic Aspects*, Rijeka, Croatia: IntechOpen <http://doi.org/10.5772/57399>.

Giller, K.E., Andersson, J.A., Corbeels, M., Kirkegaard, J., Mortensen, D., Erenstein, O. and Vanlauwe, B. (2015) 'Beyond conservation agriculture', *Frontiers in Plant Science* 6: 870 <https://doi.org/10.3389/fpls.2015.00870>.

Gnacadja, L. and Wiese, L. (2016) 'Land degradation neutrality: will Africa achieve it? Institutional solutions to land degradation and restoration in Africa', in R. Lal, D. Kraybill, D.O. Hansen, B.R. Singh, T. Mosogoya and L.O. Eil (eds), *Climate Change and Multi-Dimensional Sustainability in African Agriculture*, Cham, Switzerland: Springer.

Hathaway, M.D. (2016) 'Agroecology and permaculture: addressing key ecological problems by rethinking and redesigning agricultural systems', *Journal of Environmental Studies and Sciences* 6: 239–250 <https://doi.org/10.1007/s13412-015-0254-8>.

Kremen, C., Iles, A. and Bacon, C. (2012) 'Diversified farming systems: an agroecological, systems-based alternative to modern industrial agriculture', *Ecology and Society* 17(4): 44 <http://dx.doi.org/10.5751/ES-05103-170444>.

TEEB (2018) *TEEB for Agriculture and Food: Scientific and Economic Foundations*, Geneva: UN Environment Programme <http://teebweb.org/wp-content/uploads/2018/11/Foundations_Report_Final_October.pdf>.

Toher, D. (2018) 'Pesticide use harming key species ripples through the ecosystem', *Pesticides and You* 38(2): 17–23 <www.beyondpesticides.org/assets/media/documents/TrophicCascades-cited.pdf>.

CHAPTER 16

Conclusion: towards seed and knowledge justice for agroecology

Rachel Wynberg

The case studies, reviews, stories, analyses, and anecdotes compiled in this book thread together a remarkably consistent set of conclusions and learnings, despite their wide variations in ecologies, cultures, political conditions, and economies. Together they provide a compelling narrative of the necessity of shifting from an environmentally intolerable and inequitable agrifood system towards one that embraces a pluriverse of knowledges, cultures, and more-than-human lives, and that embeds equity, sustainability, and inclusion in its day-to-day practices. Realizing this vision has never been more important, as we head towards a polycrisis of runaway climate change, staggering biodiversity loss, growing inequality and the ongoing atrocities of war and violent conflict, hunger and human greed. This concluding chapter draws together some of the key messages emerging from the preceding sections, provides pointers for future action, and identifies gaps that require further investigation.

Agroecology works but does not receive adequate support

The first, irrefutable, conclusion is that agroecology works – as an ecologically sustainable and socially robust practice to improve the resilience and productivity of smallholder African farmers; as a transformative and forward-looking approach to restore degraded soils, landscapes, and waterways; and as a remedy for the increasingly unreliable climate future that we face (see also Bezner Kerr et al., 2023). As the contributions in this book abundantly demonstrate, agroecology provides farmers with greater agency over what they choose to grow and eat, enabling them to secure productive, diverse, and healthier food systems for themselves and their families and communities. Placing greater control over seed production, management, and distribution in the hands of those who know how to nurture and enhance the agricultural biodiversity on which all humanity depends simply makes sense.

Inspiring stories emerge throughout the book of how farmers are already working towards food and seed sovereignty: by invoking seed stewardship and collective ownership through dialogue and solidarity networks; by using local community structures to conserve and exchange seed; by reinvigorating customary practices and systems of governance for managing natural resources; and by actively restoring landscapes, soils, sacred sites, traditional crop varieties, and farming practices. As such, agroecology also forms part of a politically engaged and growing African movement for food and seed sovereignty, encapsulated in the celebration of food, culture, and diverse ways of knowing, embracing indigenous knowledge and ways of being while advocating for the right to be autonomous and to farm for the future.

Yet, despite the proven impact of such approaches, policy and financial support for transformative agroecology, indigenous knowledge, and land management systems continues to be disproportionately low, or non-existent (Moeller, 2020; Pavageau et al., 2020). Policies such as the Comprehensive Africa Agriculture Development Programme (CAADP), for example, currently serve to disincentivize the uptake of agroecology on the continent. The systematic dismantling of public-sector support for agriculture has been replaced by market-led, productivist approaches across many African countries, accompanied by a research and development agenda based on principles of uniformity, profit, and control. Despite being much less capital-intensive, more effective, and lower in cost, agroecological options are thus marginalized in policy interventions and practices, as are the voices of their proponents.

Donor aid, philanthropy, and development

Inappropriate development is a consistent theme across many chapters in the book. We have learnt how development – initially in the guise of colonization, later through structural adjustment programmes and neoliberal reforms, and more recently taking shape through public–private partnerships and seemingly benevolent philanthropic organizations – has often imposed external agendas on farmers by answering the wrong questions with the wrong tools. A seductive politics of scarcity has prevailed, promoting a 'new' African Green Revolution centred on raising the productivity of smallholder farmers through improved hybrid seeds, agrochemicals, technological interventions, and linkages to markets. Such efforts are promoted on the basis of the highly contested assumption that smallholders are the key to growth and poverty reduction (Collier and Dircon, 2014). The model is underpinned by the acquisition and enclosure of seed through comprehensive policy reforms in the name of development, while it neglects farmer-managed seed systems – even though they continue to generate most seed on the continent. Examples include the development-aid-funded CAADP under the New Partnership for Africa's Development (NEPAD), the USAID-supported New Alliance for Food Security and Nutrition, and the Gates- and Rockefeller-funded Alliance for a Green Revolution in Africa. We have also learnt about the 'side effects' of development projects (Ferguson, 1994), and how development aid, in the form of seed, might be counterproductive if it disrupts farmers' local seed systems (Ncube et al., 2023).

Supporting agrobiodiversity conservation and use, and transforming the 'maize culture'

Maize forms an integral part of such interventions and has long been promoted as a driver of modernization to propel development and foreign investment in Africa. Associated impacts of the crop that both 'feeds and robs' reveal themselves in astonishingly similar experiences across Malawi, South Africa, Kenya, Ghana, and Zimbabwe. These talk to the rich, albeit recent, cultures and foodways that have developed alongside local varieties of maize, but also to the way in which the industrialization of maize has displaced more nutritious indigenous crops such as sorghum and millet. This, alongside the fact that a shocking one-third of children in sub-Saharan Africa are stunted, is cause for concern (see, e.g., SADC, 2022). Subsidy programmes and state support for hybrid maize and associated inputs have essentially propped up multinational agrochemical and seed companies, based on a persistent yet misjudged belief that local seed systems are unproductive and that they perpetuate poverty (see, e.g., DeVries, 2019). This exclusive focus on yield and productivity has ignored other aspects of local maize that farmers value, such as taste, drought and pest resistance, climate resilience, adaptability to local conditions, and cultural heritage, and has undermined local ecological knowledge. The introduction of genetically modified (GM) maize in countries such as South Africa has added further layers of complexity, creating anxieties and psychological trauma for farmers who wish to maintain the genetic integrity of their local varieties.

How do we shift paradigms to bring into sight different values and worldviews? These might include perspectives that appreciate local agronomic conditions, that comprehend the social meaning of food, that respect cultural preferences, and that recognize the heritage importance of seed. Active support for community seed banks, and approaches such as farmer field schools and participatory plant breeding, are vital parts of a solution. So too is the need to reimagine the role and place of gene banks in sustaining African smallholders and restoring agrobiodiversity.

Bringing relationality into agriculture and recognizing diverse ways of seeing, knowing, being, and learning

An important part of this reimagining is to redefine the way in which we perceive and define agriculture, and to recognize other ways of seeing, knowing, being, and learning. For millions of smallholder African farmers, *agri/culture* is inherently a social-ecological activity, in which cultural and ecological dimensions are deeply entangled, with agriculture and seed borne out of intricate relationships between human and non-human beings, and their biophysical environments (Herrero et al., 2015; Marshak et al., 2021). This relational knowledge is context-dependent, bringing an understanding of how interconnections between animals, plants, soil, people, and weather patterns in an agroecosystem are connected to and affect one another. 'The trees and forests inspire us to a very different conversation', remarked an activist interviewed for the book, reflecting on the relationship between agriculture and food, between food and nature, and between ourselves and the more-than-human world.

Multiple examples in this volume describe how this coevolution has been altered, reoriented, and/or ruptured through the introduction of agricultural development programmes, chemical fertilizers, pesticides, and herbicides, and new seed technologies that devalue local knowledge and skills in favour of 'expert'-led innovations. Today, agriculture is framed by governments, business, and donors alike as an extractive, export-driven enterprise, designed to produce commodities, generate foreign revenue, and deliver national food security. The impacts of this mindset have been especially damaging for women, whose caregiving, agricultural, and food-gathering practices, knowledge, and activities have been unappreciated and debased. The implications of these disrupted relationships are profound for both smallholders and the agroecosystems in which they farm as they lose both their capacity to understand the complex interrelations that exist in agroecosystems and their ability to react appropriately and act autonomously. Recognizing and repairing the relationship between agriculture and food, between food and nature, and between ourselves and the more-than-human world constitutes a vital component of restoring the dignity of farming and farmers and diversifying African agroecosystems.

Realizing farmers' rights

Increasingly, international attention is turning to the rights that African smallholder farmers have to save, use, exchange, and sell farm-saved seed, given that they source most of their seed through informal channels such as local markets, own stocks, and social networks (Sperling et al., 2021). However, intellectual property rights laws which promote plant breeders' rights, and seed laws that regulate variety release and seed distribution, form part of a host of measures that prejudice the interests of smallholder farmers and restrict the legal space they have to continue customary practices (Kloppenburg, 2004; Andersen, 2017). The dramatic expansion of the rights of companies to claim ownership over biodiversity-related innovations also runs counter to practices in many traditional farming communities, where land and other natural resources are often communally owned, seed is exchanged or shared, and invention is collective. Farmers' rights thus remain under threat, especially as trade-related pressures mount for African countries to sign the restrictive 1991 UPOV Convention (establishing the International Union for the Protection of New Varieties of Plants), which includes a raft of measures that criminalize the sharing, marketing, and sale of seed and strengthen private plant breeders' rights.

In 2022, many of these concerns found expression in a campaign supported by the Seed and Knowledge Initiative and dubbed 'Our Seeds, Our Rights, Our Lives', in which small-scale farmers gathered across Southern Africa to affirm the rights they have to their traditional seed, knowledge, culture, land, and associated life systems. From Mzuzu to Chikankata, and from Chimanimani to Mtubatuba, and through fairs, markets, festivals, and dialogues, farmers from Malawi, Zambia, Zimbabwe, and South Africa celebrated, shared, and sold their diverse traditional and indigenous seeds. They exchanged views and seeds at community seed banks, set up to support household food security and conserve the agricultural diversity now lost in many countries. They shared plant breeding techniques, developed by and for farmers. And they debated with seed officers, policymakers, and gene bank representatives in their various countries.

Strong positions emerged from farmers brought together to share these experiences, culminating in a multi-country and multi-actor seminar convened in Zambia in October 2022. A foremost concern was the lack of a supportive national policy framework to recognize and support local seed systems and promote agrobiodiversity. Farmers strongly opposed laws that criminalize the sharing, marketing, and sale of traditional seed and called for an open market to sell traditional seed and crops. They requested governments to give 'maximum support' to agroecology, provide critical rural infrastructure, and protect lands, livelihoods, and rights. Demands were made for technical and financial support for community and district seed banks. Farmers asked for support for farmer-to-farmer learnings, as well as farmer-led research, training, and extension services tailored to agroecology. A strong call was made to redirect funding towards agroecology from government subsidy programmes that support fertilizers and other inputs. Farmers also expressed concern about the possible introduction of GM seed in Malawi, Zambia, and Zimbabwe. African governments were urged to avoid signing UPOV 1991, to support implementation of the International Treaty on Plant Genetic Resources for Food and Agriculture (ITPGRFA) and the UN Declaration on the Rights of Peasants and Other People Working in Rural Areas (UNDROP), and to take measures to protect and promote farmers' rights. Whether or not this powerful set of farmer-led demands will receive the policy attention it warrants remains to be seen.

Areas for future research and advocacy

Benefit sharing, digital sequence information, and reconstituting a new 'commons' for agrobiodiversity

Despite the breadth of topics covered, several important themes have not been fully explored in this book, and require further research and deliberation. One of the most contentious policy issues concerns the matter of how benefits are shared from the use of genetic resources and 'digital sequence information' (so-called DSI), which is the genetic sequence data that is uploaded onto biological databases around the world and then mined for interesting applications. Because landraces and wild relatives contain important genes for stress resistance, adaptability, and improved productivity, they are of growing interest in the context of climate change. Their commercial use in the breeding and development of new plant varieties, as well as other biotechnology applications and products, thus has direct relevance for the farmers who have stewarded, innovated, and developed interesting traits and features. Contestations about the way in which digital sequence information is used and regulated have created stumbling blocks across multiple international policy processes and have profound implications for the way in which we manage and conceptualize agrobiodiversity and its benefits (Wynberg et al., 2021).

At the same time, there is a clear need to move away from viewing genetic resources for food and agriculture as commodities that can be owned, toward a strengthened, proactive, and expansive stewardship approach that engages with the question of how we can reconstitute a new 'commons' for agrobiodiversity in the face of increasing proprietary ownership of land, seed, and now genetic sequences (Kloppenburg, 2014). Multiple open-source seed initiatives are emerging across the world to introduce these more innovative and democratic ways of working, based on collaborations to share knowledge and seed that are unencumbered by property rights and other restrictions. Nascent initiatives in Kenya and other African countries indicate the potential for such open-source seed initiatives, signifying exciting prospects for farmer-led transformations that can reclaim seed sovereignty based on norms of sharing and solidarity.

Gene editing, gene drives, and the fourth industrial revolution

Although several chapters of the book describe the impacts of so-called first-generation GM crops on smallholder farmers and their agroecosystems, we have not considered the implications of second-generation GM crops and the suite of new genetic technologies, such as CRISPR (an abbreviation of Clustered Regularly Interspaced Short Palindromic Repeats) and gene drives, nor those of the 'Fourth Industrial Revolution', which fuses physical, digital, and biological worlds with technologies that span the three. Some critics suggest that such approaches simply epitomize the dropping of 'old Green Revolution STI [science, technology, and innovation] into sustainable packaging' (Montenegro de Wit and Iles, 2021: 209). Others point out that the future of improved crop varieties in sub-Saharan Africa looks very similar to its past, characterized by 'a top-down research and development agenda that rhetorically foregrounds poor and marginalized smallholder farmers while producing technologies which benefit the most powerful and highly capitalized among them' (Schnurr and Dowd-Uribe, 2021: 384). Agroecology stalwart Michel Pimbert (2022) remarks that these so-called disruptive technologies are fundamentally corporate visions for the future that 'essentially conform with – rather than transform – the dominant agro-food regime because they are primarily based on principles of uniformity, centralisation, privatisation, concentration of power, control and coercion'. An active research and advocacy programme is clearly essential to monitor ongoing developments, to assess the impacts of new technologies on smallholder farmers, and to lobby for an approach that places smallholder farmer needs and agrobiodiversity at the centre.

Wider transitions and emerging issues

Wider transitions are essential in African agriculture if an agroecological, diverse, and seed-secure future is to be secured for the continent. Sub-Saharan Africa is one of the world's fastest-urbanizing regions, and its urban population is predicted to double over the next 25 years. It is also one of the world's regions most vulnerable to the impacts of climate change due to varying rainfall patterns, more extreme weather, low levels of adaptive capacity, and a high dependence on agroecosystems for livelihoods.

A common criticism of agroecology is that it is not sufficiently 'scalable' to address these challenges. We have seen in this volume examples both of scaling out – meaning the horizontal replication of agroecology, where many small farms and families in numerous territories produce and eat agroecologically – and scaling up, as in the case of Cuba, where transitions occurred through shifts in policies and institutions at the level of programmes, regulations, and laws. However, much more is needed. Mousseau (2015) reminds us that agroecological farmers are already active economic players involved in various forms of commercial agriculture and trade, dispelling the myth that all smallholder farmers produce only for subsistence purposes, and that a conventional, industrialized model of production is needed to achieve scalability. What is clear is that scaling goes beyond the simplistic notion of 'scale-as-yield' and cannot be achieved by a 'cookie-cutter' approach that mechanically transfers agroecological practices from one place to another. As Ferguson et al. (2019: 723) remark, scaling 'situates agroecology as one key element of broader societal transformations that challenge capitalism, colonialism, standardization, industrialization, patriarchy, and other forms of injustice'.

Further research is needed to make visible the value of agroecology, and to enhance understanding about how such scaling can be achieved in the African context, yet the availability of significant public and donor resources for agriculture signals potential opportunities for instituting programmes that can support agroecology and smallholder African farmers in meaningful and sustainable ways. Growing food sovereignty and agroecology movements can, moreover, provide the possibilities for 'scaling up' through policy leverage and inclusive, democratic change. Local seed-saving networks, the revival and use of indigenous and underutilized crops, and producers linking with users for nutritious food production point to autonomous actions that are already available, alongside solidarity networks that can reclaim seed sovereignty.

It is through seizing such opportunities, one step at a time, that a diverse, relationally robust, socially equitable, and ecologically secure future can be safeguarded for the African continent. We hope that the learnings in this book can inform this journey.

References

Andersen, R. (2017) '"Stewardship" or "ownership": how to realize farmers' rights?', in D. Hunter, L. Guarino, C. Spillan and P.C. McKeown (eds), *Routledge Handbook of Agricultural Biodiversity*, pp. 449–470, Routledge, Abingdon <https://doi.org/10.4324/9781317753285-29>.

Bezner Kerr, R., Postigo, J.C., Smith, P., Cowie, A., Singh, P.K., Rivera-Ferre, M., Tirado-von der Pahlen, M.C., Campbell, D. and Neufeldt, H. (2023) 'Agroecology as a transformative approach to tackle climatic, food, and ecosystemic crises', *Current Opinion in Environmental Sustainability* 62: 101275 <https://doi.org/10.1016/j.cosust.2023.101275>.

Collier, P. and Dircon, S. (2014) 'African agriculture in 50 years: smallholders in a rapidly changing world?', *World Development* 63: 92–101 <http://dx.doi.org/10.1016/j.worlddev.2013.10.001>

DeVries, J.D. (2019) 'The role of seed systems development in African agricultural transformation', in R. Sikora, E.R. Terry, P. Vlek and J. Chitja (eds), *Transforming Agriculture in Southern Africa*, pp. 77–85, Routledge, Abingdon <https://doi.org/10.4324/9780429401701>.

Ferguson, B.G., Maya, M.A., Giraldo, O., Terán Giménez Cacho, M.M., Morales, H. and Rosset, P. (2019) 'Special issue editorial: What do we mean by agroecological scaling?', *Agroecology and Sustainable Food Systems* 43(7–8): 722–723 <https://doi.org/10.1080/21683565.2019.1630908>.

Ferguson, J. (1994) *The Anti-politics Machine: 'Development,' Depoliticization, and Bureaucratic Power in Lesotho*, University of Minnesota Press, Minneapolis, MN.

Herrero, A., Wickson, F. and Binimelis, R. (2015) 'Seeing GMOs from a systems perspective: the need for comparative cartographies of agri/cultures for sustainability assessment', *Sustainability* 7(8): 11321–11344 <https://doi.org/10.3390/su70811321>.

Kloppenburg, J. (2004) *First the Seed: The Political Economy of Plant Biotechnology, 1492–2000*, University of Wisconsin Press, Madison, WI.

Kloppenburg, J. (2014) 'Re-purposing the master's tools: the open-source seed initiative and the struggle for seed sovereignty', *The Journal of Peasant Studies* 41: 1225–1246 <https://doi.org/10.1080/03066150.2013.875897>.

Marshak, M., Wickson, F., Herrero, A. and Wynberg, R. (2021) 'Losing practices, relationships and agency: ecological deskilling as a consequence of the uptake of modern seed varieties among South African smallholders', *Agroecology and Sustainable Food Systems* 45(8): 1189–1212 <https://doi.org/10.1080/21683565.2021.1888841>.

Moeller, N.I. (2020) *Analysis of Funding Flows to Agroecology: The Case of European Union Monetary Flows to The United Nations' Rome-Based Agencies and the Case of the Green Climate Fund*, CIDSE, Brussels, and Centre for Agroecology, Water and Resilience, Coventry <https://www.cidse.org/wp-content/uploads/2020/09/AE-Finance-background-paper-final.pdf>.

Montenegro de Wit, M. and Iles, A. (2021) 'Woke science and the 4th Industrial Revolution: inside the making of UNFSS knowledge', *Development* 64(3): 199–211 <https://doi.org/10.1057/s41301-021-00314-z>.

Mousseau, F. (2015) 'The untold success story of agroecology in Africa', *Development* 58(2–3): 341–345 <https://doi.org/10.1057/s41301-016-0026-0>.

Ncube, B.L., Wynberg, R., and McGuire, S. (2023). 'Comparing the contribution of formal and local seed systems to household seed security in eastern Zimbabwe', Frontiers in Sustainable Food Systems, DOI: 10.3389/fsufs.2023.1243722

Pavageau, C., Pondini, S. and Geck, M. (2020) *Money Flows: What Is Holding Back Investment in Agroecological Research for Africa?*, Biovision Foundation for Ecological Development and International Panel of Experts on Sustainable Food Systems <https://www.agroecology-pool.org/moneyflowsreport>.

Pimbert, M. (2022) 'Transforming food and agriculture: competing visions and major controversies', *Mondes en développement* 199–200(3–4): 361–384 <https://doi.org/10.3917/med.199.0365>.

Schnurr, M.A. and Dowd-Uribe, B. (2021) 'Anticipating farmer outcomes of three genetically modified staple crops in sub-Saharan Africa: insights from farming systems research', *Journal of Rural Studies* 88: 377–387 <https://doi.org/10.1016/j.jrurstud.2021.08.001>.

Southern African Development Community (SADC) (2022) *Synthesis Report on the State of Food and Nutrition Security and Vulnerability in Southern Africa*, SADC, Gaborone <https://www.sadc.int/sites/default/files/2022-08/SADC_RVAA_Synthesis_Report_2022-ENG.pdf>.

Sperling, L., Gallagher, P., McGuire, S. and March, J. (2021) 'Tailoring legume seed markets for smallholder farmers in Africa', *International Journal of Agricultural Sustainability* 19(1): 71–90 <http://doi.org/10.1080/14735903.2020.1822640>.

Wynberg, R., Andersen, R., Laird, S., Kusena, K., Prip, C. and Westengen, O.T. (2021) 'Farmers' rights and digital sequence information: crisis or opportunity to reclaim stewardship over agrobiodiversity?', *Frontiers in Plant Science* 12:1–16 <https://doi.org/10.3389/fpls.2021.686728>

Index

community dialogue and
  support 144
community monitoring, research,
  advocacy, and policy 145–147
crop choices 138–139
farming practices 145
issue of GM contamination
  139–142
levels of awareness among farmers
  144–145
Massive Food Production
  Programme 139
mitigation strategies 145
'Seeds of Success' programme 139
transgene flow 143
Ghana's neoliberal reforms 7
global gene bank 157
Global Justice Now 324
glyphosate-based herbicides 11, 187,
  198–199
  environmental violence 203–205
  routes of human exposure to
    201–203
  use in Africa 199–200
GM herbicide-tolerant crops 11
Gogo Qho 3, 21–23, 228–231
  about poisoned food 22–23
  agroecological farming methods
    21–22
  conservation of seeds 23
  on modern way of farming 22
  muffins of *umsuzwane* 23
GRAIN International 324
granjero 275, 282
Green Revolution (GR) 1, 5, 9, 13,
  41, 69, 133, 306, 344, 348
  criticism against 69
  food production 317–319
  new African 170–171

Harrison, Michael 76
Health Promotion Levy 190
Hebinck, Paul 4
household-level food security, study
  correlations between seed access
    and food security 111
  informal seed sources for
    smallholder farmers 110
  seed quality, effects of 111–112

socio-economic and political
  context 112
study site and methods 110
timeliness and proximity, effects
  of 111
hybrid seed 5–7
Hygrotech 173

indigenous agricultural revolution 5
industrial agriculture 10–11, 27,
  188–190, 241, 263
in situ and *ex situ* conservation 157
intellectual property (IP) 9,
  168–169
International Treaty on Plant
  Genetic Resources for Food
  and Agriculture (ITPGRFA) 16,
  157, 322, 347
*Invisible Women: Exposing Data Bias in
  a World Designed for Men* 226

Kenya Seed Co. 174
Kiaka, Richard 4
Kikuyu culture 222
Kimmerer, Robin Wall 231
Klein Karoo 173
knowledge justice 117
knowledge politics 116–117
knowledge sharing 29
knowledge systems 2, 33–34,
  345–346
KwaZulu-Natal province 105–106,
  108–109, 135, 187, 191, 276

Lake Malawi 304–305
La Via Campesina movement 28,
  316, 319–320, 324
Leyva, Dr Ángel 276
Lilimini 5
Local Agricultural Innovation
  Programme (PIAL) 14–15
Luhyaland 74
Luoland 74

Maathai, Wangari 222, 225
Mabota ritual 29
  role of women as keepers of both
    seed and rituals 30–31
macronutrients 302

www.ingramcontent.com/pod-product-compliance
Lightning Source LLC
Chambersburg PA
CBHW051255020426

42333CB00026B/3221